THERMOPLASTICS:
MATERIALS ENGINEERING

THERMOPLASTICS: MATERIALS ENGINEERING

L. MASCIA
Corporate Technology Europe,
Raychem Ltd,
Swindon SN3 5HH, UK

APPLIED SCIENCE PUBLISHERS LTD
LONDON and NEW YORK

APPLIED SCIENCE PUBLISHERS LTD
Ripple Road, Barking, Essex, England

Sole Distributor in the USA and Canada
ELSEVIER SCIENCE PUBLISHING CO., INC.
52 Vanderbilt Avenue, New York, NY 10017, USA

British Library Cataloguing in Publication Data

Mascia, L.
   Thermoplastics.
   1. Thermoplastics
   I. Title
   620.1'923   TA455.P5

ISBN 0-85334-146-X

WITH 29 TABLES AND 189 ILLUSTRATIONS

© APPLIED SCIENCE PUBLISHERS LTD 1982

Printed in Great Britain by Galliard (Printers) Ltd, Great Yarmouth

Preface

In writing this book the author set himself the task of bridging the widening gap between the texts of polymer science, polymer engineering and the long-established monographs on plastics technology.

With its ever-increasing concern for mathematical slants, polymer engineering is becoming such a specialised discipline that it is not only beyond the comprehension of practising engineers and technologists but also risks leaving wide gaps in other important areas of knowledge.

This text, therefore, seeks to overcome these limitations by attempting to provide the necessary links between knowledge relevant to industrial work and theoretical principles that have emerged from fundamental research in polymer science and engineering.

Within this context, the principles of polymerisation processes and methods of structure characterisation are not relevant, since the materials engineer would look at polymers as the starting ingredients used for the manufacture of useful products. A knowledge of the types of polymers available, their constitution and properties, and the manner in which their behaviour can be controlled or modified through the use of additives, blending with other polymers or inorganic fillers, on the other hand, is essential for work concerning the utilisation of polymers.

These concepts are dealt with in the earlier chapters of the book and are intended to provide a link between polymer chemistry and the more practical aspects of 'compounding' treated in various plastics technology books.

Two chapters are devoted to mechanical properties. These deal with basic theoretical concepts and their limitations in describing the behaviour of thermoplastics in practical situations. The methodology used for the evaluation of properties within an engineering design context is also emphasised. One chapter is devoted to the principles underlying the deformational behaviour of polymers as related to processing consideration, whilst the final three chapters describe the factors affecting the processing characteristics of thermoplastics. As a means of evaluating the relative ease with which a material can be processed, the author uses the concept of 'Processability Index' to bring together materials properties and processing parameters.

The book is aimed primarily at scientists, engineers and technologists working with thermoplastics materials, particularly in those areas falling between Product and Process Design and those concerned with materials evaluation and process optimisation. It should constitute, however, a useful text for graduate students specialising in Polymer Engineering and Technology. Whilst a basic knowledge of Chemistry and Strength of Materials is assumed, in most cases the subject has been developed from first principles and, therefore, should be within the grasp of those who have not been previously exposed to the topics treated. Many sections of the book could also satisfy the needs both of practising technicians and of students taking technological and technician-level courses sponsored by technical colleges and professional societies such as The Plastics and Rubber Institute and The Society of Plastics Engineers.

The general layout has been developed, in fact, from the writer's lecture notes while teaching at the University of Aston and several chapters were taken from the Materials and Technology curriculum developed during a leave of absence at the Plastics Engineering Program of the Institute Algerien du Petrole.

The persons who have helped in conceiving this book are far too numerous to be listed. The writer is in debt to the many authors and publishers who have granted permission to reproduce copyright material and particularly those who have provided the photographs included in the text. Individual sources are listed in the Acknowledgements. The Education Development Center in Boston (USA) is thanked for their assistance in typing some of the manuscript and drawing some of the graphs. The writer is grateful to the Advisory Committee of the Education Development Center, Professors C. G. Gogos, C. E. Rogers, S. A. Orroth and E. A. Meinecke, for reading the sections of the book derived from the curriculum material mentioned earlier, and to Mr E. B. Atkinson for his most constructive criticisms and comments on the whole text.

L. MASCIA

Contents

7 Deformation Behaviour of Thermoplastics in Relation to Processing

Acknowledgements

The following figures are reproduced by permission of the publishers and individuals named.

Figs. 1.1 and 1.4—Maclaren Publishers Ltd.

Figs. 4.1, 4.2 and 4.3—Edward Arnold (Publishers) Ltd.

Figs. 4.4, 7.24, 7.25, 7.31, 9.5, 9.6, 9.7, 9.26 and 9.27—the Society of Plastics Engineers, Inc.

Figs. 4.7, 6.17(b), 6.48 and 6.49—Chapman and Hall Ltd.

Figs. 4.12, 4.13 and 4.14—Marcel Dekker, Inc.

Fig. 5.4—Longman Group Ltd.

Figs. 5.20, 5.26, 9.3 and 9.18—IPC Industrial Press Ltd.

Figs. 5.21, 5.22, 5.23, 5.24 and 9.19—ICI Ltd, Petrochemicals and Plastics Division.

Figs. 5.27 and 5.28—Elsevier Scientific Publishing Co.

Figs. 5.29, 6.50 and 6.51—IPC Business Press Ltd.

Figs. 6.7, 6.8, 6.9, 6.46, 6.47, 7.22, 7.38, 8.4, 8.6, 9.1 and 9.29—the Plastics and Rubber Institute.

Figs. 6.11, 7.7, 7.13, 7.15, 7.21 and 9.8—John Wiley and Sons, Inc.

Fig. 6.17(a)—Mr P. I. Vincent.

Figs. 6.34 and 6.35—Dr J. Knott.

Figs. 6.43, 6.44 and 6.45—ASTM.

Fig. 7.8—Dr Dorothy E. Tobolsky.

Fig. 7.15—the Institute of Physics.

Fig. 7.26—Dr J. McKelvey.

Fig. 7.32—Elsevier Sequoia SA.

Fig. 7.35—the American Institute of Physics.

Fig. 9.9—the American Chemical Society.

Figs. 9.20 and 9.21—McGraw-Hill Publications Corp.

Figs. 9.23 and 9.24—Dr B. M. Murphy.

1

Introduction to Industrial Polymers

1.1 NOMENCLATURE

1.1.1 Plastic and Plastics

The adjective 'plastic' derives from the Greek, meaning easily shaped or deformed, e.g. like putty or clay. It was introduced in the middle of the last century with the discovery of cellulose nitrate, since it seemed to behave like clay when mixed with solvents. One prefers to use the term *plastics* as the common noun for a vast range of materials based on macromolecular organic compounds. Therefore, one would say 'plastic material' to qualify the easily deformable state of the material, but would use the term *plastics material* to indicate that the material belongs to the family of plastics.

1.1.2 Thermoplastics and Thermoset Plastics (Thermosets)

Traditionally plastics have been divided into two major classes according to their behaviour towards heat.

Thermoplastic implies that the material has acquired its 'easy deformability' by application of thermal energy (heat).

Thermoset, on the other hand, denotes that the material has 'set', i.e. become rigid, as a result of the application of heat. This term was introduced in the early stages of development of the industry in respect to phenolic materials which were '*cured*' (another term for hardening) by heat.

Nowadays one uses the term thermoset for any plastics material which does not *flow* under the influence of heat and directional forces (obviously the opposite of the behaviour of thermoplastics). In the late 1940s, plastics technologists defended the use of the term plastics for this type of material by modifying the definition of plastics to 'materials which at one stage or another of their existence will go through the plastic state'. In this case thermosets exhibit the familiar plastic (or easily deformable) behaviour prior to becoming hard, i.e. when they are still in their 'thermosetting state'. It must be borne in mind that such a definition, while practical and convenient, lacks accuracy. Although it excludes wood, for instance, from the classification, since in no stage of its life does it undergo plastic deformation, this definition suffers from the disadvantage that it also excludes a large family of materials, such as 'unsaturated polyesters', which are not normally hardened by heat and do not exhibit the classic puttylike deformability in their thermosetting state. Admittedly, such materials arrived on the scene long after the term had been introduced, but this does illustrate that definitions which are not scientifically based or properly conceived will lack generality; consequently, the terms 'thermoplastics', 'thermosetting', 'thermosets' and 'plastics' must be considered as 'jargon'.

1.1.3 Rubbers and Elastomers

These are two further examples of jargon introduced too early (Rubbers) or without sufficient consideration of denotation (Elastomers). The word 'rubber' was obviously introduced when this material was first discovered or, at least, used for practical purposes, utilising its ability to erase written marks, etc. Nowadays the term has become synonymous with the ability to stretch and quickly retract on removal of the applied forces, to the extent that any material which exhibits this type of behaviour is referred to as a 'rubbery material'. The term 'elastomer' was introduced after the war, perhaps to reflect more accurately the behaviour of this class of materials; the ending 'mer' was taken from the term 'polymer'.

Because of their cross-linked, tridimensional structure, conventional rubbers cannot undergo flow or the typical thermoplastic behaviour described earlier; consequently, one is inclined to regard them as thermosets. By analogy to thermoset plastics it would be appropriate to refer to these materials as 'thermoset rubbers'; and were it not for the fact that rubber technology developed earlier than plastics technology, this might well have been the case. The actual term used, however, is 'vulcanised rubbers', jargon used by Hancock in 1770, which presumably derives from Vulcanus (the Roman god of fire), since it involves a process of heating the rubber (vulcanisation) in order to achieve the desired properties; that is, the prevention of flow under the application of directional forces. The actual heating process to change the behaviour of the rubber was termed 'curing' to indicate, in fact, that the problem of flow (and therefore, loss of 'elasticity') was being cured. The term curing was later adopted by the plastics industry to denote the setting or hardening phenomenon produced by the formation of a cross-linked structure.

1.1.4 Polymers and Plastics

The concept of polymers as macromolecular compounds was introduced by Staudinger[1] in the 1920s when he elucidated the chemical structure of these materials which previously were believed to be 'colloidals' (another jargon term, derived from the Latin word *colla*, meaning glue). Since the molecule consisted of many repeating chemical units, the composite term 'poly-mer' was created from the fusion of two Greek words *polus* (many) and *meros* (part).

Because of the scientific origin of the word polymer, it is tempting to use it in place of more crude jargon such as 'rubbers', 'plastics', etc. However, a danger exists in using these terms synonymously insofar as the term polymer denotes certain types of chemical compounds (whether organic or inorganic); whereas the materials in question, with rare exceptions, e.g. biomedical applications, always contain a mixture of chemicals, even if in the majority of cases the main component is in fact a polymer.

If we wished, we could now modify our definition of plastics to read 'rigid materials based on organic polymers', where the term rigid may loosely differentiate them from rubbers. We would have to bear in mind, however, that by so doing, we would still not solve the problem of lack of generality insofar as the term could still be confounded with other jargon not normally associated with plastics, such as surface coatings, adhesives and fibres. So we can only conclude that since we are dealing with materials that originate in industry, where one is more concerned with simplicity and ease of identification and communication, the use of jargon is inevitable and must be accepted.

1.2 CLASSIFICATIONS

It is common to classify plastics into thermoplastics and thermosets in view of their different behaviour towards heat, which in turn puts different demands on the methods of processing. Consequently, this subdivision is primarily one based on processability, although differences in properties will also result from the differences in structure, especially with respect to ductility and resistance to solvents.

It is possible, moreover, to classify plastics on the basis of their application, tonnage, price, etc. Such a subdivision would require a knowledge of markets, structure and pricing policy within the petro-chemical/plastics industry and it would suffer the disadvantage of restricted applicability and of susceptibility to variations in time.

Hence, it is preferable to create a classification on a scientific basis in order to achieve universal acceptability. The most appropriate one is based on the chemical structure of the polymer constituent since this determines the properties and to a large extent also the price and tonnage. The classification shown in Table 1.1 should be appropriate.

Table 1.1

Thermoplastics		Thermosets
Homochains	*Heterochains*	
Polyolefins	Polyethers and cellulosics	Phenolics
Vinyls	Polyamides	Aminos
Styrene types	Polyesters	Unsaturated polyesters
Acrylics	Polyurethanes	Epoxides
Polyfluorolefins		Silicones
		Polyurethanes

1.3   DEVELOPMENT OF THE PLASTICS INDUSTRY

1.3.1 The Discovery of the First Plastics Material

The frequent wars of the 19th century created a high demand for gun powders. These are based on nitro-organic compounds and, at that time, were produced by esterification of polyhydric alcohols, so that a high concentration of nitro groups per molecule could be produced, hence achieving the required detonation characteristics. To do this on a large scale requires a readily available and cheap source of raw materials. Cellulose was, therefore, the obvious raw material to choose since it can accept three nitro groups (this was probably not realised completely at that time). When the nitration reaction does not quite reach completion (optimum at about $2 \cdot 2$ NO_2 groups/cellulose unit), the products are easily shaped, like putty and soft doughs when mixed with solvents like alcohol or ether.

This did not escape the curiosity of the 'chemists' and of the adventurous industrialists of the time. Consequently, as far back as 1832, we find a patent issued to Braconnot on the production of lacquers and films based on this new product which he called 'xyloidine'.[2] The technology was, however, far too crude and the materials so produced could not match the quality of the well-established lacquers based on natural resins (jargon derived from the Latin word *resina*, cognate with the Greek word *rhētinā*). Interest in this new man-made material remained very high, and many alchemistic attempts to improve its properties were made.

It took, however, about 30 years before plastics made their first impact on industry. This period coincided with an early realisation by man that natural resources do not last forever. When legislation was passed to preserve elephants on the continents of Asia and Africa, a new source of ivory had to be found to meet the increasing demands of 'Victorian ladies for ornaments and those of cowboys for billiard balls'. The press reported that a substantial reward (10 000 dollars) was offered in the USA to any inventor capable of producing a billiard ball from a source other than ivory. It would appear that the reward was finally collected by James Hyatt (1862) and that almost exactly to the day Alexander Parkes in England also patented a 'composition' based on cellulose nitrate.

It must be during this time that the term 'plastic' was first applied to these new materials, which became abundant and probably gave a tremendous boost to the 'billiard balls trade'.

There was, of course, interest in this type of material from other sectors. The great Paris exhibition in 1867 displayed fabrics made from 'artificial silk' or the so-called 'Chardonnet silk', named after its inventor. It would seem, however, that eventually, around the end of the last century, it was the field of varnishes and lacquers which took the lead in the use of these materials, especially in the field of fabric impregnation.

The frequent fires from the use of these highly flammable materials in such products as aeroplane wings led the industry to search for a replacement material still based on cellulose, the only source of organic raw materials at that time.

In 1865 patents had already been issued to Schützenberger[3] for materials based on cellulose tri-acetate, but these were not exploited commercially as they were not soluble in cheap solvents (e.g. acetone, or alcohols and ethers), but would only dissolve in chloroform, which is expensive and toxic. Even the efforts of Miles (USA, 1903),[4] who produced an acetone-soluble cellulose acetate (di-acetate), did not seem to produce the answer, possibly because the quality was not up to required standards. Progress was slow, partly because technologists were conditioned to thinking only in terms of solvent softening, whether for shaping of components or for coatings and impregnation of fabrics. Not until 1926[5] did development of cellulose acetate take off, when Eïchengrun (Germany) pioneered the injection-moulding machine based on the principle of the die-casting of metals developed earlier in France by Pelouse.

1.3.2 The First Synthetic Plastics Material

The production of cellulose nitrate and cellulose acetate must be considered a modification of natural products, since the polymer backbone chain is already present in the raw material. We have to wait, in fact, until 1909 to find the first patent on a process which produces a true synthetic polymer, i.e. from raw materials which are not polymeric.

The material is a thermosetting resin produced from the reaction of phenol with formaldehyde, and the invention is attributed to Leo Baekeland. The plastics materials produced were given the name 'Bakelite'.[6]

It is interesting to note that chemists in academies, e.g. Bayer around 1870,[7] were already aware of the reactions between these two materials, but because scientists are merely concerned with understanding the principles and, in this case, establishing the nature of the products, little was done towards their application. Because they could not be purified, these products were discarded. It is understood, however, that at this time industry was not geared to think in terms of innovation, and consequently, research and development were not carried out to any appreciable extent or, to say the least, in a systematic manner.

Leo Baekeland, although a chemist by training and also a brilliant practitioner, had an acute sense for business and, therefore, could see the opportunities to replace 'shellac', which was widely used both as a lacquer and for cold products. Supply was unpredictable and quality was inconsistent. All he had to do was to control the reaction (i.e. primarily the temperature, due to the exothermic nature of the reaction, and quantity of the catalyst) and find a suitable means of 'curing' the products during moulding. Heat was the obvious parameter with which to experiment since rubbers were vulcanised in this manner; while pressure was required to make the material flow into the mould and this had to be sufficiently high to suppress the formation of bubbles caused by water vapour.[6] His great achievement was undoubtedly in the use of wood flour and other fillers to overcome the inherent brittleness, a practice which has not changed since its inception. His greatest contribution, however, is in his approach to materials development, which no doubt produced the spark for the proliferation of systematic experimentation in industry.

1.3.3 The Birth of a Science

Between the discovery of cellulose nitrate and the production of phenolic resins, many chemists had described in patents and in the literature the processes of solidification of liquids (like styrene) through chemical reactions, e.g. Hoffman, 1845[8] and Ostromislensky, 1912.[9] But it was Staudinger who collected all the evidence to prove that they consisted of covalently bonded monomeric species and that their molecular weight could be estimated from measurements of the intrinsic viscosity of dilute solutions, a practice still widely used today.

Another great contribution to polymer science was made by Carothers[10] in the 1930s and 1940s with his gelation theory and introduction of step-wise condensation reactions for the production of polymers. Still in the field of polymerisation the name of Natta stands out for his original work on the production of stereoregular polymers.

Later, an impact on polymer science was made by physicists such as Tobolsky, Ferry, etc., with their work on viscoelasticity, which has become almost synonymous with polymer behaviour.

1.3.4 Technological Developments

The historical development of the technology of plastics can be regarded as having taken place in five steps or phases:

Phase I: 1850–1900 The discovery of a new material. Limited development occurred during this period owing to the lack of an established processing technology. Extrusion was the only known process at the time; it was very rudimentary (ram type) and did not receive much attention since the main concern was for varnishes and fabric impregnating products.

Phase II: 1900–1930 The birth of a technology. While it has to be recognised that the synthesis of new materials was the novelty of this period and obviously received great applause from both scientists and business-men, the great contribution to technological progress in the field of plastics came from ingenious practical men like Ëichengrun, who adapted machinery used in other sectors and brought about suitable modifications to enable these new materials to be converted into useful products. This involved the machine design element on the one hand and the tailoring of the constitution of the material on the other, which is the essence of plastics technology. Furthermore, three major processes used in the conversion of plastics were introduced in this period. The relationship between material and process development is shown in the following scheme:

Material	*Process*
Phenolics $\longrightarrow$	Compression moulding
Casein $\longrightarrow$	Screw extrusion
Cellulose acetate $\longrightarrow$	Injection moulding

Phase III: 1930–1950 The Ersatz materials. Preparations for World War II provided the impetus for revolutionary innovations in all fields. In the case of plastics the emphasis was on synthesis of new polymers, while processes became increasingly more sophisticated. This is borne out in Table 1.2, which lists the polymers and processes from this period.

The predominant objective in this period was the replacement of naturally occurring materials (e.g. metals, rubbers, ceramics, etc.). This was particularly true in Germany owing to the cut-off of supplies from the Allied countries and their colonies. It is for this reason that the term Ersatz prevailed throughout this period and was characteristic of the Nazi regime.

Table 1.2

Materials	*Processes*
Polystyrene (PS)	Thermoplastics extrusion
Polymethyl-methacrylate (PMMA)	Twin screw extruders
Polyvinyl chloride (PVC)	Large injection-moulding machines
Urea formaldehyde (UF)	Transfer moulding
Melamine formaldehyde (MF)	Vacuum forming
Low-density polyethylene (LDPE)	Cast films
Polycaprolactam (Nylon 6)	Wet-lay-up laminations
Polytetrafluoroethylene (PTFE)	
Silicones	
Thermoplastic polyesters (PETP)	
Unsaturated polyesters	
Acrylonitrile–butadiene–styrene terpolymer blends (ABS)	

Phase IV: 1950–1970 The plastic age. During this period the need for cheaper products to speed up economic recovery from the devastation of the war was a catalyst for the proliferation of technological innovation. Subsequently the economies of scale allowed industry to produce plastics materials very cheaply, which in turn created an enormous demand from the new consumer-orientated society.

On the technological front the prevailing practice was to tailor materials to the application by better control of polymerisation reactions. Processing machines became increasingly large and automated. The direct passage from material production to final product without the benefit of case histories resulted in plastics gaining the reputation of being inferior materials. Hence the slogan 'it is only plastic' to denote that the material is expected to be of poor quality.

Phase V: 1970→ The efficient use of plastics. The poor image of plastics during the economic boom of the 1960s coupled with the realisation that in order to maintain steady economic growth plastics must enter markets which call for reliable performance, has recently stimulated interest in 'engineering design'. In other words, whereas in the previous phase 'aesthetics' and 'low costs' were the major attributes of plastics; the future must be built on reliability.

1.4 THE COST OF POLYMERS

Until around 1950, chemical plants were small and processes were relatively inefficient. Unit costs were high and the feedstock element was not a determining factor when the industry was built in Europe after 1930; in fact, the feedstock was naphtha (while in the USA the main source was still natural gas), which was relatively cheap and freely available. Because of the high cost of plants and their inefficient utilisation, however, the price of plastics was quite high.

Later, as the size of the plants increased, the unit cost at full capacity fell rapidly. Consequently the price of polymers dropped accordingly owing to the feedstock remaining practically unchanged (see Fig. 1.1). Much of the reduction in the price of polymers is due to the fact that the size of the plant required for economic operation increased more rapidly than did total petro-chemical demand. Therefore, since the economies of scale can be realised only at full capacity, the base chemical manufacturers attempted to fill their new plants as rapidly as possible by reducing prices to the polymer manufacturers. This reduction in prices stimulated the consumption of polymers to such an extent that the demand for low-cost monomers (by-products of cracking plants) exceeded their availability.

In Europe the growing demand for naphtha as a feedstock both for ethylene production and for aromatic plants competed with the demand for naphtha as a petrol (gasoline) component. The US chemical industry is now following the European example of producing olefins from naphtha and gas-oil (middle distillates), which is now becoming a world-wide trend. In order to continue to attract capital investment, the polymer industry had to increase the quality of their raw materials and had to pay higher prices for chemicals than those paid by users of these chemicals as sources of energy. This automatically reversed the trend of plastics prices, a process which began towards the end of the 1960s. In the early 1970s astronomical increases in the prices of crude oil occurred. Therefore, the price of the

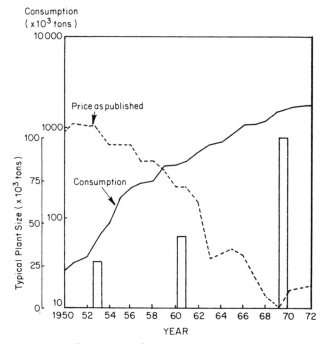

Fig. 1.1.   Consumption versus price in USA for low-density polyethylene.[11]

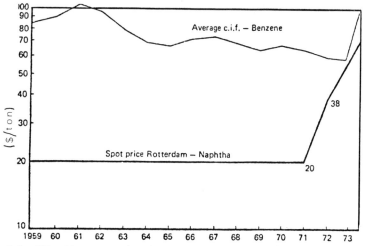

Fig. 1.2.   West European prices for naphtha and benzene.[11] (Note that between 1959 and 1971 the price of naphtha varied around a mean of $20 per ton.)

feedstock was forced up even further (see Fig. 1.2). This automatically produced an increase in the price of the high tonnage (commodity) plastics. Furthermore, the move in the USA in 1970 to reduce environmental pollution and the subsequent legislation for the abolition of lead compounds as anti-knocking additives for petrol has caused a temporary crisis in the supply of aromatics for polymer production. Aromatics (benzene) provide, in fact, a simple and effective solution to the problem of boosting the octane number of petrol in the absence of lead.

This direct competition with the gigantic petrol industry meant, therefore, that the cost of aromatic-based raw materials had to be increased tremendously, which brought about an even bigger rise in the price of certain plastics.

As a consequence, among the commodity plastics, polystyrene had to give up its lead to polyolefins in price competition. The speciality plastics, especially polycarbonates, nylons, etc. (see Fig. 1.3), suffered most. In this sector, nylons, which for a long time had been kept on a par with acetals, suddenly lost their competition. Perhaps the price of acetals was increased artificially more than necessary by the increase in feedstock prices. Furthermore, acrylics, which were already expensive owing to the complex route for the production of monomers, have increased more in price than the other commodity plastics, possibly because of the higher energy content in the monomer production. As a consequence, the price of ABS rocketed owing to its high aromatic (styrene) and acrylic content (see Fig. 1.4).

On the basis of past experience, it is predicted that in the near future

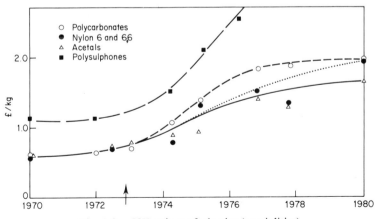

Fig. 1.3.   UK prices of plastics (specialities).

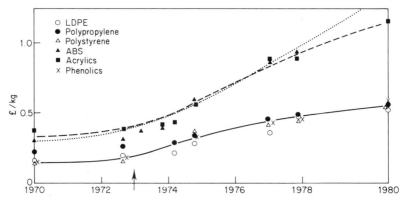

Fig. 1.4.   UK prices of plastics (commodities).

polypropylene may become the cheapest plastics material. The reason for this prediction lies once more in the value of propylene as a feedstock. Basically, modern cracking plants can now produce a greater yield of propylene than in the past and the only alternative use (the factor determining its value) is for energy production. The two competitors, ethylene (mainly) and butadiene, find many alternative uses as intermediates.

REFERENCES

1. H. STAUDINGER, *Berichte*, **57** (1924) 1203; *Kautschuk*, **63** (1927).
2. ICI Ltd, Plastics Division, *Landmarks of the Plastics Industry*, 1963, p. 7.
3. M. KAUFMAN, *The First Century of Plastics*, The Plastics Institute, London, 1963, p. 57.
4. M. KAUFMAN, *The First Century of Plastics*, The Plastics Institute, London, 1963, p. 58.
5. M. KAUFMAN, *The First Century of Plastics*, The Plastics Institute, London, 1963, p. 59.
6. US Patent 942,699.
7. A. BAYER, *Berichte*, **5** (1872) 1904.
8. A. HOFFMAN, *Annalen*, **53** (1845) 292.
9. I. OSTROMISLENSKY, *Chem. Zentral*, **1** (1912) 1980.
10. W. H. CAROTHERS, *Trans. Faraday Soc.*, **32** (1936) 39.
11. ANON., *European Rubber Journal*, **155** (1973) 1.

2

Basic Principles of Polymer Science

2.1 INTRODUCTION

It has already been mentioned that the term 'polymer' is derived from the Greek and is used to describe those organic macromolecules that contain many similar repeating units. These units can be arranged sequentially to form long chain molecules (linear polymers) or can be linked in a network fashion (cross-linked polymers) as illustrated in Fig. 2.1.

Polymers, whether naturally occurring or man made, are formed by joining small 'monomeric' molecules through so-called polymerisation reactions. It is, however, more a convenience than a requirement for polymers to be built up of similar repeating units. In other words, although the details of individual units in a polymer molecule can have a considerable influence on properties, the first order effect is achieved by virtue of molecular size.

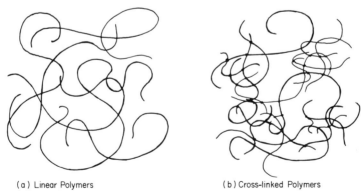

(a) Linear Polymers   (b) Cross-linked Polymers

Fig. 2.1.   Schematic representation of polymer molecules.

13

Table 2.1

Equi-intermolecular forces substances with different molecular weights			Equi-molecular weight substances exhibiting different intermolecular forces		
Chemical name	Structure	Melting point (°C)	Chemical name	Structure	Melting point (°C)
Butane	C_4H_{10}	−138	Hexanol	$C_6H_{13}OH$	−56
Octane	C_8H_{18}	−57	Hexane-1-amino	$C_6H_{13}NH_2$	−19
Dodecane	$C_{12}H_{26}$	−10	Hexanoic acid	$C_5H_{11}CO_2H$	−2
Octadecane	$C_{18}H_{38}$	+28	Hexamethylene glycol	$C_6H_{12}(OH)_2$	+41
Triacontane	$C_{30}H_{62}$	+65	Hexamethylene diamine	$C_6H_{12}(NH_2)_2$	+43
Polyethylene	C_nH_{2n+2} $(n > 500)$	+137	Adipic acid	$C_4H_8(CO_2H)_2$	+153

This is illustrated by the example in Table 2.1 where the properties of monomolecular substances are compared with compounds built up of an increasingly large number of similar units. In the first column the intermolecular forces are kept constant and the size of the molecule is increased, while in the second column the main difference between the various compounds shown is with respect to intermolecular forces. Although in both cases there is a gradual change from gas or liquid to high-melting-point solids, only the high-molecular-weight substances (e.g. polyethylene) possess adequate mechanical properties. It is this type of observation that makes one realise the importance of macromolecular (or polymer) systems in materials engineering.

2.2   DETAILS AND SIZE OF POLYMER MOLECULES

It is not necessary, for the repeating units in a polymer chain to be identical, i.e. *homopolymers*. The majority of polymers produced industrially in fact, contain more than one type of unit, which are arranged either at random, in blocks or in mixed sequences.

When there are two different units within a polymer chain or network the term *copolymer* is used to describe the macromolecule, while for the case where there are three units the system is known as a *terpolymer*. Occasionally four different units (i.e. *tetrapolymers*) are used to build up a polymer molecule but this is very rare and only occasionally found for materials used as adhesives or surface coatings.

For the simplest of all polymers, i.e. homopolymers, the molecular structure can be written as

$$*A-(A-A-A-A-A-A-A-A)-A*$$

Since the end units A* are normally either very small, e.g. hydrogen or hydroxyl groups (OH), or very similar in size to the repeating units, so that they can be actually designated as A units, it is more convenient to write the structure of the above polymer as

$$-(A)_n-$$

where n indicates the number of repeating units and is known as the *degree of polymerisation* (DPn).

From a knowledge of the atomic mass of the structural unit A and the degree of polymerisation one can obtain, therefore, an indication of the size of the polymer molecule, i.e. the '*molecular weight*'.

For the case of copolymers, say $[(A)_x(B)_y]_n$ it is necessary to know not only the molecular weight but also the relative proportions and arrangement of the two structural units in the chain.

2.2.1 Average Molecular Weights and Molecular Weight Distribution

Although under carefully controlled conditions some polymers can be produced so that all the molecules are of the same size, i.e. *monodisperse polymers*, in practically all industrial polymerisation processes the molecular chains produced can vary considerably in dimensions. Hence polymer systems have to be characterised in terms of '*average molecular weights*' and '*molecular weight distribution*'.

As far as copolymers are concerned, normally the distribution of the two comonomers varies also from one polymer chain to another. Very little can be done, however, to measure the distribution of comonomers in the polymer chains. If the polymer molecules are averaged in terms of number fractions of various lengths one obtains the so-called '*number average molecular weight*', when the average is performed on the basis of weight fractions, on the other hand, one obtains the '*weight average molecular weight*'. In other words, if the molecular weight of any single chain is M_i and the total number of such molecules is N_i, then the number fraction α_i is $N_i / \sum N_i$.

Hence the number average molecular weight becomes:

$$\bar{M}_n = \sum \alpha_i M_i = \sum \left(\frac{N_i}{\sum N_i} \right) M_i = \frac{\sum N_i M_i}{\sum N_i} \qquad (2.1)$$

Similarly if the weight fraction ϕ_i, of such molecules is $W_i / \sum W_i$ the weight average molecular weight is obtained as follows:

$$\bar{M}_w = \sum \phi_i M_i = \sum \left(\frac{W_i}{\sum W_i} \right) M_i = \frac{\sum W_i M_i}{\sum W_i} \qquad (2.2)$$

Since weights and numbers of molecules are related to each other by

$$W_i = \frac{N_i M_i}{N} \qquad (2.3)$$

where N is the Avogadro number, then substitution in (2.2) produces

$$\bar{M}_w = \frac{\sum N_i M_i^2}{\sum N_i M_i} \qquad (2.4)$$

If all the molecules were equal in size then $M_w = M_n$, because in each case (i.e. eqns. (2.2) and (2.4)) the summation would not apply.

As the distribution becomes increasingly broad the weight average assumes numerically larger values than the number average, so that the value M_w / M_n is taken as a measure of the spread of molecular chain lengths.

Until quite recently molecular weight averages were obtained indirectly from measurements of some physical properties of solutions. Solution viscosity was used to obtain an indication of the number average molecular weight, while light scattering determinations gave an indication of the weight average molecular weight. These measurements, however, are rather tedious so that for routine evaluations, e.g. quality control, simple techniques such as solution viscosity (at one concentration) and melt viscosity (at a prespecified shear rate) are often used instead. These measurements are often referred to as 'solution viscosity index' and 'melt flow index' respectively.

Nowadays the complete distribution of molecular weight can be measured directly by the so-called 'gel permeation chromatography' method, from which not only the two averages can be computed, but also the details of the distribution, e.g. degree of skewness, etc., can be determined.

2.2.2 Tacticity of Polymer Molecules

When a carbon atom in an organic molecule is asymmetric (i.e. it contains four different substituents), the molecule can adopt two different spatial configurations, each being the mirror image of the other. Since these structures cannot be converted without breaking chemical bonds, compounds having identical chemical constituents can display different properties owing to their configuration dissimilarities, i.e. they are said to exhibit stereoisomerism.

If one takes a single unit along the chain of vinyl or α-olefin polymers, i.e. $(CH_2$—$CH)_n$, it is immediately obvious that the C atom containing X and H $\hspace{1cm}|$ $\hspace{1cm}X$ substituents will exhibit isomerism because the length, and possibly the details, of the segments of the chains forming the other two substituents (R_1 and R_2) are also different from each other. In other words a unit of a poly-α-olefin or vinyl polymer can be written as R_1R_2CHX so that the H and X substituents can occupy the following positions in space[1]

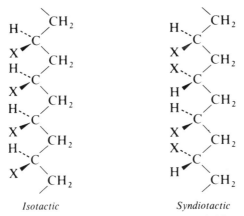

If to these two structures are arbitrarily assigned d and l configurations (initials of latin words meaning right and left respectively), then the two stereoisomeric configurations can either be distributed at random (*atactic*) or in a certain precise order (*tactic*) along the polymer chains. When each asymmetric C atom along the chain has the same configuration either l or d the polymer is said to be *isotactic*, while on alternating arrangement of l and d configurations along the chain give rise to the so-called *syndiotactic* structure (Fig. 2.2).

Isotactic   *Syndiotactic*

Fig. 2.2.   Stereo configuration of polymers exhibiting tacticity.

2.3 MORPHOLOGICAL FEATURES OF POLYMERS

2.3.1 Polymer Crystals

The bonds between the atoms in the backbone of a polymer chain are said to be 'flexible'; that is to say there is sufficient internal energy within a polymer system to enable the segments of the chains to rotate about the bonds within the backbone and create intramolecular as well as inter-molecular brownian motions. Consequently it is not difficult to envisage a situation where the polymer molecules are coiled in a completely random fashion. In such a situation there can be no structural order and the material is said to be *'amorphous'*. On the other hand, if the polymer chains were all equal in length and completely symmetrical along the chain (i.e. *stereoregular*) or free of any side group (i.e. completely uniform as in polyethylene), one would expect that the molecules would align along their length and form an array of *'chain extended'* crystals.

In view of the polydisperse nature of polymers, in practice one never achieves such a perfect order or 100 % crystallinity. Consequently polymers can only exist in amorphous or in semicrystalline states, where the level of crystallinity can vary from zero to about 90 %.

Furthermore, even when the level of crystallinity is very high the molecules are very rarely fully extended along their lengths, even when crystallisation occurs under the influence of externally applied extensional forces. The only exception to this is for polymers crystallised very slowly[2] and when the bonds are *very stiff*, e.g. aromatic types (see later). Depending on the chain stiffness, i.e. level of energy associated with bond rotations, number of imperfections in the chains, and external conditions, such as pressure and temperature, upon crystallisation polymers will form lamellae varying in thickness between about 100 and 500 nm.

An example of lamellae formed by the crystallisation of polymers is shown in Fig. 2.3.[3]

It is now normally accepted that lamellae are formed through a chain folding process starting at a *'nucleus'*. A molecule can fold within a crystal, pass through the amorphous regions and enter an adjacent lamella. This situation arises when a segment of a polymer chain is rejected from the growing lamella because imperfections such as short branches or bulky comonomer units prevent the continuation of chain folding. For this reason the more 'linear' the polymer chain is, the greater the level of crystallinity of the material.

Regions of disorder, however, can exist not only in the amorphous phase but also within the lamellae, albeit much less in number and smaller in size.

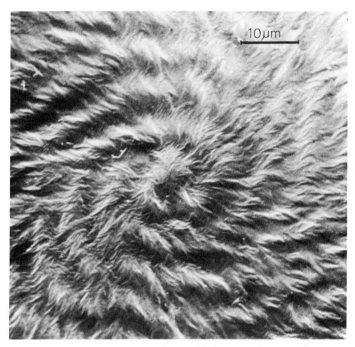

Fig. 2.3. Electron micrograph of solvent cast polycaprolactone (PCL-700).[3]

These will arise from small imperfections within the chains and, especially, at the end of the chains.

Suggestions have been made, however, that lamellae are not necessarily formed by chain folding and that molecules would simply go through a number of crystals via the amorphous regions. Nothing can be done experimentally to prove whether chain folding or simple segment alignment takes place in the crystallisation of polymers. The only rational argument that can be put forward in favour of chain folding is that plastic deformations of the material resulting from the sliding and tilting of crystals would rupture the polymer chains in the amorphous regions if all the lamellae were joined together at each polymer chain.[4]

2.3.2 Spherulites

The presence of crystalline matter in the shape of spherical domains was reported by Staudinger as early as 1937. A number of studies carried out much more recently[5-7] revealed that spherulites have the following characteristics:

(a)   A spherulite consists of a multilayer stack of crystals radiating out from a central 'single crystal'. The molecular chains in these crystals are oriented perpendicularly to the radial direction.

(b)   Viewed under an optical polarising microscope the spherulites produce a characteristic 'Maltese Cross' appearance, as shown in Fig. 2.4, which arises from the periodic fluctuations in refractive index, caused by the twist in the lamellae radiating out from the nucleus. In other words the arrangement of crystals in the lamellae causes an alternating extinction and reinforcement of the transmitted light.

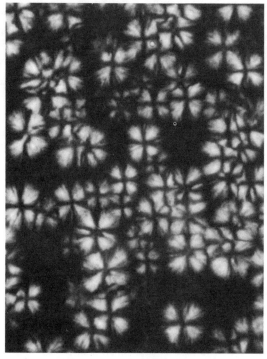

Fig. 2.4.   Optical micrograph of extruded film of short chain branching polyethylene (Sclair 11D1).[3]

2.3.3 Kinetics of Crystallisation

The formation of crystals in a material involves two steps: nucleation and growth. Nucleation is the initial process that immobilises a number of molecular chains in small regions and creates the necessary conditions for subsequent growth.

Two types of nucleations can take place in crystallisation phenomena: homogeneous nucleation (occurring spontaneously) and heterogeneous nucleation (occurring on the surface of solid impurities).

In practice, however, it is impossible to observe directly the two phenomena owing to the nuclei being too small to be detected by optical means.

Theoretical kinetic considerations for the two phenomena yield similar equations,[8] i.e.

$$\dot{N} \text{ (or } \dot{C}) = \dot{N}_0 \text{ (or } \dot{C}_0) \exp \left[-\frac{\Delta U \eta}{k(T - T_2)} \right] \exp \left[\frac{-\Delta G^*}{kT} \right] \quad (2.5)$$

where $\dot{N}$ = nucleation rate; $\dot{C}$ = crystallisation rate; $\Delta U \eta$ = activation energy for the transport of molecules across the liquid/solid interface (equivalent to the activation energy for viscous flow); ΔG^* = activation energy for the formation of nuclei (or subsequent crystal growth); k = Boltzman constant; T_2 = temperature at which the molecules are immobile (i.e. true glassy state—see later); and $\dot{N}_0$ and $\dot{C}_0$ = pre-exponential factors for the nucleation and crystal growth equations respectively, which take into account the frequency of segmental 'jumps' that lead to crystallisation.

Since viscosity and free energy have opposing temperature dependencies, the curves for $\dot{N}$ and $\dot{C}$ versus temperature will go through maxima and become zero when the temperature equals T_2 and at equilibrium conditions, i.e. at the melting point, when $\Delta G^* = 0$.

The curves corresponding to eqn. (2.5) for the two processes are shown diagrammatically in Fig. 2.5. The significant difference between the two curves is that the maximum for the nucleation rate occurs at lower temperature than that of the crystal growth. Consequently a large degree of super cooling is necessary in the absence of impurities (or nucleating agents) to induce crystallisation.

2.3.4 Secondary Crystallisation and Annealing

As crystallisation proceeds the spherulites will eventually touch each other and, as a result, the uncrystallised polymer chains at the boundaries will

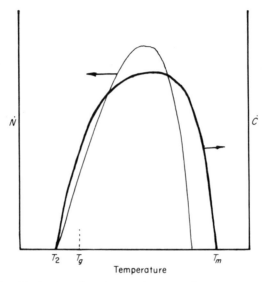

Fig. 2.5.   Nucleation $\dot{N}$ and crystal growth $\dot{C}$ rates for polymers.

find it increasingly more difficult to diffuse to the surface of existing crystals or to nucleate into new crystallites.

Consequently the crystallisation that takes place after such an event is very slow and is known as 'secondary crystallisation'.

If after crystallisation the polymer is heated to temperatures just below the 'true' melting point, some partial melting of the least perfect crystals takes place. The stable crystals, however, will continue to grow into thicker lamellae and a higher level of crystallinity is developed. This densification process carried out just below the melting point of a polymer is known as 'annealing'.

2.4 PHYSICAL TRANSITIONS IN POLYMERS

2.4.1 Primary Transitions
Classical physics recognises three fundamental states of matter, i.e. solid, liquid and gas.

The change from one state to another (transition) is brusque, i.e. it occurs at a specific temperature depending on the atmospheric pressure. Thermodynamically a transition from one fundamental state to another is known as a 'first order transition', and the transition temperature is defined as the point at which the first derivative of a primary function, e.g. volume and enthalpy, shows a sudden jump. More precisely it is the first partial derivative of the Gibbs Free Energy (F) that shows a discontinuity, i.e.

$$dF = S\,dT + V\,dP \qquad (2.6)$$

and therefore, the two functions that show a discontinuity are the entropy (S) and volume (V).

These are expressed respectively as

$$S = - \left| \frac{\delta F}{\delta T} \right|_P \quad \text{and} \quad V = \left| \frac{\delta F}{\delta P} \right|_T \qquad (2.7)$$

With polymers there can be no transition from the liquid to the gaseous state insofar as the temperature required completely to separate the molecules from one another would be far too high, and decomposition takes place instead.

For crystalline polymers the solid/liquid transition does not take place abruptly but it occurs over a range of temperatures, i.e. the primary function changes gradually between the values corresponding to solid and liquid state. This implies that the breaking of polymer crystals involves some premelting (as mentioned earlier), commencing at imperfection sites within the crystals and gradually disrupting the entire crystal lattice.

2.4.2 Secondary Transitions

With amorphous solid materials (or glasses) the change from the solid to the liquid state may also take place but the transition temperature is identified from a brusque change in the second derivative of the primary variables. For this reason a 'glass' transition is also known as the secondary transition.

In terms of the partial derivatives of the Gibbs Free Energy one obtains discontinuities in the following:[9]

$$- \left| \frac{\delta^2 F}{\delta T} \right|_P = \left| \frac{\delta S}{\delta T} \right|_P = \frac{C_p}{T} \qquad (2.8)$$

$$\left| \frac{\delta^2 F}{\delta P^2} \right|_T = \left| \frac{\delta V}{\delta P} \right|_T = -\beta V \qquad (2.9)$$

and

$$\left| \frac{\delta}{\delta T} \left(\frac{\delta F}{\delta P} \right)_T \right|_P = \left| \frac{\delta V}{\delta T} \right|_P = \alpha V \qquad (2.10)$$

where C_p = specific heat (at constant pressure), β = isothermal compressibility and α = volumetric thermal expansion coefficient.

Amorphous polymers differ, therefore, from low-molecular-weight glasses insofar as the 'glass transition' occurs over a range of temperatures. Furthermore, the behaviour of an amorphous polymer at temperatures above the glass transition is not that of an 'ordinary' liquid but resembles more that of a rubber. Hence this transition is often known as the glass/rubber transition.

The occurrence of a broad secondary transition in amorphous polymers can be likened to the premelting of crystalline polymers. In other words in passing from a frozen molecular state (the glassy state), where the molecules are immobile and their conformational entropy is zero, to the rubbery state where the molecules continually change their configuration through segmental rotations and acquire a very high entropy, there are a number of small sequential steps involving increasingly larger sections of the molecular chains.

The broad transitions observed with crystalline and amorphous polymers are represented schematically in Fig. 2.6. Note that the glass

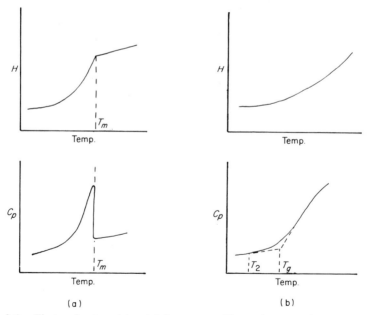

Fig. 2.6. First order transition (a) for a crystalline polymer and second order transition (b) for an amorphous polymer.

transition temperature, T_g, of a polymer occurs at temperatures above that at which the molecules are completely frozen (T_2). Consequently only T_m (true melting point) and T_2 values are to be considered as true thermodynamic (invariant) parameters. Because a transition from one state to another in polymers occurs through chain rotations (generally known as molecular relaxations) it is obvious that for each incremental step in temperature a certain amount of time has to elapse before a new equilibrium is reached (i.e. the molecules have a 'natural' relaxation time).

Hence it is to be expected that a 'fast' cooling rate will produce (in a given temperature interval) a change in volume (or enthalpy) which is smaller than the value obtained with a 'slow' cooling rate. As a result the computed T_g value as shown in Fig. 2.6, will be higher the faster the cooling rate, while the T_2 value cannot change (by definition), albeit it may be difficult to identify it with precision from practical measurements.

2.4.3 Effects of Molecular Features of Polymers on Physical Transitions
The ability of a polymer to crystallise and, therefore, to undergo a first order transition depends primarily on the factors listed below.

(a) Chain Stiffness
A high energy barrier to molecular rotations hinders the ability of the molecules to take up the required configurations to form crystals. That is to say that although the molecular structure of a polymer may be regular, the rate at which the crystals would grow could be suppressed by excessive chain stiffness and never develop a crystallographic structure.

These materials will remain amorphous irrespective of the thermal history and, therefore, will only exhibit a second order transition, i.e. the glass/rubber transition, which increases with increasing the restrictions to bond rotations within the molecular chains. Restrictions to rotations result mainly from bulky groups in the backbone and in the side branches of polymer chains.

(b) Chain Regularity or Symmetry
This determines whether the polymer molecules can come into sufficiently close proximity with one another and pack into a stable crystal lattice.

Atactic polymers and random copolymers are examples of materials that will remain amorphous because of the lack of chain regularity or symmetry.

Polymers that do not contain side groups, such as polyethylene, polytetrafluoroethylene, polymethylene oxide, etc., and stereoregular polymers have a very regular structure and can, therefore, form stable crystals.

The melting point of crystalline polymers depends, in turn, on both intermolecular and intramolecular factors. The first refers to forces between adjacent molecules and the spaces (free volumes) between molecules, while the second refers to chain flexibility. The forces between molecules become stronger when the polarity of constituent groups (or elements) increases, and reach the highest magnitude when they are capable of forming H-bonds between chains; i.e. attractions between hydrogen atoms from one chain and the unpaired electrons of an electronegative element, such as fluorine, oxygen or nitrogen, form another chain.

Side groups along the chains tend to increase the average distance between molecules and, therefore, reduce the magnitude of interactions between different chains. When such groups are bulky, on the other hand, the restrictions to rotations are increased and could have an effect on melting point that is greater than that produced by the concomitant increase in free volume.

It is understood that since crystalline polymers always contain a certain amount of amorphous matter, they will undergo a primary as well as a secondary transition. These transition temperatures are approximately proportional to each other as they are affected by the same molecular features. An approximate relationship betweeen T_g and T_m for crystalline polymers is

$$T_g \simeq \tfrac{2}{3}T_m \qquad \text{(for unsymmetrical chains)} \qquad (2.11a)$$

and

$$T_g \simeq \tfrac{1}{2}T_m \qquad \text{(for symmetrical chains)} \qquad (2.11b)$$

when both temperatures are expressed in Kelvins.[10]

2.4.4 Multiple Transitions in Polymers

The overall changes in molecular configurations that accompany the physical transitions in polymers are invariably influenced by small changes taking place within small sections (or domains) of the bulk of the material. Physical properties, therefore, do not change smoothly in a first order or second order transition but through small intermediate steps. These intermediate transitions are known as higher order transitions.

Because it is often difficult to identify a true glassy state for crystalline polymers the term glass transition is often avoided and all transitions between the melting point to absolute zero are identified as α, β and γ in order of decreasing temperature. According to the nature of the material, therefore, the glass transition could correspond to either the α or β transition and considerable controversy often exists as to which of these transitions corresponds to T_g.

No quantitative relationship exists, however, between the various second order transitions.

REFERENCES

1. N. C. BILLINGHAM and A. D. JENKINS, Chemical structure of polymers, in *Polymer Science*, Vol. 1, A. D. Jenkins (ed.), North-Holland, Amsterdam, 1972, pp. 166–70.
2. F. R. ANDERSON, *J. Appl. Phys.*, **35** (1964) 64.
3. F. KHAMBATTA, Raychem Corporation. Personal communication.
4. P. HENDRA, Southampton University. Personal communication, 1981.
5. A. KELLER, *J. Polym. Sci.*, **39** (1959) 151.
6. H. D. KEITH and F. J. PADDEN, *J. Appl. Phys.*, **35** (1964) 1286.
7. P. H. LINDENMEYER, *J. Polym. Sci.*, **C20** (1967) 145.
8. J. D. HOFFMAN and J. J. WEEKS, *J. Chem. Phys.*, **37** (1962) 1723.
9. J. J. AKLONIS, W. J. MCKNIGHT and M. SHEN, *Introduction to Polymer Viscoelasticity*, Wiley Interscience, New York, 1972, pp. 70–5.
10. J. A. BRYDSON, Glass transition, melting point and structure, in *Polymer Science*, Vol. 1, A. D. Jenkins (ed.), North-Holland, Amsterdam, 1972, p. 245.

3

Thermoplastics Materials

The purpose of this chapter is to introduce the reader to the types of polymers used for the production of thermoplastics materials. Polymers are subdivided into chemical families (as in Table 1.1) and by recalling the concepts underlying the relationship between structure and physical transitions, taking into consideration the availability of raw materials, a rationale is developed to explain why only a few polymers, in each class, have received commercial acceptance. The order in which the different families of polymers are presented reflects both the simplicity of their constitution and their industrial position in terms of production tonnage.

3.1 HOMOCHAIN POLYMERS

3.1.1 Polyolefins
The chemical composition of polyolefins is represented by the formula

$$\underset{\underset{R^1}{|}}{\overset{\overset{R}{|}}{-(CH_2-C)_n-}}$$

where R and R^1 are either hydrogen atoms or alkyl groups.

For the case of polyolefins used for the production of thermoplastics one of the side groups is always a hydrogen atom. The reason for this is that when R and R^1 are alkyl groups molecular packing becomes difficult so that even if the chain is regular, as for the case of polyisobutene

$$-(CH_2-\underset{\underset{CH_3}{|}}{\overset{\overset{CH_3}{|}}{C}})_n-$$

the distance between adjacent molecules is fairly large and the inter-molecular forces are rather low. The lack of crystallinity and the absence of polar groups in disubstituted polyolefins are responsible for their rubbery nature and ambient temperatures, i.e. T_g below room temperature, making them more suitable for rubber products or mastics than thermoplastics.

(a) Polyethylene

If one considers the case where both substituent groups are hydrogen atoms, i.e. polyethylene, the polymer chain can come close together so that sufficiently strong intermolecular forces can develop to hold the chains together into a crystal lattice. The degree of crystallinity (i.e. % crystal-line domains) in polyethylene and its melting point will depend primarily on the number of chain branches. The higher the number of side branches the greater the hindrance to the packing of molecules. In other words, during the growth of crystals the chain folding sequence will be inter-rupted at a chain branch and part of the chain will be rejected into the amorphous domains. The branches, therefore, reduce the level of crystallinity achievable and the melting point of the polymer, owing to the crystals being less perfect (or packed). As a result the overall density of the polymer is also reduced.

For ethylene homopolymers available commercially the density varies from about 0·916 when there are 2–3 branches for 100 carbon atoms ($T_m \simeq 100\,°C$) to 0·965 for an almost completely linear (branch-free) polymer ($T_m \simeq 135\,°C$). It is not necessary to say that the mechanical properties, e.g. rigidity and strength, and resistance to diffusion of soluble gases and vapours will increase with increasing density.

(b) Isotactic Poly-α-olefins

When one of the substituent groups is an alkyl radical and the other is a hydrogen atom, the polymer is known as a *poly-α-olefin*. In this case not only the nature of the side groups can be varied but also their tacticity (see Section 2.2.2). With atactic polymers the randomness of the side groups

Table 3.1: Characteristics of Polyolefins

Polyolefin	Nature of side groups	Crystals density (g/cm^3)	Density overall (g/cm^3)	Amorphous density (g/cm^3)	% Crystallinity	Melting point (°C)	Optical characteristics
Highly branched polyethylene	None	1·0	0·92	0·85	47	100	Translucent
Moderately branched polyethylene	None	1·0	0·935	0·85	57	125	Hazy
Linear polyethylene		1·0	0·96	0·85	74	135	Opaque
Polypropylene (atactic)	CH_3—		0·86	0·85			Transparent
Polypropylene (isotactic)		0·94	0·90	0·87	55	165–170	Translucent
Polybutene-1 (atactic)	CH_3—CH_2—		0·87	0·87			Transparent
Polybutene-1 (isotactic I)		0·95	0·91	0·87	55	135	Transparent
Polybutene-1 (isotactic II)		0·92	0·90	0·87	55	126	Translucent
Polybutene-1 (isotactic III)		0·89	0·88	0·87	50	106	Transparent
Polypentene-1 (isotactic)	CH_3—CH_2—CH_2—	0·87	0·86	0·84	50	75–80	Transparent
Polyheptene-1 (isotactic)	CH_3—$(CH_2)_3$—CH_2—					17	Transparent
Polynonene-1 (isotactic)	CH_3—$(CH_2)_5$—CH_2—					19	Transparent
Polydodecene-1 (isotactic)	CH_3—$(CH_2)_8$—CH_2—					49	Transparent

	Structure						
Poly-3-methyl butene-1 (isotactic)	CH_3—CH— with CH_3	0.90	0.87	0.84	50	300	Transparent
Poly-4-methyl pentene-1 (isotactic)	CH_3 / CH_3—CH—CH_2—	0.84	0.83	0.82	50	240	Transparent
Poly-4-methyl hexene-1 (isotactic)	CH_3CH_2 / CH_3—CH—CH_2—	0.84	0.83	0.82	50	188	Transparent
Poly-5-methyl hexene-1 (isotactic)	CH_3—CH—CH_2—CH_2— with CH_3	0.84	0.83	0.82	50	130	Transparent

prevents molecular packing and the polymer is amorphous and rubbery at room temperature.

Isotactic poly-α-olefins, on the other hand, can crystallise because the side groups along the chain can arrange themselves in a spiral fashion and allow adjacent molecules to come into close proximity with each other so that a stable crystal structure can develop.

In relation to polyethylene the side groups in isotactic poly-α-olefins have two opposing effects on the melting point:

(i)   By increasing the level of energy required for bond rotations along the chain the melting point increases.

(ii) The steric hindrance created by the side groups increases the average distance between molecules within the crystals and, therefore, the density decreases.

The effect of increased chain stiffness, however, predominates over the reduction in intermolecular forces and increased free volume with respect to the overall energy required to break the crystal lattice.

The overall effect of changing the nature of substituent groups in isotactic poly-α-olefins is shown in Table 3.1.

An inspection of this table reveals that when the side groups are branched the melting point increases considerably more than when they are linear. This confirms, once more, the early suggestion that restriction to molecular rotations by steric hindrance has a greater influence on melting point than average distance between molecules.

Note that although syndiotactic poly-α-olefins have been synthesised their properties are inferior to their isotactic counterparts and consequently they have not been commercialised.

Among the many different types of isotactic poly-α-olefins that have been synthesised, only polypropylene, polybutene-1 and poly-4-methyl pentene-1 are available commercially. The main reason for this is the difficulty of producing the monomers on an economic scale and their relatively unattractive properties in comparison with other polymers that can be produced more cheaply.

(c) Ethylene Copolymers

It is to be expected that by introducing at random a comonomer in the polyethylene chains the level of crystallinity will be reduced. This in turn brings about a reduction in stiffness and an increase in toughness and transparency.

The copolymers of ethylene, like the homopolymers, can be subdivided into two classes: i.e. branched, or low density polymers (produced by the high pressure process) and linear, or high density polymers (produced by the low pressure process).

With the branched types the presence of a comonomer can reduce the density from a nominal value of 0·92 to about 0·89, when the polymer becomes almost completely amorphous. The most common type of comonomers used are vinyl acetate, methyl (or ethyl) acrylate and acrylic acid.

With the linear polymers the density can be reduced from a nominal value of 0·96 to about 0·92 using comonomers such as propylene, butene or hexene.

By incorporating a termonomer, such as acrylic or methacrylic acid, in the chains and subsequently neutralising these with sodium or potassium hydroxide, to produce the so-called 'ionomers', the morphology of the polymer can change dramatically. The carboxylate ions, in fact, create a large number of nuclei in the bulk so that, upon crystallisation, the spherulites formed remain extremely small. In this way internal light scattering can be completely eliminated and the polymer remains transparent to visible light, despite the presence of crystals. This is quite a unique achievement with polymers since, normally, the spherulite structure becomes sufficiently developed to produce translucency or opacity.

(d) Copolymers of Propylene

In these copolymers the two monomers are arranged in blocks in order to minimise the reduction in degree of crystallinity so that their stiffness and strength can be preserved. By choosing a comonomer such as ethylene, the two blocks in the chains can crystallise separately forming bicomponent spherulites. It is believed that the complex morphology is largely responsible for the much increased toughness displayed by the block copolymers in comparison to homopolymers. The blocks in the chains, however, are not homogeneous but are made up of random copolymers whose composition gradually changes into homopolymer blocks and, for this reason, they are often called *tapered block copolymers.*[1]

As a result of the complexity of the morphology of these copolymers the melting range is widened and the 'ultimate' melting point is decreased by about 10 °C from that of the corresponding homopolymer.

3.1.2 Vinyl Polymers

Vinyl polymers have the following structure:

$$-(CH_2-CH)_n-$$
$$|$$
$$X$$

where X is any group other than hydrogen or alkyl/aryl groups.

Normally X is one of the following:

(a) An ester or halide group, e.g. polyvinyl acetate, polyvinyl chloride, etc.
(b) An alcohol group or its derivatives, e.g. polyvinyl alcohol, polyvinyl butyrals, etc.
(c) An ether group, e.g. polyvinylmethyl ether.
(d) Heteroaromatic rings, e.g. polyvinyl pyridine, polyvinyl carbazoles, etc.

The presence of polar and generally bulky side groups confers to vinyl polymers stronger intermolecular forces and greater restriction to rotations than occur in polyolefins.

(a) Homopolymers

The effect of the nature of the substituent X group on the glass transition temperature, melting point and density of vinyl polymers is shown in Table 3.2. To illustrate the effect of the polarity and bulkiness of the X group data are also given for polypropylene.

An analysis of these data indicates that for the atactic varieties only the polyvinyl ethers and, possibly, polyvinyl acetate would not be suitable for the production of thermoplastics in view of their rather low glass transition temperature. All the isotactic polymers, on the other hand, would be suitable, while polymers such as polyvinyl cyanide (also known as poly-acrylonitrile) and polyvinyl *tert*-butyl ethers would make excellent materials for high temperature applications.

In practice one finds that only polyvinyl chloride in the atactic form has been exploited commercially for the production of plastics materials. (Note that polyvinyl fluoride will be treated within the fluoropolymers family.)

The main reason for this is the difficulty of preventing (or significantly reducing) the decomposition of the other polymers under the influence of heat and/or ultraviolet light, which takes place according to the following scheme:

$$-(CH_2-\underset{\underset{X}{|}}{CH})_n \xrightarrow[\text{or } (h\nu)]{\text{heat}} -(CH=CH)_n + nHX$$

The conjugated polymer residue resulting from the decomposition reactions is very brittle and highly discoloured, i.e. brown or black.

A major problem in the degradation of vinyl polymers is that the reaction product HX acts as a catalyst for the decomposition reaction and, at the same time, because of its volatility it vaporises out of the system, pushing the equilibrium in favour of the decomposition reactions.

Due to their inherent thermal instability and high melt viscosity the vinyl polymers available commercially are atactic and amorphous. The high melting point of the crystalline varieties would render them virtually unprocessable since it would be difficult to control sufficiently the decomposition reactions at the temperature range where viscous flow would take place.

(b) Vinyl Chloride Copolymers

The main factors that have led to the development of vinyl chloride copolymers are first the reduction in melt viscosity, to enable the material to be processed at lower temperatures, and second the intrinsically more stable structures that are produced with the appropriate selection of comonomers.

In order to reduce the melt viscosity the comonomer has to be less polar and/or should contain fairly linear and flexible side groups. Ideally the structural changes brought about by copolymerisation must not cause a concomitant reduction in glass transition temperature, since this would reduce the creep resistance and the upper temperature range for its utilisation.

With these restrictions in mind, and with the need to keep the cost of the polymer to an acceptable level (say, not more than 20% higher than homopolymers), it is not surprising that only a few copolymers have been introduced in the market.

The earlier comonomers used were mainly vinyl acetate and, to a lesser extent, vinylidene chloride, i.e. typical comonomers that bring about a reduction in viscosity through the weakening intermolecular forces (vinylidene chloride) and increased free volumes (vinyl acetate).

To achieve a greater inherent thermal stability, on the other hand, the comonomer unit is chosen in such a way that the sequence of double bonds formed during decomposition is interrupted. Discolourations can be prevented, in fact, if the number of double bonds in the sequence is kept to less than about 20. The other advantage that results from the interrupted double bond conjugation is the reduction in the rate of evolution of hydrogen chloride, which is a catalyst for the decomposition reactions.

Neither vinyl acetate nor vinylidene chloride monomers, however, are capable of interrupting the formation of conjugated double bonds, hence they only act as viscosity depressants. Acrylate and olefinic monomers, on the other hand, have the required characteristics to produce intrinsically

Table 3.2: Structure–Properties Relationship in Vinyl Polymers[2]

Chemical name	Structure	T_g (°C)	T_m (°C)†	Density (g/cm³)	
				Amorphous	Crystalline†
Polypropylene	$-(CH-CH)_n-$ $\mid$ CH_3	−5	175	0·85	0·94
Polyvinyl bromide	$-(CH_2-CH)_n-$ $\mid$ Br	100	—	—	—
Polyvinyl chloride	$-(CH_2-CH)_n-$ $\mid$ Cl	81	215	1·38	1·52
Polyvinyl fluoride	$-(CH_2-CH)_n-$ $\mid$ F	40	175	1·37	1·44
Polyvinyl cyanide	$-(CH_2-CH)_n-$ $\mid$ CN	105	318	1·18	1·54

Polyvinyl alcohol	$-(CH_2-CH)_n-$ $\vert$ OH	90	240	1·26	1·35
Polyvinyl acetate	$-(CH_2-CH)_n-$ $\vert$ $O \cdot CO \cdot CH_3$	28	—	1·19	1·26
Polyvinyl *tert* butyl ether	$-(CH_2-CH)_n-$ $\vert$ $O-CH(CH_3)CH_3$	88	260	0·93	0·978
Polyvinyl methyl ether	$-(CH_2-CH)_n-$ $\vert$ $O \cdot CH_3$	−13	145	1·03	1·175
Polyvinyl ethyl ether	$-(CH_2-CH)_n-$ $\vert$ $O \cdot CH_2CH_3$	−19	86	0·94	0·99

† Refers to isotactic polymers.

more stable polymers. In practice mainly ethylene and propylene have been used for this purpose in view of their very low cost in comparison (say) to acrylates.

(c) Chlorinated Polyvinyl Chloride

With a glass transition temperature of 81 °C, polyvinyl chloride is unsuitable for applications where the operating temperature is close to the boiling point of water. The prospect of penetrating the hot-water pipe markets is, probably, the major factor responsible for the commercialisation of chlorinated polyvinyl chloride. By substituting hydrogen atoms along the chain with further chlorine atoms, i.e. by reacting polyvinyl chloride with chlorine gas, two possible modifications can be brought about, i.e.

(i) $\sim\!\!-CH_2-\overset{*}{C}H-CH_2-CH-CH_2-CH-\!\!\sim + Cl_2$
                  | | |
                  Cl   Cl   Cl

$$\longrightarrow \sim\!\!-CH_2-\underset{\underset{Cl}{|}}{\overset{\overset{Cl}{|}}{C}}-CH_2-\underset{\underset{Cl}{|}}{CH}-CH_2-\underset{\underset{Cl + HCl}{|}}{CH}-\!\!\sim$$

(ii) $\sim\!\!-CH_2-CH-\overset{*}{C}H_2-CH-CH_2-CH-\!\!\sim + Cl_2$
                  | | |
                  Cl   Cl   Cl

$$\longrightarrow \sim\!\!-CH_2-\underset{\underset{Cl}{|}}{CH}-\underset{\underset{Cl}{|}}{\overset{\overset{Cl}{|}}{CH}}-CH-CH_2-\underset{\underset{Cl + HCl}{|}}{CH}-\!\!\sim$$

(* indicates the position in which substitution has taken place.)

Obviously reaction (i) only leads to structures that are synonymous with vinyl chloride–vinylidene chloride copolymers, having a lower T_g than the homopolymer, in view of the cancellation of dipolar interactions (C–Cl) across the same C atom.

Reaction (ii), on the other hand, produces structures containing a larger number of both C–Cl dipoles, along the chains, and sterically hindered C–C bonds. The net effect of this is an increase in intermolecular forces and in rotational energy, both of which result in an increase in glass transition temperature. Furthermore the polymer also acquires a greater chemical resistance, which is possibly the reason why these polymers have been used mostly for waste disposal pipes.

3.1.3 Styrene-based Polymers

Styrene polymers contain large sequences of the following structural units in the chains:

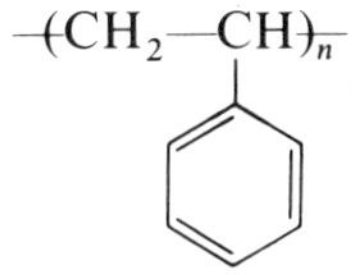

(a) Homopolymers

The main molecular features of polystyrene are:

(i)   High chain stiffness, conferred by the bulky aromatic side groups, which is responsible for the high T_g ($\simeq 80\,°C$) of the atactic (amorphous) polymer and high T_m ($\simeq 240\,°C$) for the isotactic (crystalline) polymer.

(ii)   Lack of polarity and, therefore, low intermolecular forces, which are responsible for the poor resistance of polystyrene to creep and non-polar chemicals.

Owing to the high chain stiffness, atactic polystyrene is a brittle material since molecular relaxations below T_g are relatively slow. For the case of isotactic polystyrene the ability of the molecules to dissipate energy from (say) impact stresses, is even less than for the atactic counterpart, owing to further restrictions to molecular rotations within the crystalline regions.

For this reason only atactic polystyrene is produced commercially.

(b) Copolymers

The changes in properties that one seeks to produce by copolymerisation of styrene are:

(i)   Increased inertness towards aliphatic and aromatic hydrocarbons (oils).

(ii)   Increased toughness.

(iii)   Increased resistance to creep at elevated temperature.

Random Copolymers. In order to produce the changes outlined above by means of random copolymerisation, it is necessary to choose a comonomer which alters the balance of polarity and restriction of rotation within the polymer chains:

(1) Inertness towards aliphatic solvents can be increased by means of very polar comonomers; and if water absorption is to be kept low, one would have to exclude all those comonomers which can form hydrogen bonds, e.g. acrylic acid, acryloamides, etc. It remains, therefore, that the best comonomer for this purpose is acrylonitrile and fumaronitrile.

(2) Increased toughness through random copolymerisation would be achieved only by the use of 'flexibilising' comonomers, e.g. olefins, diolefins, acrylates, vinyl ethers, vinyl esters, etc. These comonomers, however, would considerably reduce the glass transition temperature, which would in turn result in an increased deformability at high stress levels and at high temperatures. Consequently, random copolymerisation of styrene with these types of comonomers would be useful only in the field of adhesives, surface coatings, rubbers, binders, etc., but not for the production of base polymers for plastics.

(3) A reduction in the time, stress and temperature dependence of deformability parameters, such as compliance, relaxation modulus, etc. (see Chapt. 5), can also be achieved by random copolymerisation with comonomers that would restrict rotations along the C–C backbone bonds; albeit this is likely to occur at the expense of toughness, unless some very-high-molecular-weight species are simultaneously introduced in the structure. Therefore, α-methyl styrene is a suitable comonomer candidate for this purpose, as is also the incorporation of very small quantities of divinyl monomers or di-acrylic and di-methacrylic esters—enough to produce long chain branches and prevent the formation of unmeltable cross-linked species or 'gels'.

Graft Copolymers. The difficulties that arise in trying to improve the toughness of styrene polymers by the expedient of random copolymerisation can be alleviated to a large extent through the production of graft copolymers. In principle, if two incompatible polymers are brought together in the presence of copolymer species grafted on to the chains of the homopolymers, a composite (blend) morphology will result. The minor constituent would form the dispersed phase, which would be separated from the matrix by interlayers of the two polymeric species rich in graft copolymers.

If the dispersed phase consists of very-high-molecular-weight and/or possibly slightly cross-linked rubbery polymers, e.g. containing flexible links in the backbone chains that can undergo rapid relaxations, a considerable increase in toughness will result. The time, temperature and stress dependence of the deformability of such blends will be affected substantially less than in the case of random copolymers.

Block Copolymers. Block copolymerisation offers an alternative route to graft copolymerisation in the production of toughened polystyrene materials. If the two sequences of 'rigid' (styrene) units and 'flexible' units in a block copolymer are sufficiently long and incompatible, phase separation will result.

According to the type of distribution of the blocks—i.e. whether of the A–B, A–B–A or A–B–A–B–A type (where A and B denote sequences of the two comonomers)—and the relative lengths of the two sequences, different morphologies can be produced.

Naturally, the types of units which form the matrix and dispersed phases respectively depend on their relative volumetric fractions. Irrespective of the type of structures formed, there is always a substantial proportion of rubbery material, and the deformability characteristics will vary from those of a 'very soft' type of plastics material to those of a reinforced rubber, according to whether the polystyrene or the rubbery polymer sequences form the matrix.

Consequently, in order to obtain a toughened grade of polystyrene, such a block copolymer would have to be blended with a homopolymer grade of polystyrene to reduce substantially the content of the rubbery phase.

Note that the change in geometry of the dispersed rubber blocks from spherical to cylindrical and then to laminar structure takes place with increasing the proportion of rubber phase and the length of the rubbery polymer sequences in the block copolymer chains.

Terpolymer Blends. When, in addition to increased toughness, other improvements are sought, it is usual to produce random copolymer grafts onto a rubbery polymer and to disperse this into a matrix of the random copolymer free of any such grafts. In principle, however, no technical disadvantages can be anticipated in producing the same characteristics by means of block random terpolymers, e.g. sequences of random copolymer units attached to a chain of 'rubbery' homopolymer units.

For instance, if chemical resistance towards non-polar liquids (e.g. oils) is required, then the copolymer matrix and the copolymer grafts on the rubbery polymer will contain units of styrene and acrylonitrile. If ultraviolet light resistance is to be increased, then methyl methacrylate can replace much of the styrene and acrylonitrile.

It is clear that the rubbery phase may itself consist of a copolymer or a chemically modified homopolymer. Chlorinated polyethylene, for instance offers greater resistance to oils and light stability than diene rubbers, while acrylates and other copolymers can be 'tailored' to match the refractive index of the matrix so that transparent blends will result from the elimination of internal light scattering.

3.1.4 Acrylic Polymers
Within the acrylic family are included those polymers having the general chemical structure

$$-(CH_2-\underset{\underset{COOR^1}{|}}{\overset{\overset{R}{|}}{C}})_n-$$

where R and R^1 are either hydrogen atoms or alkyl groups. Two types of acrylic polymers have achieved commercial importance:

  (i)   acrylates, i.e. polymers where R is hydrogen, and
  (ii) methacrylates, i.e. polymers where R is a methyl group.

(a) Homopolymers
The nature of R and R^1 groups determines the level of the rotational energy barrier about the C–C bonds in the backbone of the molecular chains and the degree of internal plasticisation, i.e. the constraints to molecular packing and the intermolecular forces.

This effect is illustrated in Table 3.3 with respect to density and glass transition temperature of amorphous polymers.

The H bonds between adjacent molecules for the case of polyacrylic acid, polymethacrylic acid and polyacryloamide produce a large increase in T_g. The ability of these polymers to dissolve in water (through formation of H bonds with water molecules) prevents them, however, from being used as normal materials. Consequently only polymethyl methacrylate is of interest for the production of thermoplastics, while polyethyl acrylate and polybutyl acrylates are useful base materials for rubbers.

One of the most attractive properties of acrylic polymers is their resistance to ageing, which results from the ability of these polymers to

Table 3.3: Physical Characteristics of Acrylic Homopolymers[2]

Polymer	Amorphous		Crystalline	
	Density (g/cm^3)	T_g $(°C)$	Density (g/cm^3)	T_m $(°C)$
Polyacrylic acid	—	106	—	—
Polymethyl acrylate	1·22	8	—	—
Polyethyl acrylate	1·12	−22	—	—
Poly n-butyl acrylate	<1·08	−54	—	47
Poly t-butyl acrylate	1·00	43	1·06	190–200
Polymethacrylic acid	—	130	—	200
Polymethyl methacrylate	1·17	105	1·23	170–200
Polyethyl methacrylate	1·12	65	—	—
Poly n-butyl methacrylate	1·06	20	—	—
Poly n-hexyl methacrylate	1·01	−5	—	—
Polyacryloamide	1·30	165	—	—

absorb ultraviolet light in the solar spectrum and to redissipate it as harmless energy in the infrared wavelength range.

(b) Copolymers

Bearing in mind that acrylics are rather expensive, any attempt to alter the properties of acrylic polymers has to be viewed in relation to their competitiveness against cheaper polymers, with similar characteristics, e.g. polystyrene.

The main deficiencies of polymethyl methacrylate are:

(i) the susceptibility to undergoing depolymerisation when exposed to high temperatures,

(ii) the high melt viscosity due to the high chain stiffness caused by restricted rotations about the C bonds in the backbone chains,

(iii) the rather low resistance to creep at temperatures only slightly above room temperature, and

(iv) the poor resistance to solvents, causing crazing and embrittlement.

The only properties that have been improved by copolymerisation, to date, are their susceptibility to depolymerise. This is simply achieved by introducing comonomer units that will undergo chain transfer, rather than unzipping reactions. Acrylates are the preferred comonomers for poly-methyl methacrylate copolymers in view of their intrinsic resistance to thermal and ultraviolet induced oxidation, which are properties of vital commercial importance for these polymers.

It is suspected that most grades of polymethyl methacrylate available commercially contain small amounts of acrylate monomers to minimise the risks of depolymerisation during processing.

For the case of polymethyl methacrylate polymerised *in situ* in products such as sheets or rods, the comonomer is often tetrafunctional in order to introduce some cross-links in the structure, which tends to improve its resistance to crazing and hardness.

(c) Terpolymer Blends

The principle of surface grafting of polymers on cross-linked rubber particles and consequently blending these with a glassy polymer has also been exploited for the case of polymethyl methacrylate.

The earlier systems based on butadiene/styrene rubber lattices, grafted with methylmethacrylate polymer chains were known as MBS. Although transparency could be retained in view of the very similar refractive index of the rubber phase and polymethylmethacrylate matrix, the ultraviolet resistance of these polymer blends is substantially lower than that of methylmethacrylate homopolymers owing to the poor photo-oxidation resistance of the butadiene/styrene rubber. Not surprisingly, therefore, blends have subsequently been based on acrylate rubbers (i.e. ethyl acrylate/butyl acrylate copolymers), possibly containing also rigid inclusions of polymethylmethacrylate, in a manner similar to the ABS and 'impact polystyrene' systems.

3.1.5 Fluorocarbon Polymers

These are essentially polyolefins where hydrogen atoms have been replaced by fluorine atoms.

(a) Homopolymers

It is interesting to consider the effect that such substitution has on the properties of these polymers relative to those of equivalent polyolefins. This comparison is shown in Table 3.4 with respect to melting point, density and tensile yield strength at room temperature. An inspection of these data reveals the following trends:

(i) Higher melting point for fluoropolymers owing to restricted rotations about the C–C bonds along the chains, which results primarily from the repulsions between the highly electronegative fluorine atoms.

Table 3.4: Effect of Fluorine Substitution in Polyolefins

Polymer	Structure	Melting point (°C)	Density (g/cm³)	Yield strength (MPa)
Linear polyethylene	$-(CH_2-CH_2)_n-$	135	0.94	28
Polyvinyl fluoride	$-(CH_2-CHF)_n-$	175	1.4	56
Polyvinylidene fluoride	$-(CH_2-CF_2)_n-$	164	1.75	52
Polytetrafluoroethylene	$-(CF_2-CF_2)_n-$	327	2.2	13
Ethylene–propylene copolymers	$-[(CH_2-CH_2)_x-(CH-CH_2)_y]_n-$ with CH_3	Amorphous	0.92	Not applicable
Hexafluoropropylene–tetrafluoroethylene copolymers	$-[(CF_2-CF_2)_x-(CF-CF_2)_y]_n-$ with CF_3	280	2.2	10

(ii)   When only partial substitution of hydrogen by fluorine atoms is carried out the C–F dipoles produce very strong intermolecular forces, which give rise to very high yield strength values (c.f. polyvinyl and polyvinylidene fluoride).

(iii) With complete substitution of hydrogen atoms (c.f. polytetra-fluoroethylene) there are strong repulsions between fluorine atoms from different chains, which results in low interfacial strength between crystals, causing the material to yield at low stresses.

Furthermore the strong C–F bonds in fluorocarbons produce a very high resistance to thermal and photo-induced oxidation. The resistance to attack by solvents, polar and non-polar, is better than that of any other known polymers. This latter aspect is, however, rather difficult to explain.

(*b*) *Copolymers*

The copolymers that have achieved significant commercial importance within the fluorocarbon polymer family are:

(i)   Random copolymers of tetrafluoroethylene and hexafluoro-propylene (FEP). These were developed to reduce the melting point and melt viscosity so that they could be processed by conventional processing methods, which is not possible for the case of tetra-fluoroethylene homopolymers. Partially responsible for this, however, is the substantial reduction in molecular weight that can be achieved while maintaining a reasonable level of mechanical properties.

(ii)   Ethylene–tetrafluoroethylene alternating copolymers. These co-polymers have melting point, heat ageing and solvent resistance characteristics that are intermediate between those exhibited by polyvinylidene difluoride and polytetrafluoroethylene; while their mechanical properties at ambient and elevated temperatures on the other hand, are superior to those of both polymers.

The comparison of these alternating copolymers (E/TFE) with polyvinyl-idene fluoride (PVF2) is most interesting insofar as it illustrates well the importance of the details of molecular structure, other than stereo-configurations, on properties of polymers. In both cases, for instance, the total number of CH_2 and CF_2 in the chains is the same, but the E/TFE copolymers exhibit greater chain stiffness due to repulsions between neighbouring F atoms in the rotation of CF_2–CF_2 groups. This must be a major factor responsible for the higher melting point and superior mechanical properties of the E/TFE copolymers.

3.2 HETEROCHAIN POLYMERS

Heterochain polymers are those in which the covalent C–C bond sequence in the backbone of the polymer chains is interrupted by groups containing oxygen and/or nitrogen atoms. According to the details of the structure of the heterogroups in the chains, these polymers can be subdivided into: polyethers, polyesters, polyamides and polyurethanes.

The effect of in-chain substitution along the polymer backbone is to alter the chain flexibility and intermolecular forces. Rotations of the bonds in the backbone chains is, in fact, much easier at the C–O and C–N bridges than about C–C bonds. Aromatic groups, on the other hand, are completely rigid and even if they were joined by C–C bonds (as in polyparaphenylene) the melting point would be so high that decomposition would take place before a melt was formed. Consequently, a partial substitution of aliphatic C–C sequences by aromatic groups in the backbone chains produces a considerable increase in overall chain stiffness.

Furthermore, in view of the high electron density in aromatic rings the intermolecular attractions will be increased considerably over those exerted between aliphatic groups.

The effect of various substituents along the backbone chains of polymers on intermolecular forces can be inferred from the value of the molar cohesive energy of each repeating unit, as shown in Table 3.5.

One observes that there are two major increments in the values of the molar cohesive energy in introducing hetero structural units along the polymer chains. One occurs with increasing the concentration of electrons in the structural units, as in the case of ester and phenylene groups, and the other when a hydrogen atom is attached to a strongly electronegative atom as in the case of amides and urethane units. The latter is due to the formation of so-called H bonds between structural units from neighbouring polymer chains.

The possibility of developing strong intermolecular forces with heterochain polymers has been exploited commercially to produce thermoplastics with far superior mechanical strength to the more conventional materials based on homochain polymers.

3.2.1 Polyethers

Polyethers can be conveniently subdivided into aliphatic, aromatic and cellulosic. The effect of changing the nature of the units linked by ether bridges is shown in Table 3.6.

An analysis of these data reveals that there are a large number of

Table 3.5: Molar Cohesive Energy of Structural Units in Homochain and Heterochain Polymers

	Difluoromethylene	Methylene	Ether	Phenylene	Ester	Amide	Urethane
Structural unit	$-CF_2-$	$-CH_2-$	$-O-$		$-\overset{\displaystyle O}{\underset{\displaystyle \parallel}{C}}-O-$	$-\overset{\displaystyle O}{\underset{\displaystyle \parallel}{C}}-NH-$	$-O-\overset{\displaystyle O}{\underset{\displaystyle \parallel}{C}}-NH-$
Energy (kJ mole^{-1})	5·0	4·9	3·3	31·9	18·0	33·5	26·4

Table 3.6: Relationship Between Structure and Transition Temperatures of Polyethers[2,3]

Linear structures

Chemical name	Structure	T_g (°C)	T_m (°C)
Polyethylene (Reference)	$-(CH_2-CH_2)_n$	—	135
Polymethylene oxide	$-(CH_2-O)_n$	—	180
Polyethylene oxide	$-(CH_2-CH_2-O)_n$	—	66
Polytrimethylene oxide	$-(CH_2-CH_2-CH_2-O)_n$	—	35
Poly(diphenylol methane glycidyl ether)	$\left[\text{-O-}\langle\text{C}_6\text{H}_4\rangle\text{-CH}_2\text{-}\langle\text{C}_6\text{H}_4\rangle\text{-O-CH}_2\text{-CH(OH)-CH}_2\right]_n$	80	—
Poly[phenoxy-di-(4-phenylene sulphone)]	$\left[\text{-}\langle\text{C}_6\text{H}_4\rangle\text{-SO}_2\text{-}\langle\text{C}_6\text{H}_4\rangle\text{-O-}\langle\text{C}_6\text{H}_4\rangle\text{-O}\right]_n$	220	—
Poly(2-6-dimethyl-1-4-phenylene oxide)	$\left[\text{-O-}\langle\text{C}_6\text{H}_2(CH_3)_2\rangle\right]_n$	180	—

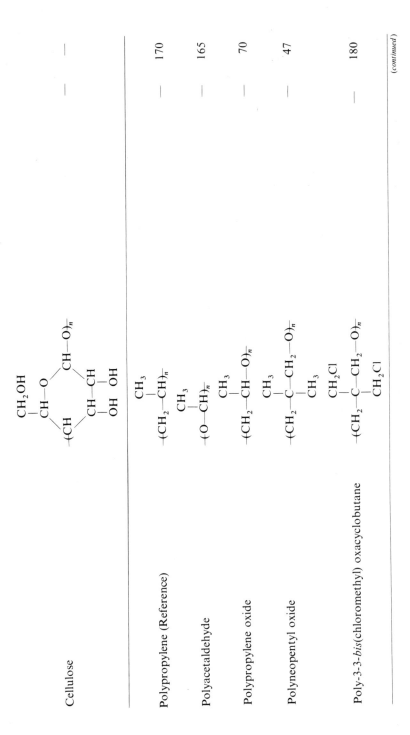

Polymer	Structure		Value
Cellulose	CH$_2$OH, $-$CH$-$O$-$, $-($CH$-$CH$-$O$-)_n$, CH$-$CH, OH OH	—	—
Polypropylene (Reference)	CH$_3$, $-($CH$_2$$-CH)_n$	—	170
Polyacetaldehyde	CH$_3$, $-($O$-$CH$)_n$	—	165
Polypropylene oxide	CH$_3$, $-($CH$_2$$-CH-O)_n$	—	70
Polyneopentyl oxide	CH$_3$, $-($CH$_2$$-C-CH_2$$-O)_n$, CH$_3$	—	47
Poly-3-3-*bis*(chloromethyl) oxacyclobutane	CH$_2$Cl, $-($CH$_2$$-C-CH_2$$-O)_n$, CH$_2$Cl	—	180

(*continued*)

Table 3.6 – contd.

Branched structures

Chemical name	Structure	T_g (°C)	T_m (°C)
Poly(diphenylolpropane glycidyl ether)		100	—
Poly[isopropylidene diphenoxy-di-(4-phenylene sulphone)]		190	—
Poly(2-6-diphenyl-1-4-phenylene oxide)		227	—
Cellulose acetate		50	—

polyethers which, in view of their high glass transition temperature and/or melting point, are attractive for commercial uses. A large number of these, however, have not been made commercially available, either for economic reasons or because of other unattractive features, e.g. solubility in water, susceptibility to degradation during processing, etc.

(a) Aliphatic Polyethers

Of the very large number of polyethers that can be synthesised, only a few have received commercial consideration because of the relatively low cost of the monomers. These are polyformaldehyde (or polyoxymethylene, POM), polyacetaldehyde (or polyoxyethylidene, POE), polyethylene oxide (PEO), polypropylene oxide (PPO), and poly 3-3 *bis*(chloromethyl) oxacyclobutane (PCMO). PEO, PPO and POE cannot be used as base polymers for the production of plastics for the following reasons:

PEO and PPO: Soluble in water and melting point too low.
POE: Depolymerises very rapidly at temperatures below the melting point, i.e. $\simeq 25\%$ weight lost at $110\,^{\circ}C$ within 2 h.

Both POM and PCMO have a high melting point and are mechanically strong. The latter, however, can exist in two crystalline forms (α and β), one of which is particularly susceptible to embrittlement. It would appear that it is rather difficult to control the morphology of PCMO to produce materials with consistent properties.

The oxymethylene polymers (also known as 'acetals'), on the other hand, have achieved enormous commercial success, owing to their very high resistance to creep.

Copolymers containing small amounts of ethylene oxide are also available commercially. These exhibit a slightly lower melting point and mechanical strength than the homopolymers, but are thermally more stable in view of the unzipping termination provided by the ethylene oxide unit during depolymerisation.

(b) Aromatic Polyethers

In producing polymer materials based on aromatic compounds there are three aspects of the molecular structure that have to be balanced in order to achieve the highest glass transition temperature, while maintaining a reasonably low melt viscosity for easy processing. These are respectively molecular weight, chain stiffness and intermolecular forces. A certain minimum molecular weight is necessary to achieve adequate toughness and solvent resistance, while the distribution of aromatic rings and polar flexible groups in the chains affects the glass transition temperature. Normally the chain stiffness and bulkiness of aromatic rings makes it difficult for these polymers to crystallise and consequently they invariably remain amorphous.

Several aromatic polyethers have been commercialised in recent years, all of which exhibit a glass transition temperature in the region of 200 °C.

Examples of commercially available aromatic polyethers and their chemical structures are shown below

(i)   poly 2-6 xylanols:

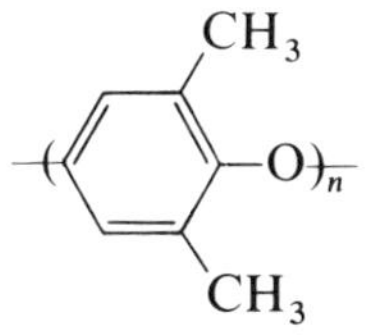

(ii)   polysulphones:

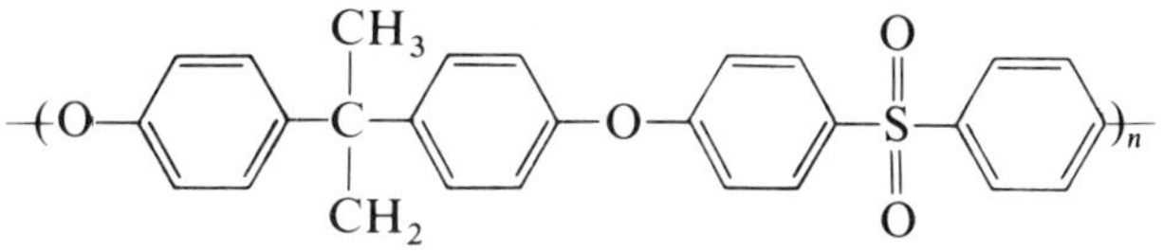

(iii) polyethersulphones:

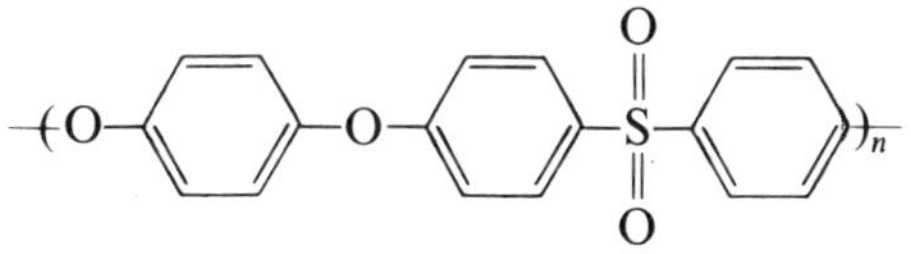

(iv)   polyetherketones:

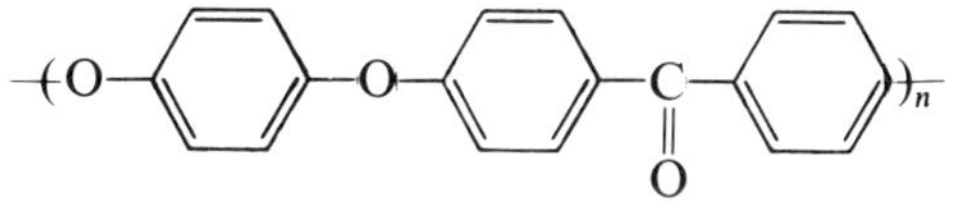

The polyethersulphones and polyetherketones offer the greatest resistance to thermal oxidation at high temperature because of the complete absence of aliphatic units in the chain.

Poly 2-6 xylanols, on the other hand, may only be exposed to high temperatures for short periods of time in view of the decomposition reactions involving hyperconjugation of the two methyl groups and subsequent chain. Hence this polymer is used more frequently for its outstanding toughness than its rigidity and strength at high temperatures. For this reason the tendency nowadays is to use it in blends with cheaper polymers, such as impact polystyrene.

(c) *Cellulosics*

As explained in the first chapter, cellulosics are the oldest family of polymers used for the production of thermoplastics. The most widely used cellulosic polymer in the plastics industry is cellulose acetate. Its structure is well balanced to provide chain flexibility about the ether links and internal plasticisation by the pendant acetate groups.

Combined with the swept free volumes due to the bulky saccharide groups these characteristics confer to the polymer a remarkable toughness.

One of the most severe drawbacks of cellulosics is their poor ageing characteristics, as they tend to hydrolyse and oxidise quite readily.

3.2.2 Polyesters

(a) *Aliphatic Polyesters*

Polyesters are generally produced from a dicarboxylic acid and a glycol. They have the following structure:

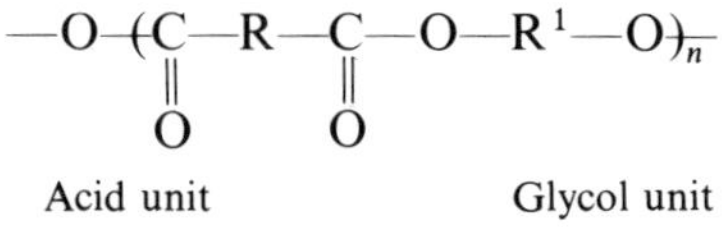

$$-O\underbrace{-(C-R-C-O}_{\substack{\| \\ O \qquad \| \\ O}}-R^1-O)_n-$$

Acid unit   Glycol unit

where R and R^1 are short olefinic chains.

In view of their linear and regular structure aliphatic polyesters are crystalline. Their melting point is determined by the orientation of the dipoles at the ester groups. This is illustrated by the data in Table 3.7 with reference to polyethylene esters.

An odd number of methylene groups in the acid unit of the chain causes a

Table 3.7: **Variation of Melting Point in Aliphatic Polyesters**

Polyethylene ester	No. of CH_2 groups in the acid unit	Melting point ($°C$)
Polyethylene oxalate	0	132
Polyethylene malonate	1	− 20
Polyethylene succinate	2	108
Polyethylene glutarate	3	10
Polyethylene adipate	4	50
Polyethylene pimelate	5	38
Polyethylene subarate	6	70
Polyethylene azeleate	7	50
Polyethylene sebacate	8	79

cancellation of dipoles, which brings about a reduction in melting point. As the number of CH_2 groups between the ester groups becomes increasingly large the dipole cancellation assumes increasingly less importance in relation to the overall intermolecular forces between polymer chains.

Aliphatic polyesters other than those shown in Table 3.7 can be produced using different glycols or from α–ω hydroxy acids. In general, however, their hydrolytic and thermal stability is rather poor while their resistance to oils can be matched by somewhat cheaper polymers, such as PVC and high density polyethylene. Consequently aliphatic polyesters find applications mainly as modifiers or auxiliary ingredients for other polymer-based compositions.

(b) Aromatic Polyesters
As already explained earlier, the introduction of aromatic rings in the chains has the effect of increasing the rotation energy and intermolecular forces of the polymer system. Completely aromatic polyesters, i.e. from condensation of aromatic diacids with aromatic dihydroxy compounds and from hydroxybenzoic acid, have been produced and commercialised. The very high melting point makes it difficult to process them by conventional melt processing techniques, hence they are more widely used as fibres. The polymer chains of these systems are so stiff that when flowing in solution through a capillary they orient themselves like rods in the flow direction and retain this conformation after removal of the solvent. Consequently they have an extremely high stiffness and strength and are, therefore, used quite widely as fibres for composites (see Chapt. 5). Amorphous aromatic polyesters, for general moulding and extrusion applications, have recently been introduced on the market. These are produced from isophthalic acid and bisphenol A. Much more important in the thermoplastics area are the mixed aliphatic/aromatic polyesters. By altering the nature of the diacid and the glycol the chain regularity and intermolecular forces can be changed considerably so that a range of properties can be achieved, ranging from high-T_g amorphous polymers to high-T_m crystalline polymers. The effect of altering chain regularity and length of aliphatic segments on the melting point is shown in Table 3.8. Of the polymers listed in the table only the ethylene and tetramethylene terephthalates have been introduced on the market.

A number of amorphous copolymers obtained from reacting glycols with a mixture of isophthalic and terephthalic acid are also available commercially for film applications.

Isophthalic and terephthalic acids are used for the production of these

Table 3.8

Glycol	Acid		
	Ortho-phthalic acid	Iso-phthalic acid	Tere-phthalic acid
Ethylene	64 °C	105 °C	256 °C
Trimethylene	—	95 °C	218 °C
Tetramethylene	18 °C	90 °C	222 °C
Pentamethylene	6–9 °C	80 °C	165 °C

polymers, despite their higher cost than the o-phthalic equivalent, because they confer a greater thermal and hydrolytic stability to the polymer.

A major drawback of the ethylene and tetramethylene terephthalate polymers is their brittleness. They retain, however, their stiffness and strength characteristics up to high temperatures. For this reason they are widely used in glass fibre reinforced grades; often blended with other polymers to improve their toughness. The best balance between retention of properties at high temperatures and toughness down to quite low temperatures has been achieved with poly diphenylol-propane carbonate, normally known simply as polycarbonates. These have the following structure

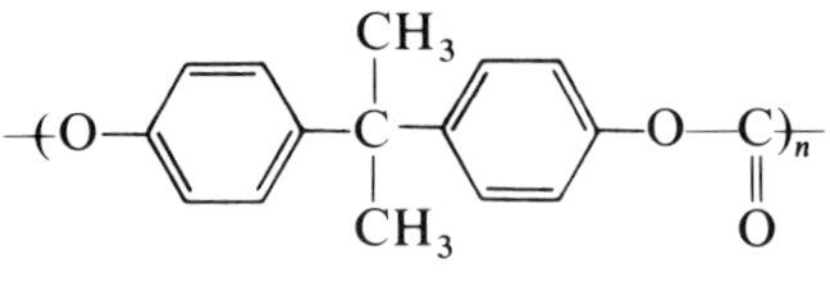

and exhibit a glass transition temperature of about 150 °C and a melting point of 257 °C.

This very high T_g value ensures that supercooling to temperatures well within the glassy state of the material is easily achieved during processing. In this way polymer chains do not achieve sufficient mobility to enable the nuclei to develop into full crystals, and consequently the material remains amorphous. As with all polyesters, however, these polymers suffer from hydrolytic decomposition during processing and on ageing at elevated temperatures.

Still in the field of mixed aromatic/aliphatic polyesters, a range of block copolymers have been introduced in recent years. These contain blocks of rubbery aliphatic polyethers (e.g. polytetramethylene glycol) and polyethylene (or tetramethylene) terephthalate. The latter blocks form fine crystallites imbedded in the rubbery polyether matrix of the adjacent blocks, hence producing self-reinforced thermoplastic rubbers.

3.2.3 Polyamides

(a) *Aliphatic Polyamides*
For historical reasons aliphatic polyamides are generally known as 'nylons'. There are two types of polyamides, i.e. those produced from α–ω amino acids and those obtained from condensation reactions between a diacid and a diamine.

The technological nomenclature for polyamides is to add to the term nylon the digit corresponding to the number of carbon atoms between the nitrogen atoms of the repeating unit. For α–ω amino acid polyamides the repeating unit is:

$$-[\underset{\underset{O}{\parallel}}{C}-(CH_2)_x-NH]_n-$$

and, therefore, the polymer will be known as Nylon $x + 1$. For polyamides produced from dicarboxylic acids and diamines the repeating unit of the polymer chain is

$$-[NH-(CH_2)_x-NH-\underset{\underset{O}{\parallel}}{C}-(CH_2)_y-\underset{\underset{O}{\parallel}}{C}]_n-$$

and the polymer will be known as Nylon $x, y + 2$.

In view of the linear and regular structure, strong molecular interactions are developed through H bonds acting between the NH group of one molecule and the C–O group of an adjacent chain. The number of H bonds that can be found for a given chain length (i.e. the chain fold length of a crystal) depends on the number of CH_2 groups separating the amide groups and on whether the number of CH_2 groups is even or odd. The manner in which this affects the melting point of the polymer is shown in Table 3.9. For the case of Nylon $x + 1$ types there will be one H bond for every repeating unit when $x + 1$ is an odd number and one H bond for every two repeating units when $x + 1$ is an even number. The opposite is true for the case of Nylon $x, y + 2$ types.

The melting point decreases when the regularity of the sequence of methylene groups in the repeating units of the chains is interrupted, as for the case of mixed polyamides (or copolyamides).

Table 3.9: Melting Point of Polyamides

Nylon x, y + 2	Melting point (°C)	Nylon x + 1	Melting point (°C)
Nylon 4, 6	278	Nylon 6†	225
6, 6†	265	7	235
8, 6	235	8	195
7, 7	205	9	210
8, 8	215	10	178
9, 9	177	11†	190
10, 10	206	12†	175

† See text.

Although there are a large number of polyamides that exhibit excellent properties, only those marked † in Table 3.9 have been commercialised in the plastics industry. The main reason for this is the availability of raw materials and their competitiveness with other engineering polymers. In the field of adhesives and fabric coatings, on the other hand, the majority of polyamides used are of mixed system so that the level of crystallinity is reduced to a minimum, hence achieving greater softness. The main attraction of polyamides is their excellent resistance to mineral oils and their good balance of toughness and strength. Because of their high melting point they exhibit good mechanical properties up to temperatures in the region of 120–150 °C.

(b) Aliphatic/Aromatic Polyamides
Although higher melting points can be achieved by replacing either the diamine or the dicarboxylic acid with aromatic types only a few of these polymers are available commercially in view of the difficulty of controlling the decomposition reactions during processing.

When, on the other hand, branches are introduced in the aliphatic portion of the chains or the aromatic units are introduced as partial substitution of the aliphatic components, so that intermolecular distances are increased and the chain regularity is destroyed, the resulting polymer will not be capable of crystallising. Hence processing can be carried out at lower temperatures and the decomposition reactions can be more easily controlled.

Polymers of this type are available commercially and are competitive with aromatic polyethers and polyesters on account of their better resistance to solvents.

(c) *Aromatic Polyamides*

When both the acid and amine components of a polyamide are aromatic the resulting polymers exhibit extremely high melting points, up to 500 °C. When the amide groups are attached to aromatic rings in asymmetric positions the chains will not be linear and, therefore, will not crystallise. The excessive chain stiffness conferred by the aromatic rings, however, does not enable these polymers to be processed by conventional melt processing methods. The amorphous nature of these polymers, on the other hand, renders them soluble in a number of solvents so that they can be spun as fibres or cast into films.

The absence of aliphatic units in the chains makes it possible for these polymers to be used up to quite high temperatures without undergoing significant decomposition.

(d) *Aromatic Polyimides*

Further improvements in high temperature properties are achieved by joining the aromatic rings by cyclic bonds. One of the most successful systems that utilises this concept is to be found among the polyimide family. Polyimides have the following structure.

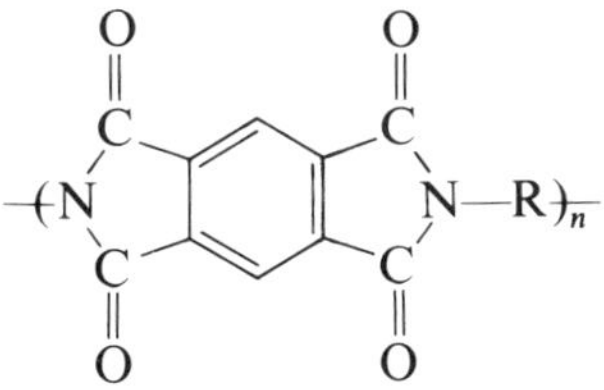

Fully aromatic systems, i.e. when R is an aromatic ring, are infusible and can, therefore, only be processed in solution to produce films. Note, in fact that the lack of possibility of H bonding in these systems renders these polymers more soluble than aromatic polyamides.

More recently thermoplastic polyimides have been introduced on the market. Sufficient chain flexibility, to make them processable in the melt state, is achieved by the introduction of ether groups or by substituting some of the aromatic units with aliphatic ones.

3.2.4 Polyurethanes

The term polyurethane is used to denote a class of polymers containing the groups $-NH-C-O-$ in the backbone of the molecular chains.
$$\underset{O}{\overset{\|}{}}$$

Unlike the rest of the polymers described so far, polyurethanes always contain other hetero groups in the chains and can all be described as mixed aliphatic/aromatic systems.

The nature and the distribution of the chain components can be changed in a variety of ways to produce a wide range of properties.

The polyurethanes that are relevant to the contents of this chapter are the thermoplastic rubber systems. The structure of these polymers is arranged in alternating blocks of rigid and flexible segments. Within the rigid blocks there are urethane groups and aromatic rings, which can form strong associations with similar blocks from adjacent chains. These form glassy domains finely distributed within a rubbery phase of flexible units, giving rise to typical properties of reinforced rubbers. At high temperatures the strong associations, mostly H bonds, are broken down to produce a low viscosity homogeneous melt.

In all thermoplastic polyurethanes the rigid blocks are made up of 4,4'-diphenylene methane–tetramethylene urethane, i.e.

$$\begin{array}{c} -\!\!\left[O-\!\!\left(CH_2\right)_4\!-\!O-\!\!\underset{\underset{O}{\parallel}}{C}\!-\!NH\!-\!\!\left\langle\bigcirc\right\rangle\!-\!CH_2\!-\!\!\left\langle\bigcirc\right\rangle\!-\!NH\!-\!\!\underset{\underset{O}{\parallel}}{C}\right]_n \end{array}$$

where n varies in the range 1–4.

These are present in amounts of approximately 30–40%. The main difference between the various thermoplastic polyurethanes available commercially is with respect to the nature of the flexible units in the chains. These are normally either polyesters (i.e. polytetramethylene adipates or polycaprolactones) or polyethers (mostly polytetramethylene oxide), with molecular weight in the region of 800–2000. Because of the highly polar nature of these polymers they exhibit a strong resistance to oils, while the strong intermolecular forces in the hard domains and the fact that these are finely dispersed through the rubbery matrix, are largely responsible for their excellent abrasion resistance.

REFERENCES

1. C. B. BUCKNALL, *Toughened Plastics*, Applied Science, London, 1977, p. 88.
2. D. W. VAN KREVELEN, *Properties of Polymers*, Elsevier, Amsterdam, 1976, pp. 574–80.
3. D. W. AUBREY, *Trans. Plast. Inst.*, **34** (1966) 291–301.

4

Additives, Blends and Composites

Since the very early stages of the development of the plastics industry, it was realised that better products could be obtained if the base polymer were mixed with additives and/or fillers. More recently different polymers have also been mixed together to obtain what are generally known as '*polymer blends*' or '*alloys*'. Within the context of thermoplastics materials technology the term additive is used to denote an auxiliary ingredient of a composition that enhances the properties of the parent polymer without appreciably altering its chemical structure. Fillers are inorganic powders, platelets or short fibres that are included into the polymer composition to increase its rigidity, strength, and hardness or to reduce the coefficient of thermal expansion. Quite often the filler is substantially cheaper than the polymer and, therefore, it is used primarily to reduce the overall cost of the composition.

Polymer alloys or blends, on the other hand, are produced primarily to enhance the toughness of the composition. Improvements in solvent resistance can also be achieved by blending polymers containing chemical groups that are substantially different in polarity.

4.1 CLASSIFICATION AND ROLE OF ADDITIVES

Additives can be conveniently classified according to their function into various classes and subdivided according to their more specific function.

This classification is shown below:[1]

Main classification	*Subdivision*
1. Processing additives	a. Processing stabilisers
	b. Lubricants
	c. Viscosity depressants
	d. Fusion promoters
2. Flexibilisers	a. Plasticisers
3. Anti-ageing additives	a. Antioxidants
	b. Ultraviolet stabilisers
4. Surface properties modifiers	a. Antistatic agents
	b. Antiblocking additives
5. Optical properties modifiers	a. Pigments and dyes
	b. Nucleating agents
6. Fire retardants	a. Ignition inhibitors
	b. Self-extinguishing additives
	c. Smoke suppressants
7. Foaming additives	a. Blowing agents

According to their mode of action in the polymer, additives can be divided also into *compatible*, *partially compatible* and *incompatible* classes.

The compatibility of additives is determined through thermodynamic consideration of the heat of mixing. Assuming that there is no volumetric change the molar Gibbs free energy (ΔG^m) is given by

$$\Delta G^m = \Delta H^m - T \Delta S^m \tag{4.1}$$

where ΔH^m and ΔS^m are the molar enthalpy and entropy respectively.

In general ΔS^m is positive since mixing involves an increase in disorder, which results from the increase in both configurational entropy and interaction entropy (i.e. that part of the entropy that is related to the freedom of the additive to form associations with itself and with the polymer molecules).[2]

Mixing occurs spontaneously when the heat of mixing is negative (i.e. exothermic), which occurs if strong bonds, such as H bonds, are formed between the additive and groups in the polymer chains. In this way a positive ΔS^m and a negative ΔH^m can only result in a decrease in free energy, i.e. the drive towards a stable system or equilibrium.

When H-bonding possibilities between the polymer and the additive are absent the compatibility is determined by the difference in solubility parameters (δ) of the two components, defined as the square root of the molar cohesive energy density (CED). The CED represents the amount of energy that is required to hold the molecules of the system together. For the case of a liquid the CED is simply the difference between the molar heat of vaporisation and the molar work of expansion, i.e.

$$CED = \frac{\Delta H^{vap} - RT}{V_m} \qquad (4.2)$$

Where V_m is the molar volume of the liquid.

The relationship between the difference in solubility parameters and the associated heat of mixing has been derived by Hildebrand[3] and is given by:

$$\Delta H^m = \varphi_1 \varphi_2 V(\delta_1 - \delta_2)^2 \qquad (4.3)$$

where V is the total volume of the mixture, φ_1 and φ_2 are the molar volume fractions and the subscripts 1 and 2 refer to the polymer and additive respectively.

Because ΔH cannot be negative on the basis of eqn. (4.3) the only way in which there can be a decrease in free energy is for ΔH to be numerically smaller than $T \Delta S$, or preferably zero.

This means, therefore, that the way to achieve compatibility is to match precisely the solubility parameter of the additive to that of the polymer. Because the product $\varphi_1 \varphi_2$ goes through a maximum when $\varphi_1 = \varphi_2$, eqn. (4.3) predicts also that for a given $(\delta_1 - \delta_2)$ value the greatest chance for an additive to be compatible with a polymer is when it is present at low concentrations. This prediction is in agreement with what is observed in practice insofar as many additives are only compatible at extremely low concentrations and will precipitate out into a separate phase when a certain critical concentration is exceeded, i.e. the additive is said to be partially compatible with the polymer.

The Hildebrand equation suffers from severe limitations in predicting the increase in compatibility with temperature. Since ΔH does not change with temperature one has to consider how the product $T \Delta S$ varies with temperature.

Furthermore, one cannot neglect the work of expansion when the additive is present in substantial amounts. It would seem for instance, that a difference in coefficient of volumetric expansion of as little as 4 % in the case of polymer blends is sufficient to cause phase separation if the molecular weight is sufficiently high.[4]

4.1.1 Processing Additives

The output from a processing equipment and the quality of the product are determined primarily by the resistance of the polymer to thermal degradation, the speed with which a solid feed can be converted into a melt and the frictional behaviour of the melt on the surface of the processing equipment.

High thermal stability in polymers enables processing to be carried out at high temperatures and to achieve higher outputs through a reduction in melt viscosity. It also improves product quality as a result of the elimination of the deleterious effects of degradation on physical properties and appearance. Fast fusion of particles is essentially to ensure that a homogeneous melt is produced before it is resolidified into the required shape. Finally, the frictional behaviour of the polymer melt on metal surfaces determines the residence time of polymer molecules in the processing equipment and, in turn, production rates and the likelihood of degradation.

(a) Processing Stabilisers

The degradation of polymers frequently involves oxidation reactions by a free radical mechanism and occurs in three consecutive steps:

(1)   The initiation step is responsible for the generation of radicals (the active species).

(2)   The propagation step consists of a series of fast chain reactions between radicals and oxygen and further generation of radicals by interactions with polymer molecules.

(3)   The termination step is the process by which the free radicals are deactivated (or quenched) through interactions with each other.

By far the most damaging species in high temperature degradation of polymers are the hydroperoxide groups, formed by the interaction of oxygen with C–H groups. These will decompose into very reactive $\cdot OH$ radicals and lead to molecular scissions.

For the case of polypropylene, for instance, degradation reactions occur as shown below:

(i)
$$\sim\!CH_2\!-\!\underset{\underset{}{|}}{\overset{\overset{CH_3}{|}}{CH}}\!-\!CH_2\!\sim + O_2 \longrightarrow \sim\!CH_2\!-\!\underset{\underset{OOH}{|}}{\overset{\overset{CH_3}{|}}{C}}\!-\!CH_2\!\sim$$

(ii)
$$\sim\!CH_2\!-\!\underset{\underset{OOH}{|}}{\overset{\overset{CH_3}{|}}{C}}\!-\!CH_2\!\sim \longrightarrow \sim\!CH_2\!-\!\underset{\underset{O\cdot\, +\, \cdot OH}{|}}{\overset{\overset{CH_3}{|}}{CH}}\!-\!CH_2\!\sim$$

$$\text{(iii)} \qquad \overset{\displaystyle CH_3}{\underset{\displaystyle O\cdot}{\sim\!\!\sim CH_2-\overset{|}{\underset{|}{C}}-CH_2\!\!\sim\!\!\sim}} \longrightarrow \overset{\displaystyle CH_3}{\underset{\displaystyle O}{\sim\!\!\sim CH_2-\overset{|}{\underset{\parallel}{C}}+\cdot CH_2\!\!\sim\!\!\sim}}$$

the $\sim\!\!\sim CH_2\cdot$ radicals can react with oxygen to give $\sim\!\!\sim CH_2\!-\!OO\cdot$, which will then abstract hydrogen from another molecule to produce more hydroperoxides and free radicals.

From the above it is obvious that, since it is practically impossible to eliminate oxygen from the system, additives have to be used to interfere with oxidation reactions.

This is accomplished by mixing, with the polymer, one or a combination of the following additives:

(i) Primary stabilisers or antioxidants. These interrupt the chain reactions by combining with the propagating free radicals, thus forming non-reactive species. Typical examples of antioxidants are hindered phenols and aromatic amines.

(ii) Secondary stabilisers or peroxide decomposers. These function by reacting with hydroperoxides as they are formed, converting them into non-radicals (inactive) species. The most common type of additives that act as peroxide decomposers are organic thio-esters, phosphites and metal thiocarbamates.

(iii) Chelating agents or metal deactivators. These protect the polymer against degradation by immobilising metal ions through co-ordination reactions. In other words they produce a cage effect around the metal ions so that they cannot interact with hydroperoxides. The most important chelating agents used for stabilisation purposes are compounds based on organic phosphites, oxamides and hydrazides.

For the case of vinyl polymers, the stabilisers must exert an additional function, i.e. they must neutralise the acidic HX species to overcome their catalytic action on the process of degradation (see Section 3.1.2(c)).

These additives are known as acid absorbers and can be either inorganic (e.g. lead, magnesium, zinc oxides and soaps) or organic compounds (e.g. dibutyl tin maleates).

Addition reactions with the double bonds formed along the molecular chains are also effective in retarding propagation reactions by preventing the formation of allyl groups ($-CH_2-CH\!=\!CH-$) which are highly reactive with oxygen. Mercaptides and maleate esters are examples of compounds that can react at the double bonds in vinyl polymers. Substitution of labile chlorine by means of metal carboxylates, mercaptides and organotin compounds is another effective way of retarding dehydro-chlorination reactions.

Finally for the case of condensation polymers such as polyesters, polyurethanes and, to a lesser extent, polyamides, additional decom-position can occur through hydrolytic reactions at the functional groups in the chains, producing scission of the chains at several points.

Although pre-drying the polymer before processing helps considerably in reducing the extent of hydrolytic reactions, it is almost impossible to remove completely the water from the system. Furthermore the carboxylic acid end groups of the chains act catalytically in the hydrolysis and, as more and more of these are formed, an autoaccelerating process for the chain scission reactions takes place.

Hence a stabiliser against hydrolysis must be capable of reacting with both water and carboxylic acids. The compounds capable of undergoing these reactions are carbodiimides. These are widely used for this purpose and are normally supplied in polymeric form to reduce health hazards resulting from their volatility.

Examples of typical processing stabilisers available commercially are shown in Table 4.1.

(b) Lubricants
There are three major friction problems encountered in polymer pro-cessing:

(i)   Interparticle friction between polymer powders or granules, which impairs their 'free-flowing' characteristics, hence creating diffi-culties in conveying operations, such as feeding into processing equipment.
(ii) Friction between the polymer melt and the metal surfaces of the processing equipment, which can seriously impair the flow of the melt and produce undesirable effects, such as low output and poor surface quality of finished product.
(iii) Friction in finishing operations, e.g. printing and packaging.

To alleviate the above difficulties two methods are available at present: *'solid layer lubrication'* for powder particles, and *'boundary lubrication'* for the melt/metal and solid polymer/metal interfaces.[5]

External lubricants used to aid processing are similar to those used for conventional boundary lubrication of metal surfaces, i.e. high-molecular-weight fatty acids, alcohols, amines, amides and metal soaps containing 12–18 C atoms in the backbone chain. More recently waxes based on high molecular weight hydrocarbons (MW = 500–5000) have also been used.

The mechanism of melt lubrication is similar to that operating in conventional boundary lubrication, i.e. the polar groups of the lubricant molecules face the surface of the metal and form a stationary layer by virtue of strong absorptive forces (Fig. 4.1). This mechanism applies to all polymer systems irrespective of whether they are highly polar or non-polar. More important than the final details of the structure of the lubricant, on the other hand, is the nature of the metal surface as it determines the intensity of the absorptive forces which it exerts towards the polar groups of the lubricant molecules, and therefore its ability to form permanent

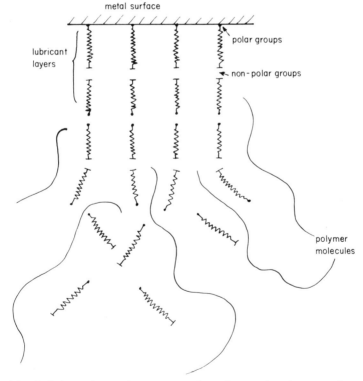

Fig. 4.1. Lubricant layers between metal surface and polymer melt. (After L. Mascia, *The Role of Additives in Plastics*, Edward Arnold, 1974.)

Table 4.1: Typical Processing Stabilisers

Chemical name		Structure
A. *Antioxidants*		
(a)	1,1,3-tris(2 methyl-4 hydroxy-5-*tert* butyl phenyl) butane	(a)
(b)	Tetra, tris methylene 3-(3′,5′-di-*tert* butyl-4′ hydroxy phenyl) propionate methane	(b)
(c)	0,*o*-di-*n*-octadecyl-3,5-di-*tert* butyl-4-hydroxybenzyl phosphonate	(c)

B. *Hydroperoxide decomposers*
(a) Dilaurylthio-dipropionate

(a) $S(CH_2CH_2-\overset{\displaystyle O}{\overset{\|}{C}}-O-C_{12}H_{25})_2$

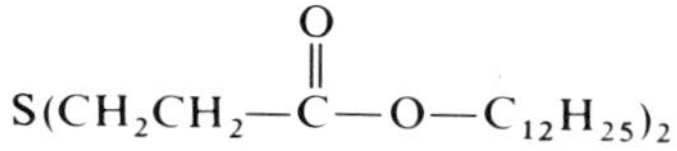

(b) Tris(nonyl phenyl) phosphite

(b) $P(O-C_9H_{19})_3$

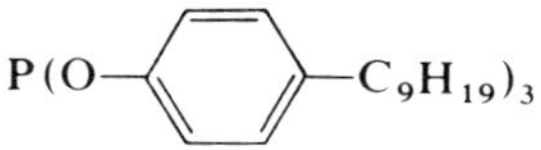

(c) Zinc dithiocarbamate

(c) $R_2N\overset{S}{\diagdown}\overset{S}{\diagup}M\overset{S}{\diagup}\overset{S}{\diagdown}CNR_2$

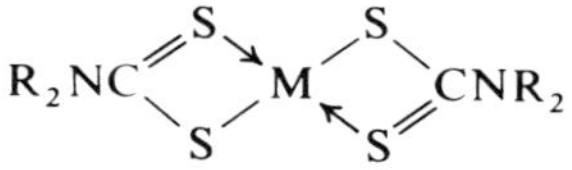

C. *Acid absorbers* (for halogenated polymers)
(a) Basic lead carbonate
(b) Co-precipitated cadmium/barium laurates
(c) Dibutyl tin, *bis*(lauryl) mercaptide

(a) $PbO \cdot PbCO_3$

(b) $\tfrac{1}{2}Cd\tfrac{1}{2}Ba(OCOC_{11}H_{23})_2$

(c) $(C_4H_9)_2Sn(SC_{12}H_{25})_2$

D. *Chelating agents*
(a) N,N'diphenyl oxamide

(a) $C_6H_5-NH-\overset{\displaystyle O}{\overset{\|}{C}}-\overset{\displaystyle O}{\overset{\|}{C}}-NH-C_6H_5$

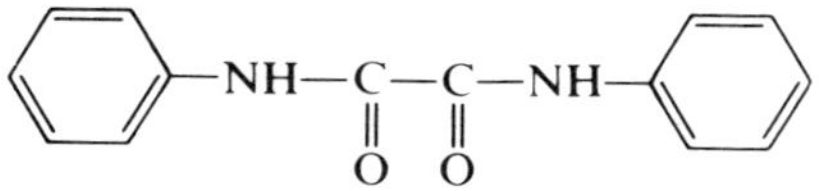

(b) *m* nitrobenzohydrazide

(b) $O_2N-C_6H_4-\overset{\displaystyle O}{\overset{\|}{C}}-NH-NH_2$

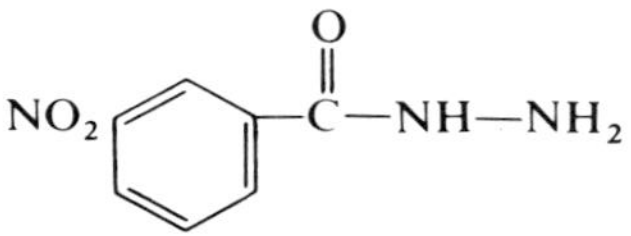

(c) 3 Amino,1,2,4 triazole

(c) benzotriazole/triazole ring structure

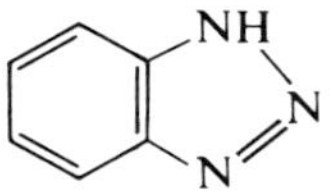

boundary layers. As already explained the nature of the metal surface can determine whether chemical reactions or physical absorption can take place at the lubricant/metal interface. When the lubricant is an acid, lauric acid for instance, it will react with metals such as copper, cadmium and zinc but not with silver, aluminium, nickel, chromium, etc. Bowden and Tabor[6] found that the coefficient of friction between metal and lubricant is considerably lower when chemical reactions take place. It is understood, however, that because the lubricant is embedded in the melt during compounding or in the early stages of the melt mixing in processing, its effectiveness depends on:

(i)   its lack of compatibility with the polymer melt so that the boundary layer at the interface is not re-dissolved by the polymer;

(ii) its diffusibility in the melt so that it can easily migrate to the interface.

The compatibility of lubricants with polymer melts is generally low and consequently the formation of a boundary layer can be achieved with very small levels of incorporation.

Examples of external lubricants are shown in Table 4.2.

(c) *Viscosity Depressants and Fusion Promoters*
Improvements in the processability of polymers can be brought about by lowering the temperature at which the polymer particles or granules will fuse together during processing and by reducing the viscosity of the melt. The ideal manner to achieve this is to use low-molecular-weight polymers whose solubility parameter is the same as that of the base polymer. A polymeric additive would not affect appreciably the T_g of the polymer but

Table 4.2: Typical External Lubricants

Lubricant type	Chemical name	Polymer
Hydrocarbon waxes	Low MW polyethylene (MW 500–2000)	Most polymers
Ester waxes	Octyl stearate, stearyl stearate, glyceryl stearate	Polyvinyl chloride
Metal soaps	Zinc stearate, calcium stearate, magnesium stearate	Most polymers
Amide waxes	Ethylene *bis* stearamide, oleamide, stearamide	Polyolefins
Silicone oils	Low MW polydimethyl siloxane	Most polymers

would simply cause a broadening of the transition and produce a concomitant increase in the deformability of the polymer at temperatures above T_g. For the majority of the high tonnage polymers, nowadays, this is achieved by producing polymers with broad molecular-weight distributions, so that the low-molecular-weight tails act as viscosity depressants.

Frequently small reductions in melt viscosity are also brought about by means of lubricants exhibiting compatibility limits greater than those displayed by conventional external lubricants, so that phase separation occurs at somewhat higher levels of addition.

In this way, the compatible amount of lubricant, known also as the 'internal lubricant' functions as fusion promoter and viscosity depressant.

For the case of PVC, sometimes a small amount of plasticiser is used to improve processability but care has to be exercised in order to avoid the onset of antiplasticisation effects (see later). To avoid excessive softening of the polymer a completely different approach is often used in PVC. Very-high-molecular-weight acrylic polymer powders are used instead so that the PVC particles are immobilised by the acrylic particles, allowing the shear stresses acting on them to effectively deform them and destroy their particular geometry.

4.1.2 Plasticisers

In order to reduce the stiffness of a polymer by external means it is necessary to reduce the intermolecular forces and increase the free volume of the system. The latter creates 'empty holes' for the free rotation of molecular segments. These effects can be produced by means of low-molecular-weight compounds that are completely compatible with the polymer, known as 'plasticisers'. An expression for the effect of plasticisation on the free volumes of the system can be simply cast in the following terms.[7]

$$f = \phi_1 f_1 + (1 - \phi_1)f_2 + k_v^1 \phi_1 \phi_2 \qquad (4.4)$$

where f = total fractional free volume of the system, ϕ_1 = volumetric fraction for the plasticiser, f_1 and f_2 = fractional free volumes for the plasticiser and the polymer respectively, and k_v^1 = interaction coefficient responsible for the deviation from the additivity rule ($k_v^1 \simeq 10^{-2}$). This equation is in accordance with experimental evidence that the total specific volume $(1/\rho)$ is related to the weight fractions of the two components w_1 and w_2 through the relationship[8]

$$\frac{1}{\rho} = \frac{w_1}{\rho_1} + \frac{1 - w_2}{\rho_2} + k_v w_1 w_2 \qquad (4.5)$$

Since the effect of adding a plasticiser can be considered to be equivalent to increasing the temperature, one can expect that the change in glass transition temperature can be calculated from eqn. (4.4).

By assuming a linear dependance of the fractional volume on temperature, one obtains[7]

$$T_g = \frac{\phi_1 \alpha_1 T_{g_1} + (1 - \phi_1)\alpha_2 T_{g_2} + k_v^1 \phi_1 \phi_2}{\phi_1 \alpha_1 + (1 - \phi_1)\alpha_2} \qquad (4.6)$$

where α_1 and α_2 are the respective coefficients of expansion, and T_{g_1} and T_{g_2} the glass transition temperatures of the plasticiser and polymer respectively.

The efficiency of a plasticiser is often measured in terms of the decrease in T_g at a given level of addition. Particularly effective in this respect are those plasticisers capable of screening the dipoles of the polymer molecules. This is accomplished when the structure of the plasticiser molecules is such that the right length of aliphatic units can be located between polar groups.

It is to be expected, however, that increasing the amounts of dipolar screening units in the plasticiser molecule will bring about a reduction in solubility parameter, causing the $(\delta_1 - \delta_2)$ factor to increase and deviate from the ideal conditions for compatibility.

The plasticiser, however, not only affects the T_g of the polymer system but also the characteristics of the polymer in both the glassy state, the rubbery state and the melt state. (The latter will be treated in Chapt. 7.)

Below the glass transition the plasticiser can form strong association with the polar groups of the polymer and, because of the small size of their molecules, they will produce an increase in the overall packing density. In other words the density can actually increase to values above those predicted by the additivity rule. This means that eqn. (4.6) is only applicable, therefore, for temperatures above T_g, or that the interaction coefficient k_v undergoes a sharp increase at the glass transition temperature, remaining constant in the temperature range corresponding to the rubbery state of the material.

The extent of densification, however, depends not only on the nature and concentration of the plasticiser but also on the thermal history of the composition (see physical ageing phenomena in Chapt. 5).

Increasing the concentration of plasticiser creates conditions in which the internal mobility can outweigh the packing effect around the polar groups, so that the density and refractive index will go through a maximum.[9] As is to be expected, the concentration at which this occurs is inversely related to the plasticising efficiency of the plasticiser.

Because the densification below T_g produces effects that are opposite to those expected by 'normal' plasticisation, the phenomenon is often known as '*antiplasticisation*'.[10]

A graph of the modulus as function of the temperature for a system exhibiting typical antiplasticisation below T_g and normal plasticisation above T_g is shown in Fig. 4.2. The qualitative relationship between

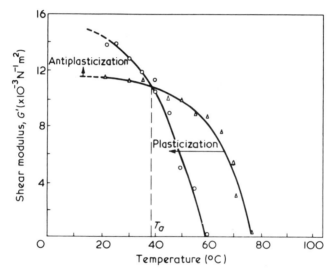

Fig. 4.2. Antiplasticisation/plasticisation transition in dynamic mechanical tests. Δ, Unplasticised PVC; ○, PVC + 10 phr tricresyl phosphate. (After L. Mascia, *Polymer*, **19** (1978) 326.)

plasticisation efficiency and critical concentration for several plasticisers for polyvinyl chloride (PVC) is shown in Fig. 4.3.

A list of typical plasticisers used for the most common thermoplastics materials is shown in Table 4.3.

For the case of crystalline polymers, plasticisation is more difficult since only the amorphous regions can readily accept the plasticiser. The inclusion of plasticiser molecules in the crystalline regions of the polymer would require a very close match of units of the two components. This happens primarily if the plasticiser is a low-molecular-weight fraction of the same polymer.

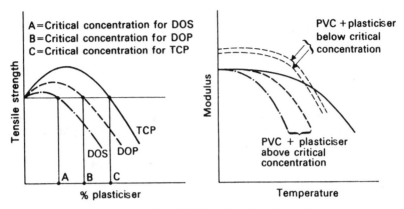

Fig. 4.3. Mechanical properties of PVC as a function of plasticiser concentration. (After L. Mascia, *The Role of Additives in Plastics*, Edward Arnold, 1974, p. 62.)

4.1.3 Anti-ageing Additives

Ageing may be defined as the process of deterioration of materials resulting from the combined effects of atmospheric radiation, temperature, oxygen, water, micro-organisms and other atmospheric agents (e.g. gases).

Although it could be argued that deterioration and failure of materials can result from the action of other external agents, such as mechanical stresses, etc., it is customary to use the term ageing to indicate that a chemical modification in the structure of the material has occurred. The manner in which each of the above factors causes deterioration of plastics is

Table 4.3: Examples of Plasticisers Used in Thermoplastics[10]

Plasticiser	Polymer
Di-octyl phthalate (DOP)	Polyvinyl chloride and copolymers
Tricresyl phosphate (TCP)	Polyvinyl chloride and copolymers
Di-octyl adipate (DOA)	Polyvinyl chloride, cellulose acetate butyrate
Di-octyl sebacate (DOS)	Polyvinyl chloride, cellulose acetate butyrate
Adipic acid polyesters (MW = 1500–3000)	Polyvinyl chloride
Sebacic acid polyesters (MW = 1500–3000)	Polyvinyl chloride
Chlorinated paraffins (%Cl = 40–70) (MW = 600–1000)	Most polymers
Bi- and terphenyls (also hydrogenated)	Aromatic polyesters
N-ethyl o,p-toluene sulphonamide	Polyamides

determined by the chemical nature of the polymer and other basic constituents. The final effect on properties, on the other hand, is invariably the same, i.e. it produces discolourations, embrittlement and increased dielectric losses.

Note, however, that stiffening and embrittlement of thermoplastics with time can also result from densification phenomena. The latter effect is often known as '*physical ageing*' and will be treated separately in Chapt. 5.

The additives that are normally used to reduce chemical decomposition of polymers caused by atmospheric agents are antioxidants and ultraviolet light (UV) stabilisers.

The most damaging environmental agent in the atmosphere is UV light. Although only 5% of the total solar energy (i.e. that portion in the wavelength range 290–400 nm) falls on earth, there are many polymers that can absorb this energy. Absorption of UV light, however, is not a sufficient condition for polymer decomposition in so far as much of this energy can be re-emitted in the form of harmless radiation, such as fluorescence or phosphorescence.

The quantum yield for bond rupture for most groups in polymer chains is rather low and only the C—Cl and C—C bonds are affected directly by the UV light in the solar spectrum. $\overset{\displaystyle \|}{O}$

For most other polymers, on the other hand, there are sufficient imperfections in the chains that will act as chromophores (i.e. centres that absorb light). These will cause localised bond rupture by a free radical mechanism, and subsequent auto-oxidative reactions. Hence antioxidants and hydroperoxide decomposers can retard considerably the photo-oxidation reactions but would soon be used up in the photo-initiation step. Consequently it is necessary for these to be used in combination with UV stabilisers. There are three classes of UV stabilisers used in polymers.

(a) UV Absorbers
These are additives capable of absorbing UV light and re-emitting it as harmless energy. Typical UV absorbers are pigments such as carbon blacks and iron oxide. When either black or brown-red pigmentation cannot be tolerated there are organic additives that will have the same function (albeit less effectively) without affecting the colour of the polymer. These are mainly derivatives of 2-hydroxybenzophenone and hydroxyphenyl benzo-triazoles.

(b) Excited State Quenchers
These are additives that are capable of transferring energy from the

activated chromophores and dissipating it through the usual photophysical processes. The most widely used excited state quenchers are nickel complexes. The structure of these compounds is often tailored in such a precise manner that in one single molecule there are also antioxidant groups.

(c) Radical Scavengers
Conventional antioxidants (alone) are not very effective in photo-oxidative processes insofar as they can absorb light, which destroys their radical trapping characteristics. Certain derivatives of piperidine, on the other hand, will not absorb UV light and, at the same time exhibit very effective radical quenching properties.

Note that carbon blacks are the most efficient additives against photo-oxidation decomposition insofar as they exhibit all three activities indicated above.

A list of typical UV stabilisers used in polymers is given in Table 4.4.

4.1.4 Surface Properties Modifiers
(a) Antiblocking Additives
The tendency of two adjacent surfaces to cause 'blocking' is determined by the following factors: (i) total area of contact, (ii) normal (compressive) forces acting at the interface, (iii) interfacial surface energy, and (iv) compliance of the material (see Chapt. 5).

A very soft (compliant) material is very prone to blocking since the protruding asperities at the interface can be readily ironed out under the influence of external (normal) forces, hence increasing the true surface area of contact. In order to eliminate blocking by means of additives, without affecting the bulk properties of the material, one can only intervene at the interface by reducing the interfacial free energy and/or promoting surface roughness.

Antiblocking properties, therefore, can be simply developed by small amounts of very fine fillers (e.g. silica flour) and/or external lubricants.

Metal soaps or other incompatible organic stearates will form a very weak boundary layer at the interface which will break very easily when the two contacting surfaces are pulled apart.

(b) Antistatic Agents
Electrification of materials results from a segregation of charges (electrons and ions) which occurs when two surfaces are parted after close initial

contact. The amount of charge build-up is determined by the rate of generation and, simultaneously, the rate of charge decay.

The rate of charge generation on the surface can be decreased to some extent by reducing the intimacy of contacts, whereas the rate of conduction can be increased by the formation of an ionic boundary layer. Consequently antistatic characteristics in polymers can be achieved by means of ionisable additives which can migrate to the surface and form conductive paths through the absorption of atmospheric moisture.

Such additives are called 'antistatic agents' and normally consist of:

(i)   Nitrogen compounds: long chain amines, amides or quaternary ammonium salts, e.g. stearamido-propyldimethyl-2-hydroxyethyl ammonium nitrate;
(ii) sulphonic acids and alkyl aryl sulphonates;
(iii) polyhydric alcohols and derivatives;
(iv) phosphoric acid derivatives, e.g. didodecyl hydrogen phosphates;
(v)   polyethylene glycol derivatives, e.g. hexadecyl ethers of poly-ethylene glycol.

Molecular weight and overall polarity considerations are important since they determine the rate of diffusion of the additive and hence their efficiency and durability. Although strongly ionic inorganic salts, e.g. LiCl, etc., would provide very effective antistatic behaviour initially, they would easily leach out of the system on subsequent ageing. Consequently a certain balance between compatibility and diffusibility must be achieved in order to obtain optimum efficiency and durability. The rate of charge decay obeys an exponential relation of the type:

$$V_t = V_0 \exp[-t/RC] \tag{4.7}$$

Hence the rate of charge decay is a logarithmic function of time, and the rate constant is defined as the time (λ) for the charge to reach $1/e$ of its original value. The rate constant can, therefore, be obtained from measurements of surface resistivity. Good antistatic properties are achieved when λ becomes of the order of a fraction of a second.

4.1.5 Optical Properties Modifiers

The optical properties of a material from a technological aspect are normally described in terms of their ability to transmit light, to exhibit colour and reflect light from the surface (i.e. gloss). The only contribution that additives can make to improve gloss is through a reduction of the surface irregularities arising during processing.

Table 4.4: Typical UV Stabilisers

Chemical name	Structure
A. *UV Absorbers*	
(a) 2-Hydroxy 4-alkoxy benzophenones	(a)
(b) 2(2H-benzotriazol-2yl) phenols	(b)
(c) Resorcinol monobenzoate	(c)

B. *Excited State Quenchers*

(a) *Bis*(2,2' thio*bis*-4-(1,1,3,3-tetra methylbutyl)phenolate) nickel

(a)

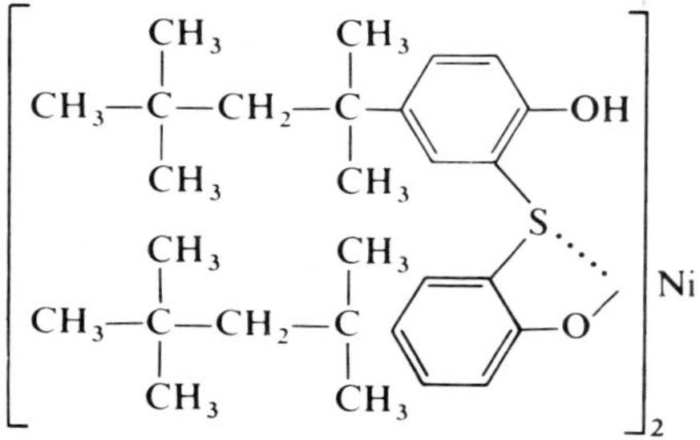

(b) 2:1 Nickel complex of 3-5 di-*tert* butyl-4-hydroxy benzyl monoethyl phosphonate

(b)

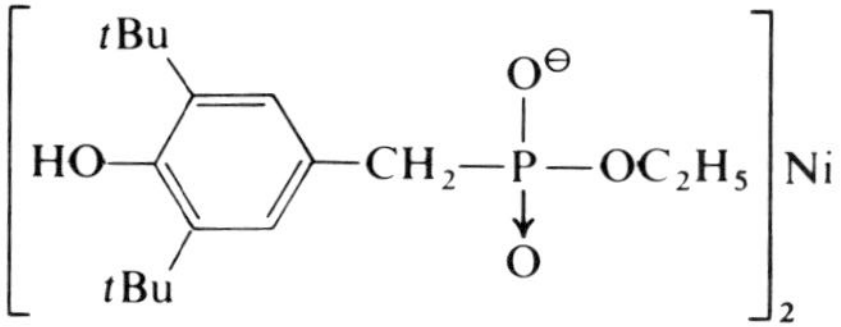

(c) Nickel dibutyl dithiocarbamate

(c)

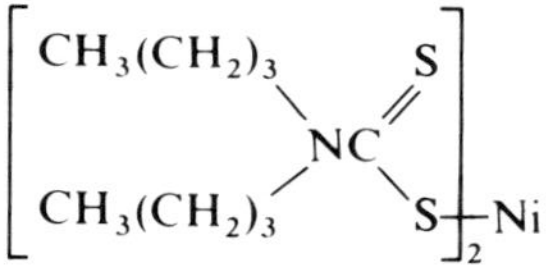

In this respect, therefore, external lubricants could also be considered as optical properties modifiers. The mechanistic aspect of external lubricants to prevent surface irregularities, normally known as *'shark skin'* will be dealt with in Chapt. 7.

(a) Dyes and Pigments

Colour in a material is developed as a result of visible light being absorbed. Absorption of light occurs through dissipation of electromagnetic energy by the electronic configuration of the material, i.e. electrons will acquire energy in excess of that in their 'ground state' when they absorb light. When all the light is absorbed by the material it appears black. If only part of the light is being absorbed (i.e. only light of certain wavelengths) and the amount of light scattered is small the material becomes 'coloured transparent' and the colour developed corresponds to the particular wavelengths of the light transmitted. If the amount of light that has not been absorbed is internally scattered, the material becomes opaque and the colour corresponds to that of the wavelengths not absorbed.

There are two types of additives used in polymers to develop colour, depending on whether the polymer is to remain transparent or to become opaque.

Additives that are compatible with the polymer or that can be finely dispersed so that they will not act as scattering centres for the transmitted light are known as *'dyes'*. The dye molecules have the function, therefore, of absorbing the unwanted wavelengths of the visible spectrum and allowing the others through.

Dyes used in polymers are normally divided into two categories; i.e. *'spirit soluble'* and *'oil soluble'* depending on whether they are soluble in aliphatic solvents (such as alcohols, ketones, etc.) or in aromatic compounds containing nitro groups, amino groups, etc.

To produce colour and opacity on the other hand, the additive must be able to absorb and scatter light. Consequently a main consideration for additives with respect to this function is their particle size, which should be greater than the wavelength of the light, and their refractive index, which should be much higher than that of the polymer. These additives are called *'pigments'* and can be either organic or inorganic in nature.

Since the number of inorganic pigments available is rather limited the majority of pigments are obtained from dyes, which are insolubilised into brittle thermosetting resins and micronised into particles of the desired dimensions. Examples of dyes and pigments used in thermoplastics are shown in Table 4.5.

(b) Nucleating Agents

Polymeric materials have a refractive index between 1·45 and 1·70. All amorphous polymers can transmit 80–90% of the incident light, while crystalline polymers tend to develop internal haze due to the difference in refractive index between the spherulites (larger than the wavelength of light) and the amorphous regions. Since the chemical constitution of the crystalline and amorphous regions is the same, the only factor that is responsible for the difference in their refractive index is the density.

The natural colour of polymers varies, therefore, from water-clear (amorphous polymers) to milky-ivory (crystalline polymers). A yellow tint is also developed in sulphur-containing polymers and those produced from phenol derivatives.

To aid transmission of white light in crystalline polymers by means of additives it is necessary that these will reduce the level of crystallinity and the size of spherulites, possibly to dimensions smaller than the wavelength of visible light.

By introducing very finely divided solid particles (nucleating agents), possessing a high surface energy, it is possible to induce nucleation of the crystals at temperatures substantially higher than in their absence. In this way a large number of nuclei are formed, while the greater chain mobility of the polymer at these temperatures will give rise to faster crystallisation rates.

Examples of nucleating agents are sodium, lithium or potassium benzoate, clays, silica flour, etc.

4.1.6 Fire Retardants

The reactions described in Section 4.1.1(a) for the thermal oxidative decomposition of polymers apply equally well to high temperatures, i.e. under fire conditions. The difference between the two cases is primarily with respect to the kinetics of the large number of competing reactions, the environmental conditions and the reactions in the gaseous phase (i.e. in the flame), which are absent in oxidative reactions below the ignition temperature of the polymer.

Although it is difficult to identify the intrinsic properties of a polymer that will describe its susceptibility to cause a fire and its propensity to sustain the flames, there are a number of empirical tests which enable the materials technologist to select additives that will make the polymer a much safer material in situations where there are fire risks.

Additives used for this purpose are known as fire retardants. Although there is no strict classification for fire retardant additives it is possible to

Table 4.5: Typical Colourants for Polymers

Chemical nature	Type	Structure
A. *DYES*		
(i) *Spirit soluble*		
(a) Salts of organic base	Zapon fast scarlet **GG**	(a) $[(C_6H_{11})_2\overset{\oplus}{N}H_2]_2$
(b) Derivatives of triphenyl methane	Spirit blue CI 689	(b)

(ii) *Oil soluble*
(a) Non-ionic aromatic dyes

(a) Oil yellow

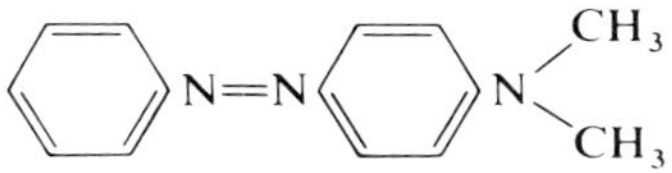

(b) Non-ionic aromatic dyes

(b) Sudan blue

B. *PIGMENTS*
(i) *Inorganic*
(a) Oxides

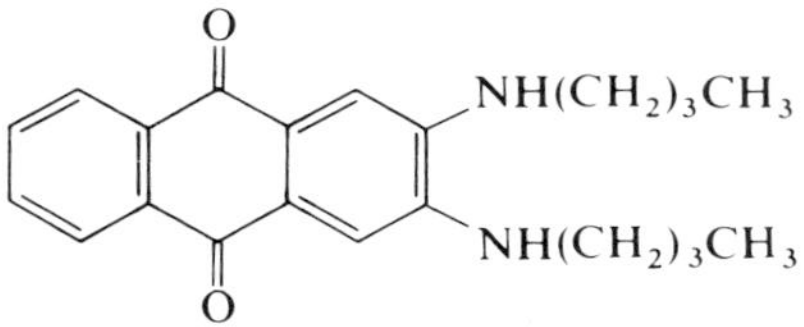

White titanium oxide   (a) TiO_2
Yellow iron oxide $Fe_2O_3 \cdot H_2O$
Red iron oxide Fe_2O_3
Black iron oxide Fe_3O_4
Green chromium oxide Cr_2O_3 and $Cr_2O_3 \cdot 2H_2O$

(b) Chromates
Moly orange   (b) $PbCrO_4 \cdot PbSO_4 \cdot PbMoO_4$
Chrome orange $(PbCrO_4)_x \cdot (PbO)_y$

(c) Sulphides
Cadmium yellow   (c) Mixed CdS/ZnS
Cadmium red   Mixed CdS/CdSe

Table 4.5—*contd.*

Chemical nature	*Type*		*Structure*
(ii) *Organic*			
(a) Yellows	Dichlorobenzidine derivatives	(a)	

(b)   Oranges

Dianisidine
derivatives

(b)

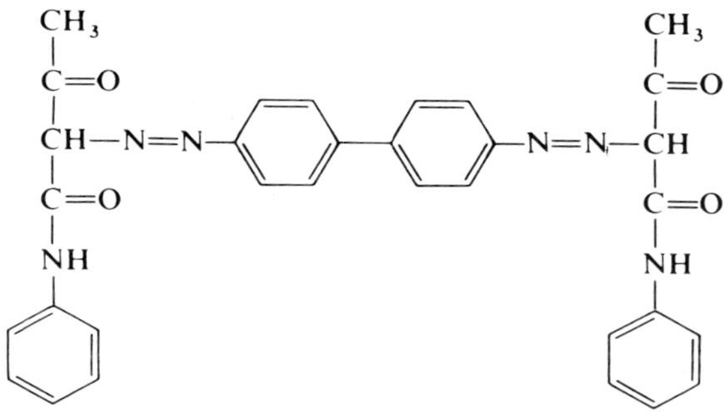

(c)   Reds

Toluidine reds

(c)

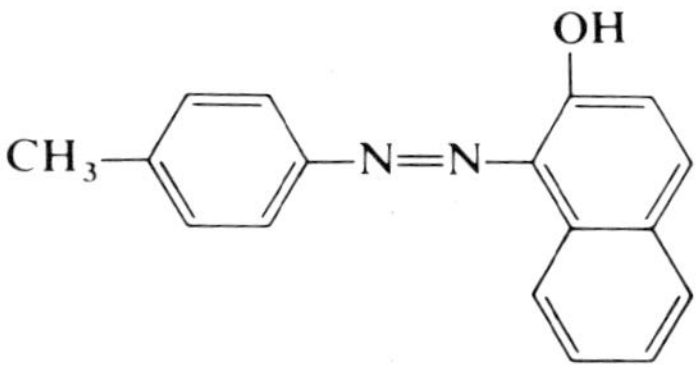

distinguish three types of systems, conveniently denoted as 'ignition inhibitors', 'self-extinguishing additives' and 'smoke suppressants'. The latter will not assist the prevention of fires but will reduce the formation of noxious by-products of a fire.

(a) Ignition Inhibitors
To ignite a material it is necessary to raise the temperature to levels at which pyrolysis of the constituents produces combustible volatiles, which will then oxidise very rapidly to form a flame. Obviously under normal atmospheric conditions (e.g. normal oxygen level) the temperature at which this takes place (i.e. the ignition temperature) depends on the relative proportions of combustible gases emitted and on the rate at which they will oxidise to CO_2 and H_2O. The interrelationship between polymer structure, nature of the gaseous products and ignition temperature is shown in Table 4.6.

An inspection of this table reveals that the formation of non-combustible gases during pyrolysis (e.g. HCl, NH_3, etc.) increases the ignition temperature of the polymer. Although the rate of evolution of combustible gases is not shown it can be inferred that materials capable of forming a char, such as polyethylene terephthalate and polycarbonates, will exhibit a high ignition temperature in view of the low amounts of gaseous products released from the pyrolysis zone. Hence for an additive to be effective in increasing the ignition temperature of the polymer it must be capable of forming large amounts of incombustible gases, which would also inhibit the free-radical oxidation reactions in the flame zone.

Table 4.6: Relationship Between Structure, Pyrolysis Products and Ignition Temperature of Polymers

Polymer type	Pyrolysis products	Flash ignition temperature (°C)
Polyolefins	Aliphatic and alicyclic hydrocarbons	343
Polystyrene	Styrene monomer and oligomers	360
Polyvinyl chloride	HCl and aromatic hydrocarbons	454
Polymethylmethacrylate	Methyl methacrylate monomer	338
Polytetrafluoroethylene	Tetrafluoroethylene monomer	ND
Polymethylene oxide	Formaldehyde	ND
Nylon 66	Ammonia, amines, carbon monoxide and dioxide	427
Polyethylene terephthalate	Aliphatic hydrocarbons and benzoic acid	460
Polycarbonate	Phenol, carbon dioxide	482

Typical additives that exhibit this function are:

(i) Mixtures of organic halogenated compounds and oxides of antimony and molybdenum. These form volatile oxychloride gases which can expel oxygen from the flame zone, and, at the same time, they produce free-radicals which, by interfering with the chain reactions in the gaseous phase, will reduce the net rate of the reactions leading to CO_2 and H_2O (Table 4.7).

(ii) Aluminium trihydrate, magnesium hydroxide, zinc borate, melamine and derivatives, etc. These function primarily by emitting incombustible gases, i.e. H_2O from the inorganic compounds, and by removing heat from the substrate through endothermic vaporisation or sublimation phenomena. Since they are used in substantial amounts in a polymer formulation (e.g. more than 40 %) they will also produce a considerable reduction in the rate of gases generated from the flame substrate.

(b) Self-extinguishing Additives

The difference between fire retardancy through ignition inhibition and self-extinction of flame is primarily with respect to the transport phenomena between the flame zone and the pyrolysis zone. Char formation is highly desirable in promoting self-extinction of the flame insofar as it provides a two-way barrier on the surface of the burning front, i.e. for the heat transfer from the flame to the substrate and for the escape of gases from the pyrolysis zone into the flame.

Since a char at the burning front will prevent dripping of molten material, this will also reduce the risk of fire spreading.

The char formation characteristics of polymers are related to their ability to form polynuclear rings during pyrolysis (i.e. chemical structures similar

Table 4.7: Typical Halogenated Compounds Used in Combination with Antimony Oxide

Compound	Suggested uses
Chlorinated paraffins	General usage
Perchlorocyclopentadiene	High melting point polymers
Decabromo diphenylene oxide	Most polymers
Tris(2,3 dibromopropyl) phosphate	General usage
Poly(β chloroethyl triphosphate)	General usage
Bromophthalimide	High melting point polymers

to graphite). In the thermoplastic family it is mainly the acrylonitrile-based polymers and those containing aromatic rings in the backbone of the chains that are capable of producing these polynuclear structures as a result of thermal decomposition reactions.

The thermograms in Fig. 4.4 illustrate the above point.[11]

It is interesting to note that the aliphatic units attached to aromatic nuclei show a negative char formation tendency. It is suggested that the reason for this is the ability of aliphatic chains to give out hydrogen, which assists the fragmentation of nuclear rings into volatile hydrocarbons.[12]

(c) Smoke Suppressants
The generation of smoke is caused by the carbonisation reactions occurring in the gas phase. This in turn is related to the ability of the polymer to generate aromatic gaseous products (polymethylene oxide, for instance, burns without smoke since it decomposes by depolymerisation) or fragments which will pyrolyse further in the flame zone in an oxygen-deficient environment. Typical of the latter are the halogenated systems, irrespective of whether the halogen is part of the polymer (e.g. polyvinyl chloride) or is contained in the fire retardant components.

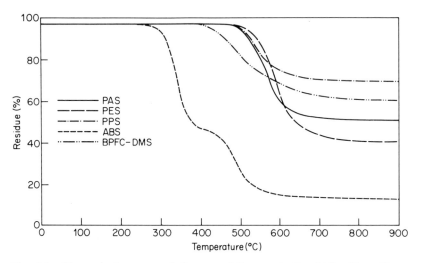

Fig. 4.4.   Dynamic thermograph for several thermoplastics. (After Kourtides *et al.*[11]) PAS = polyarylsulphones, PES = polyether sulphones, PPS = polyphenylene sulphide, ABS = acrylonitrile–butadiene–styrene terpolymer blends, BPFC–DMS = polycarbonate–silicone block copolymers.

Consequently to effectively reduce the amount of smoke it is necessary to catalyse the oxidation reactions in the gas phase. Typical additives that act in this manner are silica flour, molybdenum oxide, magnesium oxide, etc. Aluminium trihydrate and other fillers, however, can also suppress the generation of smoke by catalysing the oxidation of carbonaceous matter on the burning surface of the polymer, hence preventing them from being transported into the gas phase as very fine particles.

4.1.7 Foaming Additives

Foaming (or blowing) agents are used for the production of cellular products. In comparison with continuous (or homogeneous) materials, cellular structures can offer a number of advantages such as reduced density, lower thermal conductivity, lower dielectric losses and higher rigidity per unit weight.

Blowing agents are normally classified into physical and chemical types, according to whether the generation of gases to produce the cells takes place through a physical transition (i.e. evaporation or sublimation) or by a chemical process (i.e. decomposition reactions which result in the evolution of gases).

The physical blowing agents are normally low boiling liquids or gases which are soluble in the polymer matrix and exert their blowing action when brought to the boiling conditions by increasing the temperature and/ or reducing the pressure of the system. Some typical physical blowing agents and their boiling range are: pentane (30–38 °C), heptane (65–75 °C), methylene chloride (40 °C), trichlorofluoromethane (24 °C), trichlorofluoroethane (47 °C), etc.

Chemical blowing, on the other hand, is achieved by means of compatible or finely dispersed chemical compounds that decompose at the required rate over a fairly narrow temperature range. Typical chemical blowing agents and their decomposition temperature range are: azodicarbonamide (160–200 °C), azo *bis*-dibutyronitrile (90–115 °C), benzene sulphonylhydrazine (95–100 °C), *p*-toluene sulphonyl semicarbazide (210–270 °C).

The blowing process has to take place in three stages: a nucleation step, the expansion process and cell stabilisation.

Nucleation takes place when the amount of gas evolved exceeds the solubility limit, which is determined by the temperature and pressure according to Henry's law; i.e. at constant temperature

$$[C]_T = [S]_T P \tag{4.8}$$

where C = concentration of gas dissolved, S = solubility coefficient of the gas in the polymer at the given temperature T, P = external pressure.

This assumes, of course, that no gas escapes from the system through diffusion, which is probably true only for expansion processes in closed moulds. As in any other nucleation processes the formation of cells can be assisted by the presence of particles finely dispersed, such as pigments, lubricants, etc.

The growth rate of cells in the expansion stage is determined by the rate of increase in pressure inside the cells and by the deformability of the cell walls (i.e. tensile viscosity). Since nucleation of the cells does not occur simultaneously throughout the matrix, at any given time the cell size and pressure in the cell are subjected to the usual statistical variations.

Owing to surface tension effects there is an excess pressure within the smaller bubbles, which causes the gas to diffuse from one cell to another. Consequently some of the bubbles will grow even larger, while others will decrease in size.

If this process continued the cells would coalesce into one another to produce an open cell structure. The cells would eventually break the outer skin and the gas would diffuse into the atmosphere, causing the walls of the cells to collapse. Hence a mechanism for the stabilisation of the cells is necessary to prevent the foam from collapsing. For the case of thermoplastics this is achieved simply by balancing the rate of expansion of the cells with the rate of cooling.

4.2 POLYMER BLENDS AND ALLOYS

Economically it is obviously much more attractive to produce compositions by blending established polymers than synthesising new ones.

Despite this enormous incentive, witnessed by the large number of patents issued, the use of polymer blends in industry is not as common as might be expected. The reason for this is the deterioration in properties that invariably takes place when simply blending two polymers at random. Very few polymers, in fact, lend themselves to blending without prior modification of one or both components or tailoring their structure specifically for this purpose.

When blending polymers it is necessary to achieve one of the following morphologies:

(i)   complete homogeneity so that the two components cannot be identified as discrete phases,

(ii) partial compatibility so that although two phases may be present the boundary regions are diffuse and consist of compatible mixtures of the two components.

Deleterious effects are experienced when the two polymers are totally incompatible so that two distinct phases with sharp interfacial boundaries are obtained. The mechanical properties, for instance, are impaired because the interfacial surface energy is so low that a tear or crack can easily propagate between the boundaries. Other properties, such as solvent resistance and vapour diffusion, would also suffer as a result of the weak boundary regions between the two phases. The actual available spaces for the diffusion of a solvent into these regions are much greater than those offered by the free volumes of the individual polymer components.

4.2.1 Compatible Blends

Total compatibility between polymer pairs is fairly rare in view of the unfavourable changes in enthalpy and entropy.

The entropy of mixing, ΔS^m, of two polymers is generally very small, as can be inferred directly from the Flory–Huggins treatment,[13] i.e.

$$\Delta S^m = -k(n_1 \ln \phi_1 + n_2 \ln \phi_2) \qquad (4.9)$$

where ϕ_1 and ϕ_2 are the volume fractions, n_1 and n_2 the number of molecules of the two components and k is the Boltzman constant.

Obviously the higher the molecular weight of the two components the smaller the values of n_1 and n_2 and, therefore, the lower the value of ΔS^m.

Consequently, the main driving force for polymer mixing must come from the enthalpy term, ΔH^m, which is likely to be small unless mixing is accompanied by a reduction in volume. Favourable conditions for mixing also prevail when intermolecular forces between the two components are stronger than those existing within each individual polymer. These situations, however, are quite rare. The first seems to apply to blends of polyvinyl chloride and solution chlorinated polyethylene,[14] while examples of the latter case include blends of polyvinyl chloride with polycaprolactones[15] and with polymethyl methacrylate[15] respectively. H bonding is likely to take place between the H atom attached to the same carbon as the Cl atom in PVC and the carbonyl group of the repeating unit of the auxiliary polymer.

While the Flory–Huggins treatment predicts a favourable entropy of mixing at higher temperatures, it does not predict the possibility of phase separation, which often takes place as the temperature is increased further.

In other words, contrary to the case of mixtures with low-molecular-weight compounds, such as plasticisers, blends of high-molecular-weight polymers can exhibit an upper critical solution temperature (UCST) as well as a lower critical solution temperature (LCST). The latter can only occur if the entropy of mixing is unfavourable (negative) and the enthalpy is favourable (positive).[16] That is to say that the first derivative of the free energy with respect to concentration of the auxiliary polymer is zero for systems exhibiting only a UCST. For systems that exhibit an LCST, on the other hand, it is the second and third derivatives that will become zero at the equilibrium condition of the phase transition.

The phase diagram for compatible polymers can be represented schematically as in Fig. 4.5, where the two curves, termed respectively binodal and spinodal, represent the boundary conditions for a metastable equilibrium.

Occasionally compatible blends can be produced from mixtures of an amorphous polymer and a crystalline polymer. The most widely reported case is the blend of polyvinylidene difluoride (PVF2) and polymethyl-methacrylate (PMMA).[17] Obviously in this situation one is concerned with the effects on both first and second order transitions.

In quantifying the relative change in first order transition as a function of the concentration of the two components one is faced with the difficulty of deciding what value to assign to the volumetric fraction of the crystalline polymer, since the distribution of the two components in the crystalline and amorphous regions is not the same. The curve for T_g as function of the PVF2 content for this blend shows, in fact, a strong deviation from the theoretical curve at concentrations greater than 50 %.[17] This suggests that the underestimated T_g values are the result of the error caused by using the total volumetric fraction of PVF2, instead of that occupying the amorphous regions only.

The depression of the melting point (ΔT_m) of the PVF2 phase, caused by dilution effects from amorphous PMMA, seems to follow reasonably well Flory's thermodynamic treatment of the effect of solvents, i.e.

$$\frac{\Delta T_m}{T_m^0} = \frac{-R\bar{V}_{2\mu}}{\Delta H_{2\mu}V_{1\mu}} \chi_{12}\phi_1^2 \qquad (4.10)$$

where T_m^0 = melting point of PVF2 in isolation, $V_{1\mu}$ and $V_{2\mu}$ = molar volumes per segment for the two components, $\Delta H_{2\mu}$ = heat of fusion per segment of crystalline phase, ϕ_1 = volume function of amorphous component (PMMA), R = universal gas constant, and χ_{12} = polymer$_1$/polymer$_2$ interaction parameter.

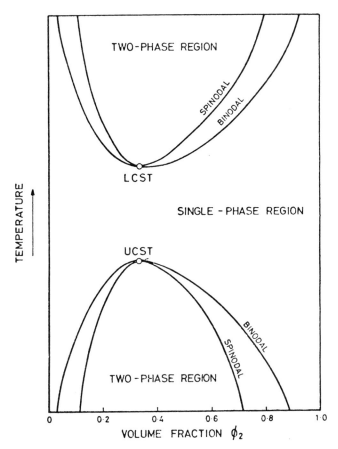

Fig. 4.5. Schematic diagram showing binodal and spinodal curves for systems exhibiting both LCST and UCST. (After C. B. Bucknall, *Toughened Plastics*, Applied Science, 1977, p. 17.)

In Flory's treatment the χ_{12} parameter determines the interactive forces between the two components and, therefore, it is related to the enthalpy of mixing, i.e.

$$\Delta H^m = kT\chi'_{12}n_2\phi_1 \tag{4.11}$$

where $kT\chi_{12}$ represents the difference between the enthalpy of one molecule of polymer$_1$ when surrounded by molecules of polymer$_2$ and that of polymer$_1$ surrounded by molecules of polymer$_1$.

The interaction parameter χ'_{12} has been found to vary linearly with temperature according to the expression[18]

$$\chi'_{12} = A_{12} + B_{12} \frac{\bar{V}_{1_\mu}}{RT} \tag{4.12}$$

From the above theoretical treatment it can be inferred that the bulk properties of compatible blends are always intermediate between those of the two components.

Consequently blending of compatible polymers is an extension of the concept of plasticisation. The latter would be used primarily to achieve a very large reduction in T_g, particularly in bringing a glassy polymer into the rubbery state at ambient temperatures. Polymer blends would be used, on the other hand, to bring about smaller changes in T_g. For the case of PVC, for instance, we have seen that using small amounts of plasticisers, to bring about a modest reduction in T_g, invariably results in antiplasticisation. The use of auxiliary compatible polymers, such as vinyl–acetate copolymers, chlorinated polyethylene, polycaprolactone, etc., is a means of getting around the problem of antiplasticisation when the extent of 'polymer softening' required is not excessive.

4.2.2 Partially Compatible Blends

Partial compatibility implies that above a certain concentration either the minor or the more viscous component remains as a dispersed phase. This differs from the concept of incompatibility insofar as the dispersing phase consists of a compatible blend of the two components.

Obviously when two phases are present the blend will exhibit the transitions of both dispersed and matrix phases. These multitransitions are particularly beneficial with respect to the mechanical properties. For instance, while the bulk properties are dominated primarily by the matrix, the events that take place at a microscopic level (e.g. crack propagation) can be influenced to a large extent by the nature of the inclusions and interfacial properties.

Systems that have been found to be particularly useful are those where the matrix is rigid, the dispersed phase is rubbery and the interfacial bond is strong. In this way the blend retains much of the rigidity of the polymer matrix but acquires a superior resistance to fracture in view of the crack stopping mechanism provided by the rubbery inclusions.

As shall be seen from more detailed discussions later, when the dispersed phase is spherical the gross external stresses acting on the bulk of the material are transmitted equally on the two phases. As a result the total strain is the algebraic average of the strain of two components, i.e. if

$$P_B = P_m = P_d$$

then

$$e_B = \phi_m e_m + \phi_d e_d \qquad (4.13)$$

and, therefore

$$\frac{1}{K_B} = \frac{\phi_m}{K_m} + \frac{\phi_d}{K_d} \qquad (4.14)$$

where P = pressure (or isotropic compression), ϕ = volumetric fraction, K = bulk modulus, and B, m and d = subscripts denoting respectively blend, matrix and dispersed phase.

Equation (4.14) applies equally well to Young's modulus and shear modulus (see Chapt. 5 for definitions). This equation, however, gives only conservative estimates for the modulus of a partially compatible blend since it represents the lower bound of the response of composite morphologies.

More accurate predictions have been obtained from the so-called 'self-consistent model'. This consists of a spherical dispersed phase particle embedded in a matrix shell which is, in turn, occluded in a continuous body having the same properties as the blend.[19] The analysis compares the mechanical response of this model to that of a homogeneous body having the same macroscopic elastic properties as the blend in question.

This analysis yields the following equation for the bulk modulus, i.e.

$$\frac{K_B}{K_m} = \frac{\phi_m K_m + (\beta_m + \phi_d) K_d}{(1 + \beta_m \phi_d) K_m + \beta_m \phi_m K_d} \qquad (4.15)$$

where $\beta_m = (1 + v_m)/2(1 - 2v_m)$, and v_m = the Poisson ratio of the matrix.

For the case of a hard glassy polymer containing soft rubbery inclusions the two equations produce the behaviour depicted in Fig. 4.6.

As can be seen from the curves in Fig. 4.6 both eqns. (4.14) and (4.15) predict that relatively low amounts of rubbery inclusions (i.e. up to about 15%) only produce a slight reduction in the rigidity of a glassy polymer.

Very little theoretical attention has been given to quantitative analysis of the enhancement of fracture resistance resulting from the incorporation of rubbery inclusions in a glassy matrix.

Bucknall,[20] however, has been able to identify two main mechanisms for the toughening of glassy polymers. One arises from multiple crazes of the

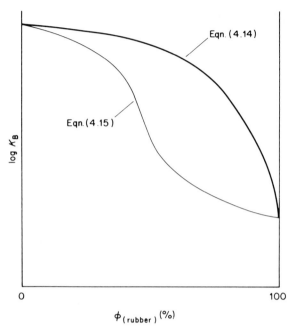

Fig. 4.6.   Prediction of the effect of increasing amounts of rubber inclusions in a glassy matrix.

matrix, which originate at the surface of the rubbery particles and dissipate the elastically stored energy over a large volume of the material. The rubbery particles provide at the same time a mechanism for stopping the crazes from developing into cracks prior to gross yielding, hence preventing the onset of brittle type fractures. This is illustrated in Fig. 4.7.

The second mechanism for the prevention of brittle fractures in partially compatible blends of glassy polymers containing rubbery inclusions is related to the formation of shear bands between rubber particles, Fig. 4.8. This simply means that the role of the rubber particles is to reduce the yield stress of the blend, so that the dimensions of the critical crack length for the onset of brittle fractures are increased (see Chapt. 6). There are two possible explanations for the reduction of yield stress in partially compatible blends of glassy polymers containing rubbery inclusions:

(i)   At the rubber particle/matrix interface any monoaxial or biaxial tensile stresses are converted into multiaxial complex stresses as a result of the difference in modulus and Poisson ratio between the two phases. Hence the compressive stresses set up at the equator will reduce the level of axial stresses required to induce yielding (see yield criteria in Chapt. 6).

(ii) Plasticisation of the matrix by the compatibilised portion of the rubber phase will result in a reduction in yield strength by the same mechanism as put forward earlier for the decrease in rigidity.

Although blending of partially compatible polymers will produce the characteristics described earlier, there are other factors that have to be controlled to achieve maximum benefits, namely the particle size and morphology of the dispersed phase.

In practice, a lightly cross-linked rubber phase is preferred in order to eliminate the possibility of breaking down the particles during processing.

The inclusion of rigid domains within the cross-linked rubber particles

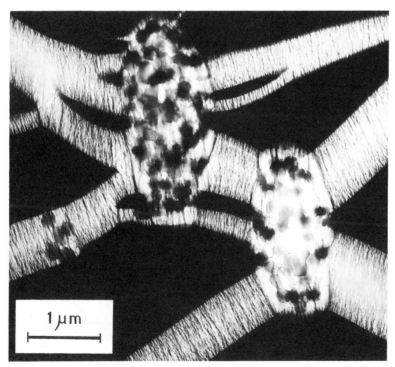

Fig. 4.7. Crazes connecting rubber particles in a cast film of high impact polystyrene. (After P. Behan *et al.*, *J. Mat. Sci.* (1976) 1207.)

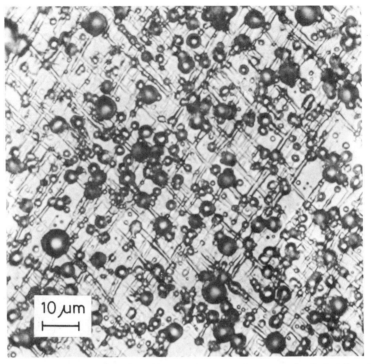

Fig. 4.8.   Shear bands in a blend of high impact polystyrene and poly(2,6 dimethyl
p-phenylene oxide). (After C. B. Bucknall, *Toughened Plastics*, Applied Science,
1977, p. 234.)

provides a mechanism for increasing the volumetric fraction of the
dispersed phase (which is a desirable condition for promoting the
formation of multiple crazes and shear bands) at the minimum level of
addition of the rubbery polymer.[21] Large amounts of the latter could
produce excessive plasticisation of the matrix.

Because of the limited choice of polymers that exhibit the right degree of
compatibility to achieve maximum toughening, it is often preferable to
chemically graft on the surface of rubbery particles a layer of polymeric
chains that are compatible with the matrix. In this way the grafted polymer
at the interface acts as an emulsifier for the dispersed phase and helps to
control the size and distribution of the rubbery particles. This practice is
widely used for the production of toughened styrene based thermoplastics
(see Section 3.1.3) and for both toughened rigid polypropylene and

thermoplastic polyolefin rubbers. In the latter case blocks of ethylene/ propylene copolymer segments attached to either linear polyethylene or polypropylene chains give rise to the formation of crystalline domains within the rubbery phase (thermoplastic rubbers) and create conditions for controlling the disperson of rubbery particles within a complex blend of polypropylene and polyethylene crystals (toughened thermoplastics).

For the case of impact grades of polystyrene, on the other hand, the formation of glassy domains within the rubber particles takes place by a phase inversion process during polymerisation.

4.3 REINFORCEMENT AND COMPOSITES

The term reinforcement is used to denote the increase in rigidity and strength from dispersing inorganic fibres or particulate fillers in the polymer matrix.

Since inorganic solids are intrinsically much stiffer and stronger than polymers a mixture of the two, known as 'a composite', is expected to exhibit properties that are intermediate between those of the constituents.

Similarly the coefficient of thermal expansion of a composite is expected to be much lower than that of the base polymer in view of the association of dilation with internal mechanical stresses.

4.3.1 Stiffness Considerations

The modulus of composites falls between two limits; the highest modulus is obtained when the two constituent phases deform isometrically (i.e. the strains are equal), while the lowest level of reinforcement occurs when the stresses acting on the two phases are identical.

(a) Upper Limit

By imposing the conditions $e_c = e_m = e_d$ it follows that the forces acting on the composite (F_c) are the sum of those acting on the individual components i.e.

$$F_c = F_m + \sum_{i=1}^{i=n} F_d \qquad (4.16)$$

where e denotes a generalised strain and the subscripts c, m and d refer to the composite, matrix and dispersed phase respectively.

It can be easily shown[22] that when the forces are converted into stresses (i.e. $S = F/A$, where A is the area over which the force F acts) and by considering the volume of the two phases in the composites one obtains

$$S_c = \phi_d S_d + (1 - \phi_d)S_m \qquad (4.17)$$

where ϕ_d = volumetric fraction of the dispersed filler (or fibres).

By dividing eqn. (4.17) by a constant strain one obtains a general expression for the modulus of the composite (known as the *law of mixtures*)

$$M_c = \phi_d M_d + (1 - \phi_d)M_m \qquad (4.18)$$

where M can be either the Young's modulus, the shear modulus or the bulk modulus.

(b) Lower Limit
Under isostress conditions, i.e. $S_c = S_m = S_d$, the deformations are additive and by similar arguments it is demonstrated that the generalised equation for the strain can be written as:

$$e_c = \phi_d e_d + (1 - \phi_d)e_m \qquad (4.19)$$

from which an inverse law of mixture can be derived for the modulus, i.e.

$$\frac{1}{M_c} = \frac{\phi_d}{M_d} + \frac{1 - \phi_d}{M_m} \qquad (4.20)$$

(c) Modulus Enhancement of Thermoplastics
The two limits shown above are strictly achieved in practice when the two phases are joined together in the form of parallel plates and the stresses are transmitted along the plane of the plates (upper limit) and in the transverse direction (lower limit) respectively. It can be shown that this also applies for the case where the reinforcing phase is present in the form of continuous fibres aligned in one direction, provided that both phases deform elastically, that their Poisson ratio is the same and that the stresses cause no debonding of the interfaces.[22]

Although in some special cases unidirectional continuous fibre thermoplastics composites have been produced, the processing methods normally used for the shaping of thermoplastics are only suitable for the production of short fibre composites.

For the case of glass fibres the final product normally contains fibres that are less than 0·3 mm long, the distribution of fibre length is skewed towards very short lengths and their orientation not only varies across the thickness but also from one position to another.

Consequently thermoplastics composites do not lend themselves to rigorous analytical treatments.

The most useful approach is the interpolation procedure of Halpin and Tsai[23,24] which leads to a general expression containing a reinforcing efficiency parameter ξ which is a function of fibre length and spatial orientation, i.e.

$$\frac{M_c}{M_m} = \frac{1 + \xi\eta\phi_f}{1 - \eta\phi_f} \qquad (4.21)$$

where

$$\eta = \frac{M_f/M_m - 1}{M_f/M_m + \xi} \qquad (4.22)$$

Note that when $\xi = 0$ eqn. (4.21) reduces to that for the lower limit, while for $\xi = \infty$ it becomes synonymous with the law of mixtures.

Hence for thermoplastics composites the value of ξ varies from approximately zero, for the case of particulate fillers, to some function of the aspect ratio of the fibres (l/d, where $l = $ length, $d = $ diameter) for fibre reinforced systems.

The value of ξ, however, varies also according to which modulus is used in eqn. (4.21).

By empirical curve fitting it has been found that the value of $\xi = 2(l/d)$ can be used to predict the Young's modulus for aligned fibre composites measured in the direction of the fibres. For particulate composites on the other hand, the shear modulus can be predicted by taking $\xi = (7 - 5v_m)/(8 - 10)$ (where $v_m = $ Poisson ratio of the matrix).

For the case of particular filler composites one can also use the self-consistent model described earlier to predict the modulus enhancement as a function of the volumetric fraction of the filler and the elastic coefficients of the two constituents (eqn. (4.15)).

With fibre reinforced composites where the fibres are aligned in one direction, the modulus decreases when the angle formed by the direction of the fibres and that of the applied stress is increased. When the stress acts in the perpendicular direction to the fibres the modulus approaches that of particulate composites. The variation of modulus with orientation angle of the fibres can be obtained from the relationships derived for orthotropic plates.[25] The effects of volumetric fraction of the reinforcing phase, the l/d ratio and orientation of the fibres on the modulus enhancement factor

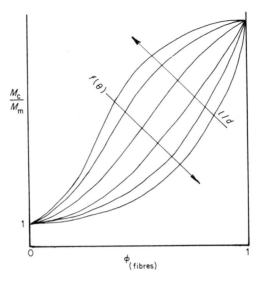

Fig. 4.9.   Modulus enhancement factor (M_c/M_m) for reinforced thermoplastics in function of the volumetric fraction of the fibres.

(M_c/M_m) are predicted by the Halpin–Tsai equation, shown in Fig. 4.9. Note that since sigmoidal curves, as in Fig. 4.9, are also obtained in the relationship between modulus enhancement factor and aspect ratio, it can be easily inferred that efficient reinforcement occurs only when both the volumetric fraction and the l/d ratio are fairly large.

In view of the processing difficulties experienced with both long fibres and high filler loadings it is to be expected that modulus enhancements for the case of reinforced thermoplastics used commercially are rather modest.

In practice the volumetric fraction of the fibres rarely exceeds the value of 0·2 and the average l/d ratio is probably of the order of 10. With these values the modulus enhancement factor obtained increases from about 2 to 8 as the M_f/M_m ratio is increased by changing the deformational characteristics of the matrix, i.e. from rubbery to glassy.

(d) *Reduction in Thermal Expansion by Reinforcement*
The dilation resulting from an increase in temperature (ΔT), is proportional to the coefficient of thermal expansion (α), i.e. for an isotropic material and in the absence of any external stress

$$e_v = \alpha_v \Delta T \qquad (4.23)$$

where e_v = volumetric strain and α_v = coefficient of volumetric expansion.

Since the volumetric strain can be related to the bulk modulus and the associated isotropic stress (S_{ii}), substituting in (4.23), one obtains

$$\frac{S_{ii}}{K} = \alpha_v \Delta T \qquad (4.24)$$

Hence for isotropic composites, i.e. for particulate fillers and short fibres distributed at random in space, one relates the reduction in volumetric expansion to the bulk modulus enhancement by the following general expression

$$\frac{K|_c S_{ii}|_m}{K_m S_{ii}|_c} = \frac{\alpha_v|_m}{\alpha_v|_c} \qquad (4.25)$$

For the case of particulate composites, where the stresses on the matrix are approximately the same as those acting on the filler (i.e. lower limit of reinforcement), the volumetric expansion reduction is approximately equal to the bulk modulus enhancement (i.e. $K|_c/K_m \simeq \alpha_v|_m/\alpha_v|_c$). This can be related to the properties of the filler and those of the matrix by means of eqn. (4.15).

The computation of the coefficient of expansion is much more complicated, on the other hand, for the case of short fibre composites, since neither isostress nor isostrain conditions can be imposed for the reinforcing phase and matrix. When the fibres are aligned in one direction the coefficient of thermal expansion in the longitudinal direction (α_L) can be calculated by the averaging procedure derived by Schapery,[26] i.e.

$$\alpha_L \simeq \bar{\alpha} + \left(\frac{\bar{E}\alpha}{\bar{E}} - \bar{\alpha} \right) \frac{\dfrac{1}{E_T} - \dfrac{1}{E_L}}{\dfrac{1}{E_T} - \dfrac{1}{E_m}} \qquad (4.26)$$

The various coefficients are calculated as follows:

$$\bar{\alpha} = \alpha_v = \phi_f \alpha_f + (1 - \phi_f)\alpha_m \qquad (4.27)$$

assuming that $v_f = v_m$

$$\frac{\bar{E}\alpha}{\bar{E}} = \frac{\phi_f \alpha_f E_f + (1 - \phi_f)\alpha_m E_m}{\phi_f E_f + (1 - \phi_f)E_m} \qquad (4.28)$$

$$\frac{1}{E_T} = \frac{\phi_f}{E_f} + \frac{(1 - \phi_f)}{E_m} \qquad (4.29)$$

$$E_L = E_m \left(\frac{1 + \xi\eta\phi_f}{1 - \eta\phi_f} \right) \qquad (4.30)$$

$$E_u = \phi_f E_f + (1 - \phi_f)E_m \qquad (4.31)$$

where f and m refer to the fibres and matrix respectively, L = longitudinal direction, T = transverse direction, E = Young's modulus, and v = Poisson ratio.

The coefficient of expansion in the transverse direction is independent of the fibre length and can be calculated as follows:

$$\alpha_T = (1 + v_f)\alpha_f v_f + (1 + v_m)\alpha_m v_m - \alpha_L^*[v_f \phi_f + v_m(1 - \phi_f)] \quad (4.32)$$

where α_L^* = coefficient of thermal expansion in the longitudinal direction.

For simplicity the value of α_L^* can be taken to be equivalent to that of a continuous fibre composite, as it would only produce a small error in the calculation of α_T, i.e.

$$\alpha_L^* = \frac{\phi_f \alpha_f E_f + (1 - \phi_f)\alpha_m E_m}{\phi_f E_f + (1 - \phi_f)E_m} \quad (4.33)$$

In doing so eqn. (4.32) simplifies to

$$\alpha_T = (1 + v_f)\alpha_f v_f + \alpha_m(1 - \phi_f) \quad (4.34)$$

Finally when the fibres are dispersed at random in space the coefficient of thermal expansion can be averaged out from the respective values obtained for unidirectional composites, e.g.

$$\langle \bar{\alpha} \rangle \simeq (\alpha_L + 2\alpha_T)/3 \quad (4.35)$$

The advantages of reducing thermal expansion in reinforced thermoplastics are often related more to processing considerations than service requirements.

Not only does the filler reduce the overall shrinkage in a mould and, therefore, make it possible to produce components with better dimensional accuracy, but it can also decrease the amount of warping in large-area mouldings. (See Chapt. 9.) These effects, however, are only partially due to reinforcing effects, since considerable reduction in volumetric shrinkage in a mould results from the increased amounts of voids (particularly with crystalline polymers) created by the constraints imposed by the fibres on the shrinkage of the polymer.[27]

4.3.2 Strength Considerations

The term strength is used to denote the maximum stress that a material can support without fracturing (brittle strength) or suffering excessive and uncontrollable deformations (yield strength) (see also Chapt. 6). As for the case of stiffness, one can obtain some estimates by identifying first the extreme limits and then introducing appropriate modifications to bring the analysis closer to the actual situation.

(a) Upper Limit

This occurs when the two phases undergo isometric deformations (isostrain conditions) up to the point where the more brittle phase fails (normally the reinforcing phase). According to the law of mixtures, when failure is caused by a tensile stress (σ), one can write:

$$\hat{\sigma}_c = \hat{\varepsilon}_f[\phi_f E_f + (1 - \phi_f)E_m] \tag{4.36}$$

where $\hat{\sigma}_c$ = tensile stress at which the composite breaks (i.e. the strength of the composite, $\hat{\varepsilon}_f$ = tensile strain at which the fibres break and the crack propagates through the matrix without undergoing further extension, and E = Young's modulus.

(b) Lower Limit

This occurs when the two phases are subjected to the same level of stress, so that the strains are additive and failure occurs when the stress reaches the strength of the weakest phase or the weakest regions of the composite (i.e. the fibre/matrix interface).

Hence for the case of unidirectional continuous fibre composites, while the upper limit is achieved when the stress acts in the direction of the fibres, the lower limit occurs when there is no adhesion at the interface and the stresses act in the direction perpendicular to the fibres. In other words this is almost equivalent to a composite where the reinforcing phase consists of voids.

If one takes the term $\eta\phi_f$ in the Halpin–Tsai equation to represent the reduced fibre volumetric fraction it is obvious that the lower limit of η must occur for systems where the fibres are replaced by voids, which should produce the lowest level of reinforcement for a given geometric factor.

In this case[28] $\eta = 1/\xi$, hence the Halpin–Tsai equation reduces to

$$\frac{E_c}{E_m} = \frac{1 - \phi_f}{1 + \phi_f/\xi} \tag{4.37}$$

and, therefore,

$$\frac{\sigma_c}{\sigma_m} = \frac{\varepsilon_c}{\varepsilon_m}\left(\frac{1 - \phi_f}{1 - \phi_f/\xi}\right) \tag{4.38}$$

Since in such instances the condition $\varepsilon_c = \varepsilon_m$ must prevail the strength of the composite can be described by the equation

$$\hat{\sigma}_c = \hat{\sigma}_m^* \left(\frac{1 - \phi_f}{1 + \phi_f/\xi} \right) \tag{4.39}$$

So that the lower limits are obtained by substituting $\xi = 2l/d$ for short fibre composites with the stress acting in the direction of the fibres, and $\xi = 1\cdot1{-}1\cdot5$ for particulate composites when $v_m = 0\cdot3{-}0\cdot5$.

Note, however, that the value of $\hat{\sigma}_m^*$ is less than the strength of the matrix in isolation since the presence of voids introduces notches in the system, which can reduce considerably the stress to failure (see Chapt. 6). If the matrix fails in a brittle fashion (and the maximum principal stress failure criterion is obeyed), the term $\hat{\sigma}_m^*$ could be replaced by $\hat{\sigma}_m'/K$, where K is the elastic stress concentration factor and $\hat{\sigma}_m'$ is the strength of the matrix in isolation. The value allotted to K would depend on the geometry of the voids (i.e. non-bonded filler particles) and would be greater for cylindrical and platelet shapes than spherical holes.

Alternatively one can assume that when there is no adhesion between the filler and the matrix, failure will occur at the minimum cross-section area of the continuous phase (i.e. the matrix).

For the case of spherical particles a unit volume of the matrix will contain n^3 particles of radius r, and, therefore, the minimum cross-section area A_m is given by

$$A_m = 1 - \pi(nr)^2 \tag{4.40}$$

while the volume fraction of the filler particles ϕ_f is

$$\phi_f = \tfrac{4}{3}\pi(nr)^3 \tag{4.41}$$

For the balance of the forces to be maintained it will be required that

$$F|_{A_{max}} = F|_{A_m} \tag{4.42}$$

and, therefore,

$$\hat{\sigma}_c = \hat{\sigma}_m A_m \tag{4.43}$$

where $\hat{\sigma}_c$ = global failure stress, and $\hat{\sigma}_m$ = failure stress at the minimum cross-section area.

Substituting eqn. (4.41) in eqn. (4.40) and then in eqn. (4.43) one obtains[29]

$$\hat{\sigma}_c = \hat{\sigma}_m \left[1 - \pi \left(\frac{3}{4\pi} \right)^{2/3} \phi^{2/3} \right] = \hat{\sigma}_m (1 - 1\cdot21\phi_f^{2/3}) \tag{4.44}$$

Note that for volume fractions up to 0·30, eqns. (4.43) and (4.38) give the same results when $v = 0·5$ and a stress concentration factor (K) of 1·045 is used.

(*c*) *Strength of Short Fibre Composites*
When the fibres are aligned in one direction and a tensile stress is applied in the direction of the fibres, the middle regions will deform uniformly, i.e. the matrix and the fibres are under isostrain conditions.

At the fibre ends, on the other hand, the stress distribution is more akin to that prevailing with particulate fillers and, therefore, the fibres and the matrix in these regions are very close to the isostress conditions. This means that the matrix will deform much more than the fibres (owing to its lower modulus) and that shear stresses are set up at the interface. The distribution of tensile stresses (σ) and shear stresses (τ) along the fibres is shown in Fig. 4.10.

To relate the applied stress (σ_c) to the stresses acting on the fibres and matrix one can make use of the simple law of mixtures by taking the average tensile stress values, i.e.

$$\sigma_c = \phi_f \bar{\sigma}_f + (1 - \phi_f)\bar{\sigma}_m \qquad (4.45)$$

If the fibres are more brittle than the matrix it is likely that fracture takes place in the middle of the fibres and propagates through the matrix.

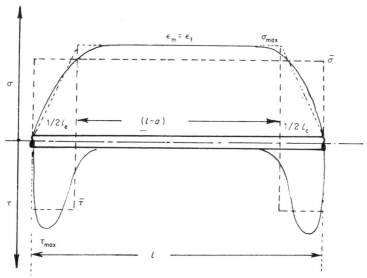

Fig. 4.10.   Stress distribution along short fibres when the external stress acts along the direction of the fibres.

Consequently to calculate the strength of the composite, $\hat{\sigma}_c$, it is necessary to know the relationship between the strength of the fibres, $\hat{\sigma}_f$, and the average stress acting on the fibres in the composite. It can be assumed that the difference between the maximum and average stress in the matrix is too small to have any significant effect on the calculated value of $\hat{\sigma}_c$.

It is more important, however, to note that the matrix will fail at a lower strain in a composite than it would in the absence of the fibres, i.e. it will fracture at a stress equal to $\hat{\varepsilon}_f E_m = \hat{\sigma}'_m$, where $\hat{\sigma}'_m < \hat{\sigma}_m$. One can calculate the average tensile stress along the fibres from the stress distribution shown in Fig. 4.8. In fact, approximating this to a trapezium, the average stress can be easily obtained by finding the height of the rectangle having the same area, i.e.

$$\hat{\sigma}_f \left[\frac{(l-a)+l}{2} \right] = \bar{\sigma}_f l \qquad (4.46)$$

and, therefore,

$$\bar{\sigma}_f = \hat{\sigma}_f \left(1 - \frac{l_c}{2l} \right) \qquad (4.47)$$

Hence eqn. (4.40) can be re-written as

$$\hat{\sigma}_c = \phi_f \hat{\sigma}_f \left(1 - \frac{l_c}{2l} \right) + (1 - \phi_f)\hat{\sigma}'_m \qquad (4.48)$$

where the factor $l_c/2l$ is the amount by which the strength of short fibre composites differs from that of a continuous fibre composite.

To calculate l_c, known as the critical fibre length (for reasons that will become obvious later), one considers the balance of forces at the fibre ends, i.e. the tensile forces along the fibre are opposed by the shear forces at the interface, i.e. at the point of fracture $\hat{\sigma}_f A = \bar{\tau}_{m/f} S_{m/f}$ or

$$\hat{\sigma}_f \frac{\pi d^2}{4} = \bar{\tau}_{m/f} \frac{\pi d l_c}{2} \qquad (4.49)$$

and, therefore,

$$l_c = \frac{\hat{\sigma}_f d}{2\bar{\tau}_{m/f}} \qquad (4.50)$$

where $\bar{\tau}_{m/f}$ = interfacial shear strength (bond strength), and d = diameter of the fibres.

It is obvious, therefore, that the maximum enhancement in strength for short fibre composites is achieved by using very fine fibres and by promoting strong adhesion between the fibres and the polymer matrix.

From an inspection of the stress distribution in Fig. 4.8, it is apparent that when the length of the fibres is shorter than l_c the maximum tensile stress in the middle of portion will never reach the ultimate strength of the fibres, $\hat{\sigma}_f$. Hence fracture cannot initiate within the fibre but either at the matrix/fibre interface or within the matrix.

In either case the composite is likely to fail under the influence of shear stresses at the interface, since these can cause debonding and yielding of the matrix.

The average tensile stress acting on the fibres can be calculated directly from the balance of forces, as in eqn. (4.49), i.e.

$$\bar{\sigma}_f = 2\bar{\tau}_{m/f} \frac{l}{d} \tag{4.51}$$

and consequently the strength of the composite for fibres shorter than the critical fibre length becomes

$$\hat{\sigma}_c = 2\phi_f \bar{\tau}_{m/f} \frac{l}{d} + (1 - \phi_f)\hat{\sigma}_m \tag{4.52}$$

Once more the diameter of the fibres and the adhesion between fibres and matrix play a major role in reinforcement by means of short fibres.

Obviously the accuracy of the predictions in both eqn. (4.47) and (4.51) increase with increasing the difference between the actual fibre length and the critical fibre length.

When the external stress forms an angle θ with the direction of the fibres the matrix and interfacial layers are subjected to shear stresses, owing to the decomposition of forces. The maximum shear stress in the plane of the fibres occurs when θ is 45° and, therefore, the composite is likely to fail by a shear–slip mode along the direction of the fibres (see Chapt. 6).

At angles just above or below 45° the normal stress components do not alter the failure mechanism and, consequently, the strength of the composite measured at an angle θ (where $45° > \theta < 90°$) can be obtained by relating the applied stress, σ_θ, to the critical resolved shear stress, $\bar{\tau}_{m/f}$ (i.e. the stress that causes failure on the microscale)

$$\hat{\sigma}_c|_\theta = \frac{\hat{\tau}_{m/f}}{\sin\theta\cos\theta} \tag{4.53}$$

When the stress, on the other hand, acts in the direction transverse to the direction of the fibres, failure is likely to occur through a crack opening mechanism, i.e. the crack is likely to originate at the fibre ends and propagate through the matrix and along the fibres. Consequently the failure conditions in the transverse directions can be written as

$$\hat{\sigma}_c|_\perp = \hat{\sigma}_m^* \qquad (4.54)$$

where σ_m^* is the stress at which the matrix fails in the composite and *not* the intrinsic strength of the matrix.

Unless one enters into fracture mechanics arguments (see Chapt. 6) the failure stress σ_m^* may be related to the intrinsic strength of the matrix by means of an arbitrary stress concentration factor. When the stress acts at an angle just below 90 °C, the failure mechanism is likely to remain unaltered and, therefore, the failure stress can be related to the strength in the transverse direction by the simple relationship

$$\hat{\sigma}_c|_\theta = \frac{\hat{\sigma}_c|_\perp}{\cos^2 \theta} \qquad (4.55)$$

Since in practice the fibres are invariably organised at random or only slightly biased towards one preferred orientation, none of the relationships described earlier can be applied directly. They are helpful, however, in making some rational estimates of the true strength of composites through the usual averaging procedures. For instance, if one considers the case where the fibres lie in one plane (as is frequently the case in practice) then one could make estimates for various levels of fibre alignments by assuming that the composite behaves in a manner similar to that of a laminate consisting of a number of layers, each containing fibres aligned at specific angles varying from 0 to 90°. Assuming that the three failure mechanisms described earlier operate simultaneously, one can assign different thickness values to each of these laminae according to the prevailing level of fibre alignment. The simplest procedures would consist of taking the distribution curve and dividing the total population in three intervals, as in Fig. 4.11; each of these would contain the number of fibres that form average angles of 0, 45 and 90° to the direction considered. Hence the relative thickness of each lamina will be taken as $t_1 = N_{||}/N$, $t_2 = N_{\sharp}/N$ and $t_3 = N_\perp/N$ and the strength of the composite can be estimated as

$$\langle \hat{\sigma}_c \rangle = t_1 \hat{\sigma}_c|_{||} + t_2 \hat{\sigma}_c|_{\sharp} + |t_3 \hat{\sigma}_c|_\perp \qquad (4.56)$$

For more accurate estimates, obviously, one would have to take into account also the fibre length distribution. This does not present any

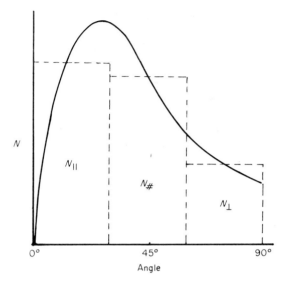

Fig. 4.11. Decomposition of the orientation of fibres in three laminae, with average orientation angles of $0°$, $45°$ and $90°$ respectively.

particular problem insofar as one can use a procedure similar to the one described above. More important is to establish the critical fibre length since this would determine the formula to be used for the calculation of $\hat{\sigma}_c|_{||}$. In using eqn. (4.53) one would have to decide on the value to assign to the term $\hat{\tau}_{m/f}$. If the interfacial bond between the fibres and the matrix is very strong so that failure originates within the matrix, $\hat{\tau}_{m/f}$ can be taken to be equal to half the value of the yield strength, on the basis of Tresca criterion (see Chapt. 6). When failure, on the other hand, occurs at the fibre/matrix interface the value of $\hat{\tau}_{m/f}$ would have to be estimated from practical experiments, i.e. from pull-out tests of fibres immersed in a block of the matrix, which are rather tedious and not very reliable.

4.3.3 Packing of Fibres and Fillers in the Matrix

As can be inferred from the previous discussions, the matrix plays a predominant role in determining the strength of short fibre and particulate filler composites. Particularly important is the minimum distance between filler particles since this determines the maximum stress acting within the matrix. For a given filler loading, therefore, particles or fibres that give the maximum packing density will exhibit the largest interparticle distance and, therefore, will produce the highest enhancement of strength by minimising stress concentrations within the matrix.

For the packing of spherical particles of different diameter ratios, R, the replacement of large spheres by small ones produces an increase in 'free' volume that is proportional to volume fraction of the smaller spheres, i.e.

$$V = C\phi_s \qquad (4.57)$$

where C is a proportionality constant that increases with the value of R.

When large spheres replace small ones the increase in 'free' volume follows the relationship

$$V = \phi_s + M\phi_l \qquad (4.58)$$

where ϕ_l is the volume fraction of larger spheres and M is the proportionality constant.

The values of C and M have been measured experimentally by

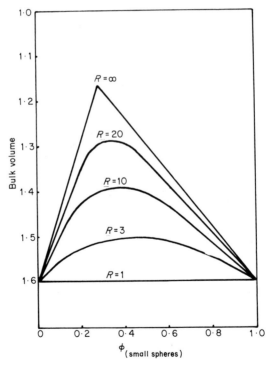

Fig. 4.12. Packing of small and large spherical particles. (After Milewski.[31])

McGeary[30] for a number of R values, ranging from 3·4 to 16·5. Their effect on the bulk volume (or occupied volume) is shown in Fig. 4.12. A similar approach was taken by Milewski[31] to determine the bulk volume of short fibres mixed with spherical particles, for R values ranging from 4 to 70 and ratios of the fibres varying from 7 to 50 (Fig. 4.13).

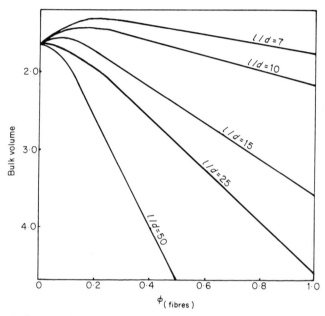

Fig. 4.13.   Packing of fibres and spheres for $R = 10$. (After Milewski.[31])

By defining the packing efficiency as the ratio of the deviations of bulk volume for $R = 1$ and $R = \infty$ respectively, a series of curves were obtained for a range of fibre aspect ratios as a function of the diameter ratio R (Fig. 4.14).

These data are helpful in determining the maximum amount of fillers and fibres that can be used in a matrix to avoid contiguity.

4.3.4 Enhancement of Filler/Matrix Adhesion

Earlier discussions have clearly pointed out the need to have a strong interfacial adhesion in order to maximise strength enhancement.

Since inorganic fillers invariably contain hydroxyl groups on the surface, owing to inevitable reactions with atmospheric water or simply due to

Table 4.8: Typical Silane and Titanate Coupling Agents

Chemical name	Chemical structure	Applicable polymer
Vinyl triethoxysilane	$CH_2{=}CHSi(OC_2H_5)_3$	Polyolefins, ABS, impact polystyrene, etc
γ-Methacryloxypropyltrimeth-oxysilane	$CH_2{=}\overset{\underset{\textstyle CH_3}{\mid}}{C}{-}\overset{\underset{}{\overset{\textstyle O}{\parallel}}}{C}{-}OCH_2CH_2CH_2Si(OC_2H_5)_3$	
γ-Aminopropyltrimethoxysilane	$H_2NCH_2CH_2CH_2Si(OC_2H_5)_3$	Polyamides, polyesters, polyurethanes, etc.
γ-Glycidoxylpropyltrimethoxysilane	$\underset{\textstyle O}{CH_2{-}CH{-}CH_2}{-}O{-}CH_2{-}CH_2{-}CH_2{-}(SiOCH_3)_3$	Polyvinyl chloride, polyesters, polyamides, polyurethanes, etc.
γ-Mercaptopropyltrimethoxysilane	$HSCH_2CH_2CH_2Si(OCH_3)_3$	Polyvinyl chloride, ABS, polyurethanes, etc.
Monoalkoxy isostearoyl dimethacrylate titanate	$RO{-}Ti\left(\underset{\textstyle O}{\overset{\textstyle O}{\parallel}}C(CH_2)_{14}{-}\underset{\underset{\textstyle CH_3}{\mid}}{CH}{-}CH_3\right)\left(\overset{\underset{\textstyle CH_3}{\mid}}{O}CC{=}CH_2\right)_2 \atop {\scriptstyle 2} {\scriptstyle O}$	Polyolefins

Monoalkoxy tri(dioctyl pyro phosphato)titanate

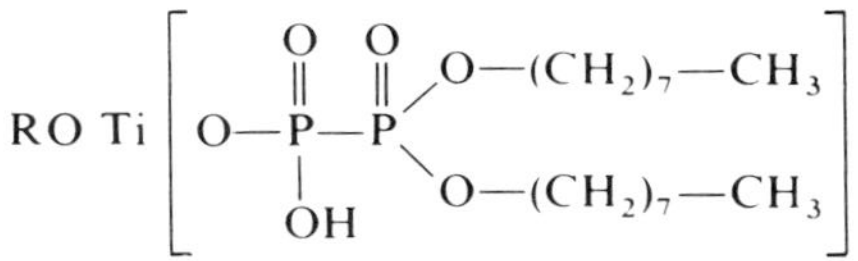

$$RO\,Ti\left[O-\underset{\underset{OH}{\parallel}}{\overset{\overset{O}{\parallel}}{P}}-\underset{}{\overset{O}{\parallel}}{P}\begin{matrix}O-(CH_2)_7-CH_3\\O-(CH_2)_7-CH_3\end{matrix}\right]$$

Vinyl polymers

Monoalkoxy 4-amino benzine sulphonyl di(dodecylbenzene) sulphonyl titanate

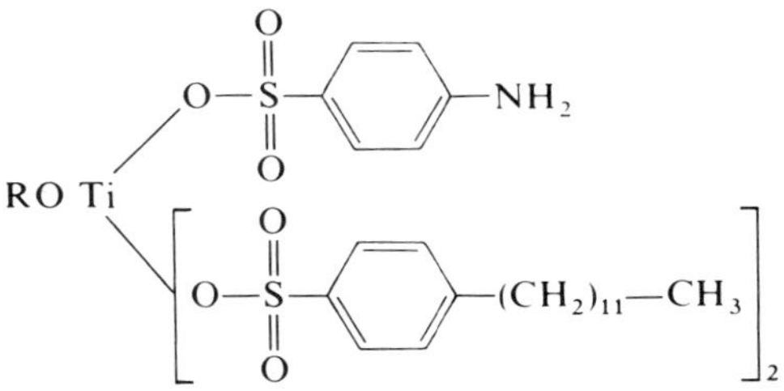

Polyamides

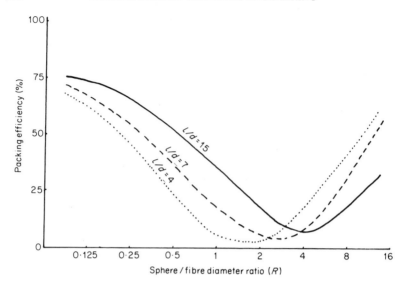

Fig. 4.14.   Packing efficiency of fibres and spherical particles. (After Milewski.[31])

strong adsorption forces resulting from the high surface energy, there are very few polymers that are intrinsically capable of forming strong interfacial bonds, i.e. only those that can form hydrogen bonds.

Furthermore the hydrophilic nature of inorganic fillers makes it easy for atmospheric water to accumulate at the interface by diffusing through the matrix. As a result, interfacial bonds of thermoplastics/filler systems are either intrinsically weak or deteriorate on ageing, when the composite is exposed to humid environments. For this reason the majority of fibrous and particulate fillers are precoated with so-called 'coupling agents' prior to being blended with the polymer.

The most widely used coupling agents are 'silane' and 'titanate' based compounds, whose chemical composition allows them to react with both the surface of the filler and the polymer matrix.

The manner in which the coupling agent may react with the filler surface is shown in Fig. 4.15.

The constitution of the organic part of the molecules of a coupling agent varies according to the nature of the polymer matrix. Examples of various silane and titanate coupling agents used for thermoplastics are shown in Table 4.8.

The surface of particulate fillers, such as clay, calcium, carbonate and talc, is often modified by completely different treatments from those

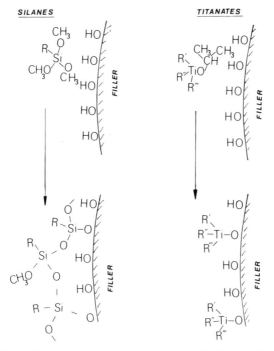

Fig. 4.15.   Reaction of coupling agent with filler surface.

described earlier. Typical modifications consist of depositing on the surface of the filler long chain fatty acids or amines in order to reduce interparticle friction, to minimise agglomeration and improve the dispersion in the polymer matrix.[32]

Other treatments include the deposition of polymeric coating on the

Table 4.9: Typical Properties of Inorganic Fibres Used for Reinforcement of Thermoplastics

Fibres	Density (g/cm^3)	Young's modulus $\times 10^{-9}$ (N/m^2) (average values)	Tensile strength $\times 10^{-9}$ (N/m^2) (average values)
Glass E type	2·55	75	2·0
Glass S type	2·49	75	5·5
Carbon	2·00	170–200	0·5–1·0
Chrysotile asbestos	2·50	160	2·0
Silicone carbide	3·15	220–300	4·0–10·0

Table 4.10: Comparison of Mechanical Properties of Typical Thermoplastics and Glass Fibre Reinforced Grades

Polymer	Non-reinforced grade		Reinforced grade	
	Young's modulus $\times 10^{-9}$ (N/m^2)	Yield strength $\times 10^{-6}$ (N/m^2)	Young's modulus $\times 10^{-9}$ (N/m^2)	Tensile strength $\times 10$ (N/m^2)
Polypropylene (carboxylated type)	1·3	35	5·0	80
Nylon 6,6	2·8	80	8·0	160
Polyethersulphone	2·6	85	8·4	140

surface either directly, such as low-molecular-weight (liquid) carboxy-terminated-polybutadiene or by depositing a liquid monomer which is then polymerised *in situ*. Upon drying the polymer can oxidise to produce carboxylic acid groups, which will form strong bonds with the surface of the filler.

REFERENCES

1. L. MASCIA, *The Role of Additives in Plastics*, Edward Arnold, London, 1974, p. 2.
2. P. J. FLORY, *Principles of Polymer Chemistry*, Cornell University Press, 1953, p. 509.
3. J. H. HILDEBRAND, *J. Am. Chem. Soc.*, **51** (1929) 66.
4. C. B. BUCKNALL, *Toughened Plastics*, Applied Science, London, 1977, p. 19.
5. A. S. AKHMATS, *Molecular Physics of Boundary Friction*, Israel Program for Scientific Translations, 1966, p. 283.
6. F. P. BOWDEN and D. TABOR, *The Friction and Lubrication of Solids*, Oxford University Press, London, 1950.
7. G. BROWN and A. J. KOVACS, *Proc. Conf. on Physics of Non-crystalline Solids*, J. A. Prins (ed.), North-Holland, Amsterdam, 1965, p. 303.
8. E. JENCKEL and R. HEUSCH, *Kolloid-Z.*, **89** (1953) 130.
9. V. JACOBSON, *Br. Plast.*, **32** (1959) 152.
10. L. MASCIA, *Polymer*, **19** (1978) 325.
11. D. A. KOURTIDES, W. J. GILWEE JR and J. A. PARKER, *SPE Regional Tech. Conf. Connecticut Sect.*, 1977, p. 113.
12. D. W. VAN KREVELEN, *Polymer*, **16** (1975) 615.
13. P. J. FLORY, *Principles of Polymer Chemistry*, Cornell University Press, 1953, p. 503.
14. C. DOUBE and D. J. WALSH, *Polymer*, **20** (1979) 1115.

15. J. W. SCHURER, A. DE BOER and G. CHALLA, *Polymer*, **16** (1975) 201.
16. D. J. WALSH and J. G. MCKEOWN, *Polymer*, **20** (1980) 1330.
17. Y. HIRATA, S. UEMURA and T. KOTAKA, *Report on Progress in Polymer Physics in Japan*, **22** (1979) 180.
18. T. K. KWEI, *Contemporary Topics in Polymer Science*, Vol. 2, E. M. Pearce and J. R. Schaefgen (eds.), Plenum Press, London, 1977, p. 165.
19. R. A. DICKIE, *Polymer Blends*, Vol. 1, D. R. Paul and S. Newman (eds.), Academic Press, London, 1978, p. 359.
20. C. B. BUCKNALL, *Toughened Plastics*, Applied Science, London, 1977, p. 189.
21. C. B. BUCKNALL, Cranfield Institute of Technology, Personal communication, 1981.
22. L. MASCIA, *The Role of Additives in Plastics*, Edward Arnold, London, 1974, p. 66.
23. J. E. ASHTON, J. C. HALPIN and P. H. PETIT, *Primer on Composite Analysis*, Technicon Publishing Co., Stamford, Conn., 1969, Chapt. 5.
24. S. W. TSAI, *Formulas for Elastic Properties of Fibre Reinforced Composites*, AD 834851, June 1958.
25. R. M. OGORKIEWICZ, *Composites*, **3** (1972) 232.
26. R. A. SCHAPERY, *J. Comp. Mat.*, **2** (1968) 380.
27. L. MASCIA and F. A. SPECK, *Reinforced Plastics Congress*, Brighton, 12–14 Nov. 1974, Sect. 3, pp. 91–100.
28. R. M. JONES, *Mechanics of Composite Materials*, McGraw-Hill, New York, 1975, p. 120.
29. L. NICOLAIS and N. NARKIS, *Polym. Eng. Sci.*, **11** (1971) 194.
30. R. K. MCGEARY, *J. Am. Ceram. Soc.*, **44** (1961) 513.
31. J. V. MILEWSKI, *Polym. Plast. Technol. Eng.*, **3** (1974) 101.
32. M. MOTOYOSHI, *Japan Plastics Age*, **13** (1975) 33.

5

Deformation Behaviour of Thermoplastics in Relation to Product Design

The transition from the 'plastic age' to that of the 'efficient use of plastics', which began to take effect in the 1960s and is expected to gather momentum within the next decade, has created a need for a rational approach to designing with plastics. Whereas in the past, with few exceptions, designers have taken advantage of the ease of processing and colouring of plastics to produce aesthetically desirable (or acceptable) products, they must now turn their attention to the penetration of markets where reliability of performance is a prime criterion for material selection. Transportation and building are examples of such markets. Economic advantages once enjoyed by plastics have been reduced by the increase in raw material costs and, therefore, designers must concentrate increasingly on desk calculations in order to establish optimum geometry, size and materials for the production of a given component.

The purpose of this chapter is (a) to establish the requirements for rational engineering design, (b) to determine what data are necessary, (c) to determine how the data can be obtained experimentally and presented to designers, and (d) how the engineering properties (as opposed to bulk properties, i.e. T_g, T_m, etc.) are related to the formulation and structure parameters of plastics compositions.

On the question of materials evaluation, one must distinguish three basic types of test:

(i)   Tests for quality control. These are generally simple and involve standardised procedures, e.g. ISO, ASTM, BS, DIN, etc. They are generally known as 'one-point' tests, for they measure the value of a particular function under one set of conditions only. For this reason the procedures must be standardised to eliminate discrepancies in results.

(ii) Tests for research purposes. These are often elaborate and somewhat subjective since they are exploratory in nature and are aimed only at advancing knowledge.

(iii) Tests to obtain data for engineering design. These may or may not be covered by standard procedures.

5.1 INTRODUCTORY CONSIDERATIONS FOR DESIGNING WITH POLYMERS

In engineering design, calculations are normally based on the following analyses:

(i) Stiffness or equilibrium considerations.

(ii) Failure criteria and estimation of the maximum (or safe) working stresses or strains.

(iii) Friction and wear considerations.

5.1.1 Design Based on Stiffness Considerations

Engineering formulae for the relationships between deformations, loads and geometry factors can be generalised as follows:

$$\Delta = \frac{P}{E} f(v, d) \tag{5.1}$$

where Δ = deformation (or deflection) observed, P = load applied. $f(v, d)$ = function of the Poisson ratio and geometry, E = Young's modulus.

Alternatively, the compliance (D) may be used instead of the Young's modulus; so that the general expression for the deformation becomes

$$\Delta = PDf(v, d)$$

For instance, in a centre-loaded beam supported at the two ends, the maximum deflection is given by:

$$\Delta = \frac{L^3 P}{4Ebd^3} = \frac{DL^3 P}{4bd^3}$$

where L = distance between supports, d = thickness of the beam, b = width of the beam.

The Poisson ratio does not enter into the calculations, since the only normal stresses acting on the beam are along its length (perpendicular to

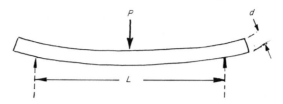

Fig. 5.1. Three point bending of a rectangular beam.

the applied load). Note also that the deflection due to shear stresses is very small and can be neglected (Fig. 5.1).

If a simply supported circular plate is subjected to a centre load P, the plate will be subjected to biaxial stresses (normal stresses) in the plane of the plate (Fig. 5.2). Consequently, the formula for the maximum deflection will contain the Poisson ratio term, i.e.

$$\Delta = \left[\frac{3(3 + v)(1 - v^2)}{4\pi(1 + v)E}\right]\left(\frac{Pr^2}{d^3}\right)$$

The general expression above holds true also for beams subjected to torsion, where the angle of twist θ can be related to the torque T and the geometric parameter through the shear modulus of the material, i.e.

$$\theta = \frac{T}{G}\left(\frac{l}{\beta bd^3}\right)$$

where l is the length of the beam from the clamped end and β is a shape factor which depends on the aspect ratio b/d. Other examples involve the bulk modulus K, but occur less frequently than those involving the Young's modulus and shear modulus. Since all of these constants are related to one

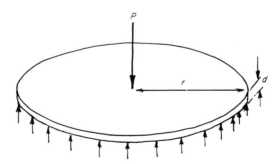

Fig. 5.2. Simply supported circular plate loaded at the centre.

another, it is normally unnecessary to measure all of them for a given material; i.e.

$$G = \frac{E}{2(1 + v)} \qquad G = \frac{3(1 - 2v)K}{2(1 + v)} \qquad v = \frac{1 - 2G/3K}{2 + 2G/3K} \qquad (5.2)$$

However, several assumptions must be made in the application of the equations for deflections, deformations or angles of distortion. These are:

(i)   The deflections or deformations are very small.
(ii) The material is isotropic.
(iii) Each of the stiffness parameters (K, G, v and E) is constant and, therefore, independent of the levels of deflection, deformation, volumetric changes, etc., and of the direction of the stresses, i.e. whether in tension or compression.

The fact that all of these conditions are satisfied by the traditional engineering materials (e.g. metals, ceramics, concretes, etc.) has enabled materials technologists to make the necessary measurements from simple test procedures and to present the data to the design engineer in tabulated form. Furthermore, the accuracy of the predictions that can be made by use of these simple constants has helped the practising engineer to acquire a certain degree of confidence in the theoretical approach to design.

Even when the geometry of the component and the loading distribution have been much more complex than stipulated in a given analysis, the engineer has been able, by making some simplifying assumptions, at least to deduce orders of magnitude which could subsequently be checked by means of service simulated tests or prototype evaluations.

The introduction on the market of cold-rolled metal sheets, which possess a degree of anisotropy, may have produced some uncertainty since this material does not conform to the second assumption. For metals, however, the degree of anisotropy is often small and can be neglected. When the degree of anisotropy is larger, especially with wood, the analysis can be modified (albeit only for the most simple loading situations) through the use of the 'orthotropic' plate assumption; i.e. the material is assumed to be isotropic across the thickness, while the anisotropy ratio is constant along the plane of the sheet. From the materials evaluation point of view the only complication that such a modification introduces is that 5 constants must be measured instead of 3. These are the Young's modulus and the Poisson ratio in the two principal directions of the plane (i.e. $E_{\parallel}$, $E_{\perp}$, $v_{\parallel}$ and $v_{\perp}$) and the shear modulus (G_{12}).

It is noteworthy that from the values of these constants in the principal

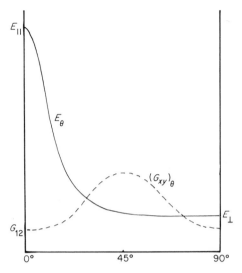

Fig. 5.3.   Modulus variation in function of the angle for an anisotropic (orthotropic) plate.

directions of the plane, it is possible to predict their variation as a function of the angle formed by the direction of the load and the principal direction along which the modulus is highest, i.e.

$$\frac{E_{\parallel}}{E_{\theta}} = \cos^4\theta + \frac{E_{\parallel}}{E_{\perp}}\sin^4\theta + \left[\frac{E_{\parallel}}{G_{12}} - 2v_{12}\right]\sin^2\theta\cos^2\theta$$

and

$$\frac{1}{(G_{xy})_{\theta}} = \frac{1}{E_{\parallel}}\left[1 + 2v_{12} + \frac{E_{\parallel}}{E_{\perp}}\right] - \left[1 + 2v_{12} + \frac{E_{\parallel}}{E_{\perp}} - \frac{E_{\parallel}}{G_{12}}\right]\cos^2 2\theta$$

(see Fig. 5.3). The consequence of the anisotropy can be immediately seen for the cases of the beam and plate mentioned earlier. The central load would cause the two opposite edges (corresponding to the diagonal line along which the modulus is highest) to rise from the supports, and it may well be necessary to predict this in addition to the deflection.

5.1.2 Stiffness Approach to Designing with Plastics
The biggest problem that arises from attempts to apply the classical elasticity formulae to designing for plastics materials is that none of the three conditions are valid. The isotropy assumption, probably, presents the

least difficulty in most cases, while attempts to apply the other assumptions encounter insurmountable difficulties. Furthermore, the properties can be quite sensitive to environmental conditions, particularly temperature, and may change with ageing through morphological readjustments; absorption or environmental agents, such as water (or even changes in equilibrium concentration of absorbed matter): and chemical reactions with atmospheric agents, such as oxygen, ultraviolet light, etc.

(a) *Analysis of Limitations Arising from Large Deflections*
Owing to the intrinsic low rigidity of plastics materials, with the exception of some high performance composites, deflections and deformations are always likely to be high and may well exceed the limits of the applicability of the theory. Modifications of the standard type of equations to take into account the larger deflections are sometimes possible through a more rigorous analysis of the deflections or simply by empirical curve-fitting procedures.

For a centrally loaded (freely supported) rectangular beam, a theoretical analysis produces the following relationship between the applied load and the geometry of the beam.[1]

$$\frac{\Delta P}{L} \left(\frac{PL^2}{EI} \frac{1}{\cos \alpha} \right)^{1/2} = - \int_0^\alpha \sin \phi \sin^{-1/2} (\alpha - \phi) \, d\phi \qquad (5.3)$$

where α = angle of deflection formed by the axis of the beam at the supports, ϕ = angle of deflection formed by the beam at a distance from the support, I = second moment of area (i.e. $I = \int_A y^2 \, dA$, where A = cross-section area of an element of the beam at distance y from the central axis).

The complete solution of eqn. (5.3) is shown in Fig. 5.4, from which it can be observed that a linear relationship between the applied load (P) and the deflection (Δ) exists only for very small angles of deflection. It is important to note also that the beam will collapse unless the load is reduced for $\Delta/L > 0.24$.

(b) *Analysis of the Limitations Resulting from the Assumption that the Stiffness Coefficients are Constant*
The major problems that have arisen when classical elasticity formulae have been used are due to the failure of both polymer technologists and design engineers to recognise that plastics do not behave like metals and other traditional engineering materials, such as ceramics, etc. This meant that the test procedures and testing equipment developed were akin to those already established for metals. Because of their simplicity, rapidity

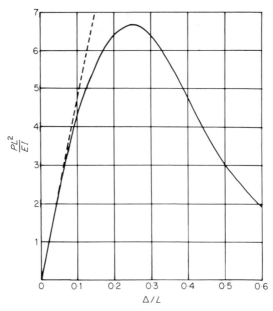

Fig. 5.4.   Deflection of a freely supported centrally loaded rectangular beam, freely supported at the edges. (After J. G. Williams.[1])

and low cost, these tests have become very popular and widely accepted as evaluation procedures for plastics. Although industry now defends this practice on the grounds that these tests were developed merely for quality control purposes and as a basis for setting up materials specifications between suppliers and users, their intrinsic limitations are still not fully appreciated.

5.2   VISCOELASTICITY APPLIED TO PLASTICS

In the case of plastics, stresses and strains are not related by simple proportionality constants or elastic coefficients, but depend on time, temperature, level of external excitation (e.g. stress or strain level acting on the material) and indirectly also on changes in structure resulting from interactions with environmental agents. Therefore, the use of classical elasticity formulae for plastics would entail replacing these coefficients (e.g. E, G and v) with some appropriate functions that take into account the above effects; e.g. $E = f(t, \varepsilon, T, A)$; $D = F(t, \sigma, T, A)$, etc., where A is an arbitrarily defined structural parameter which is likely to change as a result of the interactions of the material with environmental agents.

If the effect of each variable could be separated, the above functions could be evaluated experimentally by examining the effect of each variable in turn while keeping the others constant. From a practical point of view the evaluation of these functions should not be too time-consuming, nor should it require the use of expensive and elaborate equipment.

Since one is dealing with functions of time, without a reliable theory to make long-term predictions, the task of obtaining these functions is impossible. The only reasonably accurate theory which is currently available is that of 'linear viscoelasticity', which is applicable to plastics only at low stress or strain levels (normally strains less than 0.5%) and at temperatures not too far from the glass transition temperature.

5.2.1 Linear Viscoelasticity Principles

The time and temperature dependence of the compliance and modulus of polymers arises from the low intermolecular forces and the flexibility of polymer chains.

An external force applied at time $t = 0$ causes deformations which result from internal displacements of molecular chains and bond distortions. Thereafter the stored potential energy is gradually redistributed through rotations of segments of molecular chains, producing further deformations until a new equilibrium between the external forces and the intermolecular forces is established.

It is to be expected that if the temperature is raised the internal expansion would cause both a reduction in the initial modulus and a more rapid decrease in modulus with time, as a result of the faster molecular rotations accompanying the redistribution of internal stress.

A convenient way of quantifying the decrease in modulus (or increase in compliance) with time and temperature is to use mechanical analogues, made up of combinations of springs and dash-pots. The term 'visco-elasticity', in fact, derives from these models where the spring represents the elastic response while the dash-pot simulates the viscous component.

The relationship between stress and strain and, therefore, the modulus or compliance as function of time for these analogues is obtained from solutions of general linear differential equations of the type:

$$a_n \frac{d^n \sigma}{dt^n} + a_{n-1} \frac{d^{n-1} \sigma}{dt^{n-1}} + \cdots + a_0 \sigma = b_m \frac{d\varepsilon^m}{dt^m} + b_{m-1} \frac{d\varepsilon^{m-1}}{dt^{m-1}} + \cdots + b_0 \varepsilon$$

(5.4)

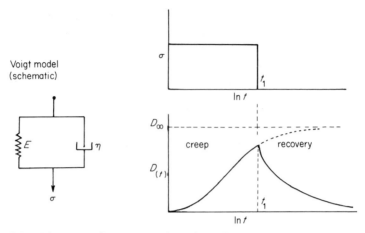

Fig. 5.5. Diagrammatic representation of the time-dependent compliance for a Voigt model during creep and recovery.

(a) The Voigt Model

This simple model consists of a spring and dash-pot in parallel, as in Fig. 5.5, and is used to describe the time-dependent deformation resulting from a constant stress input, i.e. *creep conditions*. In this situation the deformation of the two elements is isometric while the stresses are additive. In other words one can write

$$\sigma = \sigma_E + \sigma_D \qquad (5.5)$$

where $\sigma_E = \varepsilon_E E$ is stress acting on the spring, $\sigma_D = \eta(d\varepsilon/dt)$ is stress acting on the dash-pot, $E =$ modulus of the spring element, $\eta =$ viscosity of the dash-pot, $\varepsilon_E =$ strain in the spring, and $d\varepsilon/dt =$ strain rate of the dash-pot.

Substitution in (5.5) gives

$$\sigma = \varepsilon_E E + \eta \frac{d\varepsilon}{dt} \qquad (5.6)$$

The solution of eqn. (5.6) is

$$\varepsilon \exp\left[\frac{Et}{\eta}\right] = \frac{\sigma}{E} \exp\left[\frac{Et}{\eta}\right] + \text{const.}$$

and, therefore,

$$\varepsilon = \frac{\sigma}{E}\left[1 - \exp\left(-\frac{Et}{\eta}\right)\right] \qquad (5.7)$$

Note that the subscript E for the strain has been omitted, since at any given time t the strain on the spring and dash-pot is the same (isometric conditions). The parameter η/E determines the rate at which the strain ε reaches its equilibrium value (see Fig. 5.5) and is known as the 'retardation time', since it has the dimensions of time.

One can also replace the term $1/E$ with D_∞, i.e. the equilibrium compliance and, dividing eqn. (5.7) by the stress input σ, an equation for the time-dependent compliance is obtained:

$$D_{(t)} = D_\infty \left[1 - \exp \left(-\frac{t}{\lambda_c} \right) \right] \tag{5.8}$$

Note that the solution for the Voigt model is a specific case of the general constitutive equation, (5.4), where all the high order derivatives and a_1 are zero, i.e.

$$a_0 \sigma = b_0 \varepsilon + b_1 \frac{d\varepsilon}{dt}$$

where b_0/a_0 is equivalent to E and b_1/a_0 is synonymous with η in eqn. (5.6).

When the stress is removed after time t_1, the strain will recover to zero at a rate depending only on the retardation time and on the strain at time t_1. It can be shown that the strain recovery equation is

$$\varepsilon_{(t)} = \varepsilon_{(t_1)} \exp \left[-\frac{(t - t_1)}{\lambda_c} \right]$$

If written in terms of the compliance the above equation becomes

$$D_{(t)} = D_\infty \left[1 - \exp \left(-\frac{t_1}{\lambda_c} \right) \right] \exp \left[-\frac{(t - t_1)}{\lambda_c} \right] \tag{5.9}$$

(b) The Maxwell Model
The Maxwell model consists of a spring and a dash-pot connected in series and can be used to represent the viscoelastic response for a constant strain input, i.e. stress relaxation conditions (Fig. 5.6).

For this model the stress on the spring is the same as that acting on the dash-pot and, therefore, the strains are additive, i.e.

$$\varepsilon = \varepsilon_E + \varepsilon_D$$

through the usual substitutions one can also write

$$\frac{d\varepsilon}{dt} = \frac{1}{E} \frac{d\sigma}{dt} + \frac{\sigma}{\eta}$$

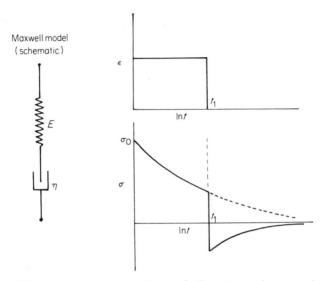

Fig. 5.6. Diagrammatic representation of the stress decay under stress relaxation conditions at constant strain for the Maxwell model.

which is a specific case of the general linear differential equation, (5.4), when all the higher order derivatives and b_0 are zero, i.e.

$$a_0 \sigma + a_1 \frac{d\sigma}{dt} = b_1 \frac{d\varepsilon}{dt}$$

where b_1/a_1 is equivalent to E and b_1/a_0 is equal to η.
At constant strain $d\varepsilon/dt = 0$ and, therefore,

$$\frac{1}{E} \frac{d\sigma}{dt} + \frac{\sigma}{\eta} = 0$$

or

$$\ln \sigma = -\frac{E}{\eta} t + \text{const.}$$

Since at $t = 0$, $\sigma = \sigma_0$ the solution of the above equation becomes

$$\sigma_{(t)} = \sigma_0 \exp \left(-\frac{t}{\lambda_R} \right) \tag{5.10}$$

where $\lambda_R = \eta/E$ is known as the 'relaxation time' and determines the rate at which the stress will decay to zero.

Written in terms of the modulus, i.e. dividing by the strain ε, eqn. (5.10) becomes

$$E_{(t)} = E_0 \exp\left(-\frac{t}{\lambda_R}\right) \tag{5.11}$$

where E_0 is the modulus at $t = 0$, known also as the 'instantaneous modulus'.

If at time t_1 the strain is made to go to zero the stress function takes the form

$$\sigma_{(t)} = \sigma_0 \left[\exp(-t/\lambda_R) - \exp\left(-\frac{(t - t_1)}{\lambda_R}\right) \right] \tag{5.12}$$

In other words for the strain to become zero after time t_1, the stress has to assume negative values, from which it will recover to zero as the time t approaches infinity (see Fig. 5.6).

Since in the deformation of polymers an increase in temperature produces a steeper variation of the strain with time and a more rapid relaxation of the stress, one must infer that the retardation and relaxation time are inversely related to temperature. It can also be deduced that the retardation and relaxation time must be associated with segmental rotations imposed by the external stress and, consequently, their variation with temperature must be viewed as a rate process.

In other words the retardation and relaxation time can be related to temperature by means of the Arrhenius equation, i.e.

$$\frac{1}{\lambda} = \frac{1}{\lambda_0} \exp\left(-\frac{\Delta H}{RT}\right) \quad \text{or} \quad \lambda = \lambda_0 \exp\left(\frac{\Delta H}{RT}\right) \tag{5.13}$$

where ΔH is the activation energy and λ_0 is the retardation (or relaxation time) at very high temperatures, i.e. when $\Delta H/RT \simeq 0$.

(c) The Standard Linear Solid

The models described earlier fail to satisfy the 'materials objectivity' requirements insofar as they cannot be used indiscriminately to represent both the creep and stress relaxation behaviour of polymers. The Voigt model does not exhibit stress relaxation under contant strain conditions, i.e. eqn. (5.6) has an elastic solution when $d\varepsilon/dt = 0$, while the Maxwell model does not display full strain recovery when the stress is removed, i.e. when the stress goes to zero at time t_1, the strain in the dash-pot will not recover and remains at the value $\varepsilon_D = \sigma t_1/\eta$.

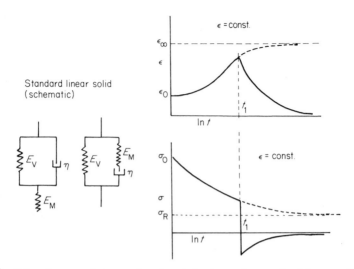

Fig. 5.7.   Creep and stress relaxation behaviour of a standard linear solid.

Materials objectivity is satisfied, however, by a three-element model as in Fig. 5.7, which is known as the standard linear solid, since it contains the features of both Voigt and Maxwell models.

The constitutive equation for the standard linear solid is a specific case of the general equation, (5.4), i.e.

$$a_0 \sigma + a_1 \frac{\mathrm{d}\sigma}{\mathrm{d}t} = b_0 \varepsilon + b_1 \frac{\mathrm{d}\varepsilon}{\mathrm{d}t}$$

or, if derived from the model in Fig. 5.7, it becomes

$$\frac{\mathrm{d}\sigma}{\mathrm{d}t} + \frac{E_M}{\eta}\,\sigma = (E_V + E_M)\,\frac{\mathrm{d}\varepsilon}{\mathrm{d}t} + \frac{E_V E_M}{\eta}\,\varepsilon$$

The solutions of this constitutive equation are respectively:

(i) *Stress relaxation* (i.e. constant strain input $\varepsilon = \varepsilon_0$):

$$\sigma_{(t)} = \frac{E_M E_V}{(E_M + E_V)}\,\varepsilon_0 + c_1 \exp\left[-\frac{(E_M + E_V)}{\eta}\,t\right] \qquad (5.14)$$

where

$$\frac{\eta}{(E_M + E_V)} = \lambda_R \qquad c_1 = \sigma_0 \qquad \text{(i.e. the stress at } t = 0)$$

and

$$\frac{E_M E_V}{(E_M + E_V)} \varepsilon_0 = \sigma_R \quad \text{(i.e. the relaxed stress at } t \to \infty)$$

Equation (5.14) can be written in the form of a relaxation modulus, i.e.

$$E_{(t)} = E_R + E_0 \exp(-t/\lambda_R) \quad (5.15)$$

where E_R = Relaxed modulus and E_0 = Elastic (or instantaneous) modulus.

(ii) *Creep conditions* (i.e. constant stress input, $\sigma = p_0$):

$$\varepsilon_{(t)} = \sigma_0 \left\{ \frac{1}{E_M} + \frac{1}{E_V} [1 - \exp(-t/\lambda_c)] \right\} \quad (5.16)$$

where $\lambda_c = \eta/E_V$.

This can be written in the form of two terms containing respectively an elastic compliance (D_0) and an equilibrium compliance (D_∞), i.e.

$$D_{(t)} = D_0 + D_\infty [1 - \exp(-t/\lambda_c)] \quad (5.17)$$

(d) Retardation and Relaxation Spectra

Although the standard linear solid gives a more accurate representation of the deformational behaviour of polymers than either the Voigt or the Maxwell models, it displays a rather steep change in the respective deformational parameters $D_{(t)}$ and $E_{(t)}$ when the excitation time approaches the value of λ. The curves obtained experimentally are much flatter and not necessarily symmetrical at values of $t > \lambda$ and $t < \lambda$ respectively, as depicted in Fig. 5.7.

This situation can be corrected, however, by using a large number of the same models connected either in series or in parallel. By doing so the solutions to the constitutive equations can be added up for each model and become

$$E_{(t)} = E_R + \sum_{i=1}^{n} E_0|_i \exp(-t/\lambda_R|_i)$$

and

$$D_{(t)} = D_0 + \sum_{i=1}^{n} D_\infty|_i [1 - \exp(-t/\lambda_c|_i)]$$

where E_R and D_0 are the sum of the relaxed and elastic components and, therefore, can be written as constants.

Obviously if an infinite number of such models are considered the above equations can be written as integrals, i.e.

$$E_{(t)} = E_R + \int_0^\infty E_0 \exp\left(-t/\lambda_R\right) d\lambda_R \qquad (5.18)$$

and

$$D_{(t)} = D_0 + \int_0^\infty D_\infty [1 - \exp(-t/\lambda_c)] d\lambda_c \qquad (5.19)$$

Since the true functions are nominally presented on a logarithmic scale, one can replace $d\lambda$ with $\lambda\, d(\ln \lambda)$. At the same time the equilibrium functions of the deformational parameters can be replaced by the so-called 'distribution spectra', i.e.

$$E_0 \lambda = H(\ln \lambda) \qquad \text{(distribution of relaxation times)}$$

and

$$D_0 \lambda = L(\ln \lambda) \qquad \text{(distribution of retardation times)}$$

Consequently the above equations will be rewritten as

$$E_{(t)} = E_R + \int_{-\infty}^\infty H(\ln \lambda) \exp\left(-t/\lambda\right) d(\ln \lambda) \qquad (5.20)$$

and

$$D_{(t)} = D_0 + \int_{-\infty}^\infty L(\ln \lambda)[1 - \exp\left(-t/\lambda\right)] d(\ln \lambda) \qquad (5.21)$$

(e) Viscoelastic Functions in Complex Notation

When a viscoelastic material is subjected to dynamic stresses it is convenient to express the modulus and compliance functions with complex notations.

This convenience arises from the fact that stress and strain, in a given cycle, are out of phase and can be separated, therefore, into in-phase (real part) and out-of-phase (imaginary part) components.

To illustrate this approach one can take the two extreme cases, i.e. an elastic material and a viscous material; the viscoelastic case can be considered as being intermediate between these two extremes. For an elastic material the strain is proportional to the stress at any given time in the cycle, i.e.

$$\varepsilon_{(t)} = \frac{\sigma_{(t)}}{E}$$

so that, if the stress varies sinusoidally with time, the strain will follow exactly the same variation, i.e.

$$\sigma_{(t)} = \sigma_0 \sin \omega t$$

and

$$\varepsilon_{(t)} = \frac{\sigma_0}{E} \sin \omega t$$

For a viscous material, on the other hand, the stress input is related to the strain rate through the viscosity, i.e.

$$\frac{d\varepsilon}{dt} = \frac{\sigma_{(t)}}{\eta}$$

In this case, therefore, the strain rate goes through a maximum when the stress reaches the maximum value, and vice versa. Hence if the stress is a sinusoidal function, the maximum strain will occur when the stress is at the minimum and, therefore, stress and strain will be out of phase by 90°, i.e.

$$\frac{d\varepsilon}{dt} = \frac{\sigma_0}{\eta} \sin \omega t$$

and

$$\varepsilon = \frac{\sigma_0}{\eta\omega} - \frac{\sigma_0}{\eta\omega} \sin \omega t$$

It follows, therefore, that for a viscoelastic material the stress and strain will be out of phase by an angle between 0 and 90°, called the loss angle δ.

This means that the stress and strain vectors can be decomposed into in-phase and out-of-phase components as in Fig. 5.8.

In the decomposition of the stress vector the projection of σ on ε yields σ' (the component of σ in phase with ε), while the projection of σ on the axis perpendicular to ε yields σ'' (the component 90° out of phase with ε). From these one can define the in-phase and out-of-phase moduli, E' and E'', as

$$E' = \frac{\sigma'}{\varepsilon} \quad \text{and} \quad E'' = \frac{\sigma''}{\varepsilon} \tag{5.22}$$

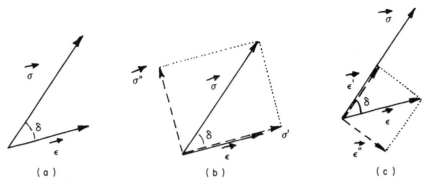

Fig. 5.8.   Real and imaginary parts of stress and strain in dynamic tests. (a) Phase angle between stress and strain vector. (b) Decomposition of the stress vector. (c) Decomposition of the strain vector.

Similarly, the projection of ε on σ yields ε' (the strain component in-phase with the stress), while the projection of ε onto a vector axis perpendicular to σ yields ε'' (the out-of-phase component of the strain). Hence the compliance functions D' and D'' will be defined as

$$D' = \frac{\varepsilon'}{\sigma} \quad \text{and} \quad D'' = \frac{-\varepsilon''}{\sigma}$$

Note that the negative sign in front of ε'' denotes its placement in the negative quadrant in relation to ε'.

Because of the vector nature of stress and strain, their decomposition into in-phase and out-of-phase components can be written in complex number notation, i.e.

$$\sigma^* = \sigma' + j\sigma'' \quad \text{and} \quad \varepsilon^* = \varepsilon' - j\varepsilon''$$

Consequently the corresponding modulus and compliance become

$$E^* = E' + jE'' \quad \text{and} \quad D^* = D' - jD'' \quad (5.23)$$

where E^* and D^* are the complex components, E' and D' are the elastic components, and E'' and D'' are the loss components.

The loss components, therefore, determine the extent of deviation of the deformability parameters E^* and D^* from the elastic behaviour.

This can also be inferred from the observation that the loss angle δ is the arc tangent of the ratio of the loss (out-of-phase) component to the elastic (in-phase) component, i.e.

$$\frac{E''}{E'} = \tan \delta \quad \text{and} \quad \frac{D''}{D'} = \tan \delta \quad (5.24)$$

Obviously the constitutive equations for the Maxwell, Voigt and standard linear solid models can also be solved for conditions where the stress and strain vary sinusoidally with time.

For the case where spectra, rather than discrete (single models), retardation and relaxation times are considered the solutions become

$$E' = \int_0^\infty \frac{H(\lambda_R)\omega^2\lambda_R^2}{1+\omega^2\lambda_R^2}\, d\lambda_R \tag{5.25}$$

$$E'' = \int_0^\infty \frac{H(\lambda_R)\omega\lambda_R}{1+\omega^2\lambda_R^2}\, d\lambda_R \tag{5.26}$$

and

$$D' = \int_0^\infty \frac{L(\lambda_c)}{1+\omega^2\lambda_c^2}\, d\lambda_c \tag{5.27}$$

$$D'' = \int_0^\infty \frac{L(\lambda_c)\omega\lambda_c}{1+\omega^2\lambda_c^2}\, d\lambda_c \tag{5.28}$$

Because these equations have a sound theoretical foundation, the functions obtained from one set of experimental conditions (say, at constant stress) can be transformed into corresponding functions for other conditions.

While for time-independent (elastic) materials the modulus is the reciprocal of the compliance, for linear viscoelastic materials the same relationship applies only when transforms that eliminate the time variation are considered, i.e.

$$p\bar{E}(p) = \frac{1}{p\bar{D}(p)}$$

where $p\bar{E}(p)$ and $p\bar{D}(p)$ are the Carson transforms of $E(t)$ and $D(t)$ functions respectively.

When the approximation is made that a power-law type of relationship exists between the modulus and the elapsed time and m is taken as the slope of the log/log plot, the relationship between modulus and compliance becomes

$$D(t) = \frac{\sin \pi m}{m\pi}\frac{1}{E(t)} \tag{5.29}$$

The values of m for typical thermoplastics are shown in Table 5.1.

Table 5.1[2]

Polymer	m Values
Polyoxymethylene (copolymer)	0·095
Polyvinyl chloride	0·07 (rising to 0·10 after 2 months)
Polypropylene	0·125
High density polyethylene	0·17

Furthermore, the linear theory permits the calculation of the material response to any arbitrary stress input through the so-called 'Boltzmann superposition principle'.

In other words, an arbitrary stress input (Fig. 5.9) can be subdivided into a series of elementary step inputs; so the total strain response can be obtained by adding up each elemental strain response (i.e. each of these is independent of the previous history). If each step taken is infinitesimal, the strain response can be calculated through the use of the so-called convolution integrals; i.e.

$$\varepsilon(t) = D_0\sigma(t) + \int_{\theta=-\infty}^{\theta=t} \sigma \frac{d\sigma}{d\theta} \int_{\lambda=-\infty}^{\lambda=+\infty} [L(\ln \lambda)(1 - \exp^{-(t-\theta)/\lambda})\,d\theta]\,d(\ln \lambda)$$

which can also be written as

$$\varepsilon(t) = D_0\sigma(t) + \int_{-\infty}^{t} \dot{D}(t-\theta)\sigma_{(\theta)}\,d\theta$$

The same approach would be used with an arbitrary strain input variable. The relaxation of the stress would be calculated by means of the equation

$$\sigma(t) = E_\infty\varepsilon(t) + \int_{-\infty}^{t} \dot{E}(t-\theta)\varepsilon_{(\theta)}\,d\theta$$

By taking Laplace transforms, all the viscoelastic coefficients can be used in all calculations involving stress/strain relationships in exactly the same manner as for elastic materials; i.e.

$$\bar{\sigma}_x(p) = \bar{E}(p)\bar{\varepsilon}_x(p)$$

or

$$\bar{\varepsilon}_x(p) = \bar{D}(p)\bar{\sigma}_x(p)$$

Under triaxial stresses, for instance, the principal strains would be related to principal stresses in the following manner:

$$\bar{\varepsilon}_1(p) = \bar{D}\{(p)\bar{\sigma}_1(p) - \bar{v}(p)[(\bar{\sigma}_2(p) + \bar{\sigma}_3(p)]\}$$

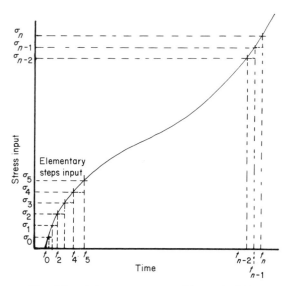

Fig. 5.9.   Boltzmann superposition principle.

(*f*) *Effects of Temperature on Linear Viscoelastic Behaviour*
The effects of temperature on the modulus or compliance can be taken into
account by letting the time-independent component be a function of density
and by applying the time/temperature superposition for the time-
dependent component. So, if the effect of temperature is considered in
relation to a conveniently chosen temperature, the ratio of the relaxed
modulus (or instantaneous compliance) at any temperature relative to a
reference temperature is simply:

$$\frac{[D_0]_{T_R}}{[D_0]_T} = f\left(\frac{\rho_T}{\rho_{T_R}}\right) \quad \text{or} \quad \frac{[E_\infty]_{T_R}}{[E_\infty]_T} = \varphi\left(\frac{\rho_{T_R}}{\rho_T}\right) \tag{5.30}$$

The same applies for the time-independent components of the shear and
bulk modulus and for the Poisson ratio, so that the relationships between
these coefficients do not change with temperature.

The strength of these theoretical arguments lies in the experimental
evidence that the time-independent coefficients are related only to the
intermolecular forces and the internal bond 'strain' energy of the polymer
molecules.

The time-dependent component, on the other hand, is related to the bond
rotational energy along the polymer backbones and to the steric hindrance

exerted by neighbouring molecules, which in turn affect the retardation and relaxation times of the individual components of the spectrum.

Increasing the temperature reduces both retardation and relaxation times by increasing the free volume and reducing the potential energy barriers for the molecular rotations. This is normally taken into account by utilising the concept of 'shift factor' a_T, defined as

$$a_T = \frac{[E(t)]_{T_R}}{[E(t)]_T} \quad \text{or} \quad a_T = \frac{[D(t)]_T}{[D(t)]_{T_R}} \tag{5.31}$$

Where $T > T_R$.

In other words, while the decrease in density resulting from an increase in temperature corresponds to a vertical shift in the modulus and compliance curves as functions of log (time), the above shift factor a_T determines the amount of horizontal shift, i.e. towards shorter times, when the temperature is increased (Fig. 5.10).

For temperatures around the main glass transition temperature, the shift factor, a_T, can be calculated by means of the WLF equation; i.e.

$$\log a_T = \frac{-c_1(T - T_g)}{c_2 + T - T_g} \tag{5.32}$$

The constants c_1 and c_2 depend on the nature of the material and are related to factors, such as free volumes, rotational energy barriers of segments within the polymer chains and the activation energy of the relaxation processes involved. In other words, each relaxation (or

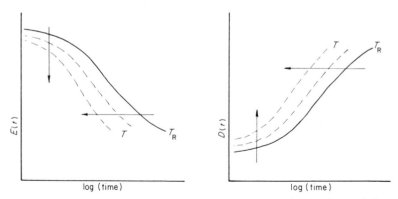

Fig. 5.10.   Effects of temperature on the deformational behaviour of linear viscoelastic materials.

retardation) time, λ, can be assumed to change with temperature according to the Arrhenius rate equation; i.e. if

$$K = A \exp\left(-\frac{\Delta H}{RT}\right)$$

where K = rate of molecular rotations, then

$$\frac{1}{\lambda} = \frac{1}{\lambda_0} \exp\left(-\frac{\Delta H}{RT}\right) \quad \text{or} \quad \lambda = \lambda_0 \exp\left(\frac{\Delta H}{RT}\right) \tag{5.13}$$

The shift along the time axis which results from an increase in temperature, ΔT, is associated with a reduction in the values of λ.† This means that the distribution functions $L(\ln \lambda)$ and $H(\ln \lambda)$ are shifted along the λ axis by an amount equivalent to the shift factor a_T; so that from eqn. (5.13), above, one can write

and

$$a_T = \lambda_1/\lambda_0 = \exp\frac{\Delta H}{RT} \quad \text{or} \quad a_T = \frac{\lambda_T}{\lambda_{T_R}} = \exp\left[\frac{\Delta H}{R}\left(\frac{1}{T} - \frac{1}{T_R}\right)\right]$$

hence

$$\log a_T = \frac{\Delta H}{2 \cdot 303 RT} \quad \text{or} \quad \log a_T = \frac{\Delta H}{2 \cdot 303 R}\left(\frac{1}{T} - \frac{1}{T_R}\right)$$

$$\frac{d(\log a_T)}{dT} = -\frac{\Delta H}{2 \cdot 303 RT^2} \quad \text{or} \quad \frac{d(\log a_T)}{dT} = -\frac{\Delta H}{2 \cdot 303 RT(T_R)}$$

Consequently, the activation energy of the retardation or relaxation processes can be obtained from the slope of the plot of $\log a_T$ versus $1/T$. Because, in effect, this is the global activation energy for all the individual relaxation processes involved, a curve, rather than a straight line, is obtained from the plot, i.e. ΔH varies with temperature.

From these discussions, it can be concluded that a linear viscoelastic material is fully characterised when the following parameters are known:

(a) The elastic components of $E(t)$, $G(t)$ and $v(t)$ at a reference temperature, and the change of density with temperature.

† Note, for instance, that a master curve implies single values of D or E at a specific elapsed time, hence a change in λ arising from a change in temperature must be compensated by an exactly equal change in t, i.e. $\log \lambda_T - \log \lambda_{T_R} = \log t_T - \log t_{T_R}$.

(b)   The distribution of retardation times of the creep compliance and the distribution of retardation times over a wide range of relaxation and retardation times. These can be obtained from the slope of the plots of the modulus or compliance as a function of the logarithm of the time from experiments carried out over a range of temperatures and by constructing master curves at some reference temperature.

The above distributions are obtained by the Tschoegl method:[3] i.e.

$$L(\ln \lambda) \simeq \frac{1}{2 \cdot 303} \frac{d[D(t)]}{d(\log t)} \qquad \text{(constant stress experiments)} \qquad (5.33)$$

$$H(\ln \lambda) \simeq \frac{1}{2 \cdot 303} \frac{d[E(t)]}{d(\log t)} \qquad \text{(constant strain experiments)} \qquad (5.34)$$

For dynamic (sinusoidal input) experiments, the distributions are obtained from the plots of modulus or compliance as a function of $\log \omega$.

(c)   The values of m, i.e.

$$m = \frac{d[\log D(t)]}{d(\log t)} \qquad (5.35)$$

(d)   The distribution functions for shear stress and hydrostatic stress inputs.

(e)   The distribution function in respect to the Poisson ratio.

(f)   The shift factor through the coefficients c_1 and c_2 or the activation energy of the shift factor.

A full experimental evaluation of the linear viscoelastic functions is, therefore, a formidable task. However, certain assumptions can reduce considerably the amount of work required. Within the glassy state and rubbery state, in fact, it can be assumed that the bulk modulus is time-independent, while within the glass/rubber transitional state, which is the region of application of most plastics materials, the bulk modulus is only a mild function of time. It seems, for instance, that while the shear modulus may change by a factor of 1000 over a given time range, the bulk modulus changes by a factor of only 10 over the same time range.[2] This indicates that the errors arising from the assumption of a time-independent bulk modulus are likely to be quite small. From eqns. (5.2) one can write:

$$D(t) = \tfrac{1}{3}J(t) + \frac{B_0}{9} \quad \text{and} \quad E(t) = 3K_0(1 - 2v)$$

where B_0 is the time-independent bulk compliance, K_0 is the time-independent bulk modulus and $J(t)$ is the time-dependent shear compliance.

An alternative simplifying assumption is that the Poisson ratio is constant.[4] This would simplify considerably the problem of design, since it would make all the other viscoelastic functions have the same time dependence, i.e.

$$E(t) = 2(1 + v_0)G(t)$$

and

$$E(t) = 3(1 - 2v_0)K(t)$$

Although at variance with experimental observations, small adjustments in the assigned values for v may be sufficient to correct this situation.

In fact, since the Poisson ratio varies between approximately $\frac{1}{3}$ for a short-time glassy state behaviour to $\frac{1}{2}$ for a long-time glassy/rubbery state and for a rubbery state behaviour at any elapsed time, the maximum error that will result from the assumption of a constant v value of $0 \cdot 4$ is likely to be approximately 5% in estimations using the above equations.

When a third simplification is made, that is, the assumption of incompressibility, the relationship

$$\frac{G}{K} = \frac{3}{2} \frac{(1 - 2v)}{(1 + v)}$$

predicts a rapid change in K as the value of v approaches $0 \cdot 5$. This assumption would be realistic only for materials in the rubbery state and particularly for cross-linked (unfilled) amorphous polymers. However, even if the above assumptions can be made in various situations, it must be remembered that the linear viscoelastic analysis is applicable only to designs involving small strains and fairly short elapsed times, consequently it would be useful only for the design of very rigid structures, such as those involving high performance composites.

5.2.2 Non-linear Viscoelastic Behaviour

A linear viscoelastic behaviour implies that the coefficients that relate stresses and strains are functions of time only; hence, at any given elapsed time (i.e. under isochronous conditions) these coefficients are constant. In effect, this is why the relationships among the various elastic coefficients can be applied to linear viscoelastic materials through the Laplace or Carson transforms.

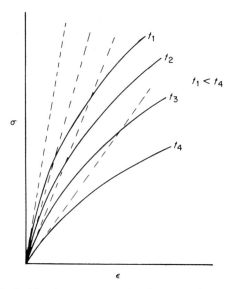

Fig. 5.11.　Typical isochronous stress/strain curves for thermoplastics.

In practice, the isochronous stress/strain relationships of most plastics are not linear; the relaxation modulus decays with an increase in the level of strain, while the compliance increases with stress level (see Fig. 5.11).

From Fig. 5.11 it can be seen that the isochronous stresses and strain values are related by simple proportionality coefficients only at small levels of σ and ε and that the deviations from linearity increase with an increase in the level of the excitation variable (input) and with the elapsed time. Outside the linear regions the compliance and modulus functions can be written as

$$D(t) = f(t, \varepsilon) \quad \text{and} \quad E(t) = E(t, \sigma)$$

Therefore, while the strain response to a constant stress input (σ_0) for a linear viscoelastic material can be written as

$$\varepsilon(t) = \varepsilon_0 + \phi(t)$$

or

$$\varepsilon(t) = D_0\sigma_0 + \sigma_0\phi^*(t)$$

For a non-linear material this becomes

$$\varepsilon(t, \sigma) = g[\varepsilon_0, \sigma, \phi^*(t)] \quad (5.36)$$

which, if divided into time-independent and time-dependent components, yields

$$\varepsilon(t, \sigma) = f_1(\varepsilon_0, \sigma) + g_1(\phi^*(t), \sigma) \qquad (5.37)$$

For instance, Findley[5] found a good fit for his experimental data with the following equation:

$$\varepsilon = \varepsilon_0^* \sinh a\sigma + bt^n \sinh c\sigma \qquad (5.38)$$

where ε_0^*, a, b, c and n are empirical constants.

For very small values of σ the Findley equation is typical of a linear viscoelastic material, in view of the fact that the hyperbolic sine function becomes equal to its argument; i.e.

$$\sinh x \simeq x|_{x \to 0}$$

Hence

$$\varepsilon(t) = \varepsilon_0^* a\sigma + bc\sigma t^n$$

or

$$D(t) = D_0 + At^n$$

where the power expression for the time-dependent component is approximately equal to the exponential function for a certain narrow range of t values.

However, there are several indications that the Findley equation and other similar ones that have been proposed do not have general applicability.[6] The suggestion is often made that the non-linear behaviour of the compliance (presumably, also the relaxation modulus) can be taken into account from isochronous compliance curves plotted as a function of the stress input (Fig. 5.12). The problems arising from such an approach are that (a) the material has to be tested fully, and (b) the Boltzmann superposition principle to estimate the response to a variable stress input is no longer applicable.[2]

Although some modified forms of the Boltzmann superposition principle have been suggested,[7] the equations involved are intractable, and the functions themselves are difficult to measure experimentally. Ward and Onant[8] have proposed that the following convolution integrals could be used:

$$\varepsilon(t) = \int_{-\infty}^{t} \frac{d\sigma(u_1)}{du_1} D_1(t - u_1)\, du_1 + \int_{-\infty}^{t} \int_{-\infty}^{t} \frac{d\sigma(u_1)}{du_1} \frac{d\sigma}{du_2}$$

$$\{D_2[(t - u_1), (t - u_2)]\, du_1\}\, du_2 + \cdots$$

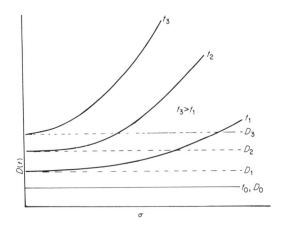

Fig. 5.12. Empirical description of the non-linear viscoelastic behaviour of plastics. $D_{(t)} = D_0 + D_1|_{t_1}\sigma + D_2|_{t_2}\sigma^2 + D_3|_{t_3}\sigma^3 + \cdots$.

in which the terms $D_1(t - u_1)$, $D_2[(t - u_1), (t - u_2)]$, etc., are new materials constants.

A more practical suggestion has also been made that it should be possible to construct time/stress superposition curves in a similar fashion to the time/temperature master curves. In other words, the isochronous compliance curves in Fig. 5.12 should yield a master curve through shifts along the σ axis and, possibly, also through vertical movements if the first term is also affected by σ. This suggestion is theoretically sound on the basis that rate processes are stress activated and, therefore, Eyring theories of the effects of stress on viscosity should be applicable. It has been argued, in fact, that the presentation of creep curves through the hyperbolic sine functions by Findley is a manifestation of the applicability of the stress activation concept.[9] The author has suggested that if the time-independent component of the compliance is also independent of the stress level (the justification for this assumption is that, by definition, a time-independent parameter cannot enter a rate process), increasing the stress level has the effect of amplifying the shift of the 'distribution' to shorter values of retardation times, arising from an increase in temperature.

In other words, if the time/temperature shift factor is

$$a_T = \frac{\lambda|_{T_R}}{\lambda|_T} \quad \text{where } \lambda = \lambda_0 \exp\left(\frac{\Delta H}{RT}\right)$$

the time/temperature/stress shift factor can be written as

$$a_{T,\sigma} = \frac{\lambda|_{T_R,\sigma_R}}{\lambda|_{T,\sigma}}$$

where λ can be related to temperature and excitation stress through the Eyring rate equation, i.e.

$$1/\lambda = 1/\lambda_0 \exp\left(-\frac{\Delta H}{RT}\right) \sinh\left(\frac{v\sigma}{RT}\right)$$

or

$$\lambda = \lambda_0 \exp\left(\frac{\Delta H}{RT}\right) \bigg/ \sinh\left(\frac{v\sigma}{RT}\right) \qquad (5.39)$$

where v is the so-called 'activation volume', which is defined thermodynamically as the coefficient of the change in free energy as a function of the stress at constant temperature and pressure. For example,

$$v = \left|\frac{\delta F}{\delta \sigma}\right|_{T,P} \qquad \text{or} \qquad v \equiv kT\left[\frac{\delta L(\lambda)}{\delta \sigma}\right]_{T,P} \qquad (5.40)$$

Consequently, while the thermal activation energy of the deformation process is obtained by measuring the slope of the plot of either $\log \lambda$, $\log L(\ln \lambda)|_\lambda$ or $\log a_T$ versus $1/T$, the combined effect of the stress and temperature (i.e. the lowering of thermal activation energy by the stress) should be obtainable from a plot of $\log [a_T/\sinh (v\sigma/RT)]$ versus $1/T$, where the value of a_T is that obtained from measurements of the compliance (or modulus) at very low stress levels, such as in dynamic tests (i.e. in the linear viscoelastic region). This requires a pre-knowledge of the activation volume, which according to the thermodynamic definition given above should be obtainable from plots of $L(\lambda)|_\lambda$ versus σ at constant temperature and pressure. These hypotheses, however, have not yet been verified experimentally, and there are some indications that the assumption of a stress-independent component of the compliance may not be valid.[10]

In any case, even if through exploitation of these concepts the non-linear theories of viscoelasticity can be rationalised and the relevant data on materials gathered, it is improbable that exact analytical solutions will ever be used in designing with plastics. The major obstacles will always be:[11]

(i) The difficulty of finding suitable superpositions (like the Boltzmann superposition principle) to predict the response to variable inputs.

(ii)   The lack of adequate transforms (like Laplace transforms for the linear viscoelasticity case) which would permit the use of established relationships between the various coefficients which relate stress to strain. This is complicated, in fact, by the practical observation that the modulus in compression becomes greater than that in tension above certain strain levels, depending on the nature of the material.[12]

(iii)   Ageing factors cannot be incorporated into a generalised theory.

(iv)   The anisotropy often encountered in plastics components, if neglected in order to make the analysis easier, may produce errors of order of magnitude greater than those arising from the other sources already mentioned.

5.3   STIFFNESS COEFFICIENTS MEASUREMENTS BY STANDARD TEST METHODS

The majority of standard tests are carried out on equipment that operates at constant rate of clamp separation, even though most instruments are capable of operating at different speeds. Occasionally, test procedures make use of purpose-built 'jigs', or equipment that is recommended for the testing of specific materials.

In the following sections are described some standard test methods to demonstrate that they have been developed without any real consideration of the viscoelastic nature of polymers.

5.3.1 Tensile Modulus Measurements

ASTM D638: This test recommends the use of dumbell-shaped specimens with a parallel waist (2 in long × $\frac{1}{2}$ in wide × $\frac{1}{8}$ in thick), which are pulled at a speed of 0·2–0·25 in/min, and a load/extension curve is recorded (Fig. 5.13).

BS 2782 method 302A: This test recommends the use of rectangular specimens which are clamped. An initial small load is then applied to straighten the specimen in the clamps. The specimen is then pulled at a speed of 1 mm/min, and the load/extension curve is recorded (Fig. 5.13).

In the ASTM method the modulus is calculated from the tangent of the curve to the origin by taking the slope after the load and extension are converted into nominal stress and engineering strain. In the BS method a secant from the origin is drawn through the point on the curve corresponding to 0·2 % extension (or strain). The ASTM test gives the so-

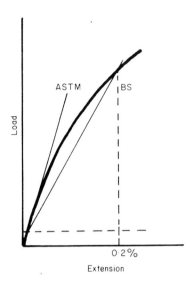

Fig. 5.13.   Examples of standard methods for the calculation of the modulus from load/extension curves.

called 'tangent modulus', while the BS method measures the 'secant modulus'. Hence the values obtained by the two methods are different.

The limitations of the values obtained for engineering design are quite obvious in relation to the theoretical principles outlined earlier. The ASTM method not only gives rise to errors in the drawing of the tangent to the curve but also neglects completely the non-linearity aspects. The BS method overcomes, to some extent, these errors by specifying the strain level and gives at least a measure of reliability as far as reproducibility is concerned. Both of them are, however, 'one-point' tests; i.e. they measure only one value of the stress/strain ratio of the functions described earlier.

5.3.2 Flexural Modulus Measurements

Flexural tests, usually in the form of three-point bending of rectangular specimens, are used more for convenience and simplicity than for fundamental reasons. Since deflections are much easier to measure than tensile extensions, these tests do not require the use of expensive extensometers. (See later.) From bending theory the normal stress σ at a distance y from the neutral axis, either on the tension or compression side, is related to the bending moment M and the second moment area I of the cross-section, defined as $I = \int_A y^3 \, dA$ (Fig. 5.14).

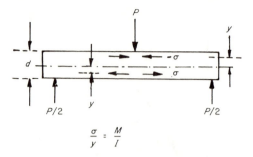

$$\frac{\sigma}{y} = \frac{M}{I}$$

Fig. 5.14. Centrally loaded freely supported beam (three-point bending).

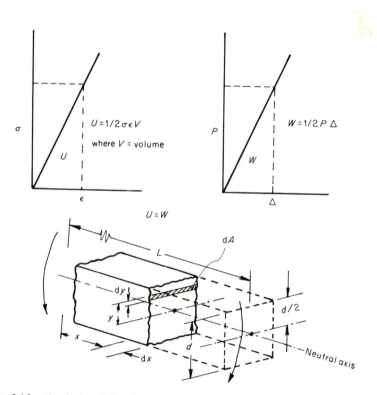

Fig. 5.15. Analysis of the three-point bending test for the calculation of the flexural modulus.

To obtain an expression for the 'flexural modulus' (which corresponds to the Young's modulus if the modulus in tension and compression for the material at the particular level of deflection are the same), one can equate the strain energy (i.e. the internal response of the material) to the work done on the beam by the external load P (Fig. 5.15).

Now $U = \frac{1}{2}\sigma\varepsilon V = (\sigma^2/2E)V$; where for an infinitesimal element at distance y from the axis, $dV = dA\, dx$.

Hence

$$U = \int \frac{\sigma^2}{2E}\, dV$$

and since $\sigma = (M/I)y$

$$U = \int \frac{dx}{2E} \int \frac{M^2 y^2}{I^2}\, dA$$

Substituting $\int_A y^2\, dA = I$, one obtains

$$U = \int \frac{M^2}{2EI}\, dx$$

Now the bending moment taken from one support at the end of the beam is $M = (P/2)X$ (Fig. 5.16), which reaches a maximum at the mid-point where $X = L/2$. Thus,

$$U = 2 \int_0^{L/2} \frac{P^2 x^2}{8EI}\, dx$$

Therefore

$$U = 2\left[\frac{P^2}{8EI} \int_0^{L/2} x^2\, dx \right] = \frac{P^2 L^3}{96EI}$$

For a rectangular cross-section $I = bd^3/12$, therefore

$$U = \frac{P^2 L^3}{8Ebd^3} = W = \frac{1}{2}P\Delta$$

Hence,

$$E = \frac{PL^3}{4bd^3 \Delta} \quad \text{or} \quad D = \frac{4bd^3 \Delta}{PL^3} \qquad (5.41)$$

Note that a certain amount of deflection occurs due to the shear stresses (maximum at the neutral axis); hence by similar arguments one can

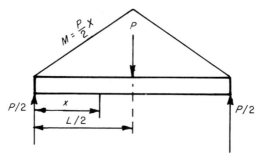

Fig. 5.16.   Variation of bending moment along the length of specimen in a three-point bending test.

calculate the shear strain energy in the beam and equate this to the work done by the load in deflecting the beam through shear deformations. Hence if the shear stress τ at a distance y from the neutral axis is[13]

$$\tau = \frac{6P}{bd^3}\left(\frac{d^2}{4} - y^2\right)$$

and the strain energy in the strip of volume $(b\,dx\,dy)$ is $\frac{1}{2}(\tau^2/G)b\,dx\,dy$, the total shear strain energy becomes

$$U = \frac{3P^2L}{5bdG}$$

Therefore,

$$\frac{3P^2}{5bdG} = \frac{1}{2}P\,\Delta s \qquad \text{(work done by shear)}$$

and

$$\Delta s = \frac{6PL}{5dbG}$$

The ratio of the deflection due to normal stresses to that due to shear stresses is

$$\frac{\Delta}{\Delta s} = \frac{5L^2 G}{24d^2 E} \simeq \frac{5L^2}{72d^2}$$

Therefore, if the thickness of the specimen (d) is very small and the length of the beam (L) is very large, the deflection due to shear becomes insignificant and can be neglected.

Standard test methods normally recommend that L/d should be at least 15, which makes the deflection due to normal stresses approximately 16 times greater than that due to shear. In order to remain within the applicability of the bending theory, one would not exceed (for the particular $L/d = 15$) deflections of about 4 mm; therefore, the deflection due to shear would be at the most of the order of $\frac{1}{4}$ mm, which probably is not even detectable by the usual methods of measurements, and the error would only be of the order of 1 %.

(a) Standard Tests, ASTM D790 and BS 2782 (Method 302D)
The rate of cross-head motion (i.e. deflection rate) and the span vary according to the thickness and width of the specimen. The recommended values are normally given in tables but can be calculated. The span-to-thickness ratio is usually in the range 15–20. The rate of cross-head motion is determined from the desired strain rate at the outer skin of the specimen and can be calculated from the relationship between maximum strain and deflection at the midpoint of the bar.

For example,

$$\Delta = \frac{\varepsilon}{6} \frac{L^2}{d}$$

Therefore

$\dot{\Delta} \simeq \dfrac{\dot{\varepsilon}L^2}{6d}$ (deflection rate or approximate rate of cross-head motion)

The ASTM recommends that the maximum strain in the outer skin should not exceed the value of 0·05 (i.e. 5 %), while the rate of strain should not be more than 0·01/min. With an L/d ratio of 18, this means that the deflection rate should be between 0·05 and 0·24 in/min.

The flexural modulus can be expressed as a 'tangent' value or 'secant' value (Fig. 5.17), at a specified maximum skin strain, using the formula

$$E = \frac{L^3 P}{4bd^3 \Delta}$$

where P/Δ is the slope, i.e. the tangent and secant respectively through the origin. While the tangent values should be the same as those obtained in tension (if the strain rates selected are the same), the secant values may differ in view of the compressibility effects in the flexural measurements. It

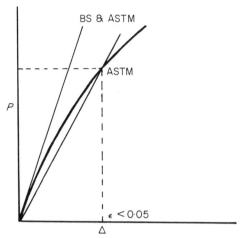

Fig. 5.17.   Tangent and secant modulus determination in a three-point bending test.

is expected, therefore, that the flexural modulus will be somewhat higher than the tensile modulus.

5.3.3 Heat-distortion-temperature (HDT) or Deflection-under-load Tests

The HDT test is carried out to obtain an indication of the modulus of the material at high temperatures. Measurements are made on a three-point bending fixture (ASTM D648, BS 2782 methods 102G–102H) by submitting the specimen to a constant load and recording the increase in deflection while the temperature is raised at a constant rate. The temperature at which a certain pre-specified deflection is obtained is recorded and the value is taken as the 'heat distortion temperature' of the material. Two loads (corresponding to 0.45–1.7×10^6 N/m² skin stress) are normally used according to the rigidity of the material and have to be specified when quoting the results.

Similar tests, often carried out by means of a cantilever beam (BS 2782), are sometimes used to obtain the same information. The values obtained from HDT tests lack physical meaning in terms of material properties for design calculations; in addition, there is a fundamental reason why such tests cannot provide even reproducible results: thermoplastics can undergo further crystallisation during the test, while thermosets can undergo further curing. Even if these problems could be avoided the procedure still could not evaluate the complex function

$$D = f(t, T, \sigma)$$

since the stress level is fixed and the temperature and time variables are changed simultaneously.

5.3.4 Shear Modulus Measurements

There are only two standard tests which use a shear mode of deformation. The first is the dynamic shear modulus using a torsion pendulum (ASTM 2236) and the second is the 'Clash and Berg' test for the measurement of the 'cold flex temperature' of flexible materials (ASTM D1043, BS 2782 method 104B). The Clash and Berg test operates on the same principle as the heat-distortion-temperature test, i.e. the temperature is raised at constant rate from very low temperatures in a refrigerated medium and the angular deflection resulting from the application of a torque at one end of the specimen is monitored at regular temperature intervals. In the BS test the temperature at which a certain pre-specified angular deflection (200°) is obtained on a standard size specimen, and is recorded and quoted as the 'cold flex temperature' of the material: the ASTM method stipulates that the angular deflection should fall in the range of 5–100° (in order to remain within the applicability of torsion theory) and the shear modulus at each temperature is calculated using the following formula:

$$G = \frac{917TL}{bd^3\phi\beta}$$

where T = torque applied, L = specimen length (between clamps), b = width of specimen, d = thickness of specimen, ϕ = angle of deflection, and β = geometric factor which depends on the b/d ratio.

The results for the ASTM test are either presented in the form of a plot of log G versus T or as the temperature at which $G \simeq 1 \times 10^9$ N/m^2. Note that although the ASTM test is by far the more accurate of the two, neither of them takes into account the viscoelastic nature of the material.

The torsion pendulum (ASTM) test method records the frequency and the decay of the oscillations of the specimen subjected to a known moment of inertia, from which the elastic component of the shear modulus and the loss tangent are calculated. This method takes into account the viscoelastic characteristics of the material but only in the linear viscoelastic range, since the shear stress to which the specimen is subjected is normally very small.

5.4 RATIONALISATION OF TEST PROCEDURES FOR THE MEASUREMENT OF THE STIFFNESS PARAMETERS OF PLASTICS MATERIALS

5.4.1 Standard Tests for Routine Evaluations and Quality Control Purposes

Test methods for routine evaluations have to be simple, speedy and reliable (i.e. operator errors must be minimal). Furthermore, because of the very large number of tests to be carried out, it is necessary that the equipment be relatively inexpensive and possibly built as multi-stage systems or 'battery units'. By far the simplest method for loading the specimens is by applying a constant weight and monitoring the deformations as a function of time, i.e. a 'creep' loading system.

For experiments in tension and compression, it is normally necessary to provide load amplifications through mechanical leverage, a procedure which normally requires eliminating friction and damping arrangements to avoid impact loading. (See later discussions.) Furthermore, compression presents enormous difficulties with respect to specimen alignment in relation to the surfaces on which the load is applied. This has to be accomplished by means of very accurate axial bearings and hemispherical thrust plates.[14] Extensions and/or contractions of specimens are very low and require for their measurement the use of special 'extensometers' which amplify the output; consequently standard tests are not suitable for the purpose outlined.

Workers at ICI Ltd have developed two fairly simple test methods and apparatus for the accurate measurement of the tensile and shear compliances by three-point bending and torsion of rectangular specimens respectively.[14] Apart from the simplicity of the apparatus, tests in three-point bending and in torsion on a plastics material obviate the difficulty of measuring the Poisson ratio function and can detect with good approximation the extent of anisotropy present in the specimen. The only disadvantage is that, in order to remain within the limits of applicability of the theories of bending and torsion, the beam deflection and the angle of twist of the bar have to be fairly small. This means that the material can be tested only in the region of its linear viscoelastic behaviour. For this reason, these tests are more suitable for routine evaluations and quality control purposes than for producing extensive data for design purposes.

From measurements of the compliances as a function of time, the Poisson ratio function can be obtained, with some approximation, from the relationship:

$$[1 + v(t)] \simeq \frac{E(t)}{2G(t)}$$

Strictly speaking, one should use Carson transforms in the relationship above. However, if one takes isochronous values (which, in effect, eliminate the time function), the relationship is analogous to that for elastic materials and, therefore, can be used to obtain approximate estimates.

Since in creep experiments one measures the compliance and not the modulus, it would be necessary to make measurements at several time intervals in order to obtain the m value so that the modulus can be calculated using eqn. (5.29). However, for very low stress levels and short times, which are normal conditions for these tests, a good approximation can be obtained by taking the 'creep modulus' values (i.e. the inverse of the compliance) instead of relaxation modulus; i.e.

$$[1 + v(t)] = \frac{E_c(t)}{2G_c(t)}$$

where the subscript c stands for creep.

Typical values for the stiffness parameters of several plastics materials are shown in Table 5.2.

Note in the table the anomalous values (marked †) of the Poisson ratio for anisotropic materials. The values of E_c/G_c ratio greater than 3 and the Poisson ratio greater than 0·5 can be explained by reference to Fig. 5.3. When the specimen is cut parallel to the direction of the orientation, whether fibre orientation or molecular orientation, the modulus in tension (and compression) corresponds to $E_{\parallel}$ (i.e. the highest value), while the shear modulus corresponds to, approximately, the lowest value.[15]

If specimens are cut at an angle from the direction of the orientation, the ratio E_c/G_c decreases and reaches a minimum when specimens are cut at 45°. The discrepancy in the calculated values for the Poisson ratio arises because

Table 5.2: Shear and Tensile Creep Moduli and Poisson Ratio at 100 s, 20 °C and Small Strains[15]

Material	Creep moduli (N/m^2)		$\dfrac{E_c(100)}{G_c(100)}$	$v(100)$
	Shear $G_c(100)$	Tensile $E_c(100)$		
Polymethyl methacrylate (sheet)	$1\cdot17 \times 10^9$	$3\cdot12 \times 10^9$	$2\cdot66$	$0\cdot33$
Rigid PVC (sheet)	$1\cdot17 \times 10^9$	$3\cdot21 \times 10^9$	$2\cdot74$	$0\cdot37$
Polymethylene oxide	$1\cdot15 \times 10^9$	$3\cdot24 \times 10^9$	$2\cdot82$	$0\cdot41$
Polypropylene (injection moulded)	$0\cdot54 \times 10^9$	$1\cdot64 \times 10^9$	$3\cdot02$	$(0\cdot51†)$
Glass reinforced Nylon 6,6 (injection moulded) and at 65% RH	$0\cdot884 \times 10^9$	$6\cdot26 \times 10^9$	$7\cdot08$	$(2\cdot54†)$
Carbon fibre laminate	$3\cdot88 \times 10^9$	176×10^9	$45\cdot3$	$(21\cdot7†)$

† See text.

the above formula is valid only for isotropic materials. Consequently when anisotropy is suspected from such measurements, a thorough investigation must be carried out. Measurements must be made at all angles, and a check of the applicability of the orthotropic plate theory must be made. Reasonable agreement with the above theory gives at least some hope that the results can be handled analytically.

For instance, in such a case the dependence of the Poisson ratio on angle of orientation is given by:[16]

$$v_{(xy)} = E_x \left[\frac{v_{12}}{E_1} (\sin^4 \theta + \cos^4 \theta) - \left(\frac{1}{E_1} + \frac{1}{E_2} - \frac{1}{G_{12}} \right) \sin^4 \theta \cos^4 \theta \right]$$

Note that the $v_{(xy)}$ value at $0°$ corresponds to v_{12} which can be taken to lie between the value of the polymer matrix and that of the fibre for composite materials; typically for glass fibres $v \simeq 0.25$, while for glassy polymers $v \simeq 0.35$ and for crystalline polymers $v \simeq 0.42$. When anisotropy results from molecular orientation, the v_{12} value can be taken to equal approximately that of the unoriented material.

Note that the above equation for $v_{(xy)}$ shows a maximum value at about $30°$ and a minimum value at $90°$. This corresponds to the v_{21} value. It is understood that if isochronous values are taken, the above relationships should be applicable since one is still in the regime of linearity where it is implicit that the behaviour at any elapsed time is independent of the previous deformation history of the material.

5.4.2 Generation of Data for Design Purposes

The low modulus (or high compliance) of plastics necessitates designing components at levels of strain that are much greater than those possible with traditional materials, such as metals. In addition, the onset of non-linearity can take place at strain levels which are too low to enable designers to specify working stresses within the linear regions, since this would result in the use of excessive amounts of materials.

On the other hand, non-linear theories are not sufficiently well established to enable a full characterisation of materials to be made from accelerated tests through the use of superposition rules to predict the behaviour under variable stress conditions. It would seem necessary, therefore, to carry out a large number of tests over a wide range of conditions and for long time periods. However, since this is neither economically attractive nor practically feasible, it is necessary to rationalise the evaluation procedures in order to obtain the maximum amount of useful information in the minimum amount of time.

5.5 CHOICE OF TESTING PROCEDURE

5.5.1 Stress Relaxation and Creep Testing Procedures

In view of the non-linear behaviour of plastics, it is clear that neither the usual constant-strain-rate nor the dynamic tests are suitable for the prediction of long term behaviour. The only obvious alternatives are creep procedures and stress relaxation methods. This does not imply, of course, that dynamic tests are irrelevant; they must be considered as complementary to long term tests and an absolute necessity for the generation of design data in situations where the component is subjected to vibrations. In many cases a design problem reduces to one involving either stress relaxation or creep. However, to carry out both tests is not economically feasible, except in critical cases. Consequently it is necessary to decide which of these two tests is likely to produce the most valuable information. The conversion of compliance into modulus values (and vice versa) can be carried out by means of relationships similar to eqn. (5.29) or simply by taking the reciprocals and making adequate allowance for the errors that may arise from any deviation from the rule used for linearity.

(a) Stress Relaxation Tests

Stress relaxation tests seem to offer a distinct advantage over creep evaluations insofar as the non-linearity of the deformational behaviour of plastics and other important phenomena, such as absorption of environmental agents, crazing, etc. (to be discussed later), seem to be governed by the level of strain rather than stress.[17] This is particularly true for the case of composites whose fibre/matrix bond seems to break at a certain critical strain level. These factors have, in fact, led to the stipulation of design procedures based on limiting strain rather than working stress.[18-20]

Stress relaxation procedures would provide, therefore, the means of identifying quickly the strain levels at which the above phenomena occur and the rate at which they affect the behaviour of the material as a function of time, temperature, etc. However, they present some formidable experimental difficulties; in addition, the data obtained are not easy to handle in situations involving variable stress and strains.

The experimental difficulties arise mainly from the complexity of the apparatus required to produce an 'instantaneous' or step input of the required strain level, while to monitor the decay of the stress as a function of time it would be necessary to use load cells. This means that the capital investment to build a multi-stage or a 'battery' of suitable apparatus to test a material under various conditions (e.g. strain levels, temperature, etc.) and to evaluate the interspecimen variability would be excessive. From a fundamental point of view, it would be necessary to measure the behaviour under intermittent strain conditions in the form of alternating step functions (Fig. 5.18) since there is no applicable superposition principle to enable the designer to make the predictions from continuous step functions.

Step input is obviously preferable to any other variable type of input in empirical evaluations of complex functions, when response (in this case the stress) as a function of time is being monitored. This means that in order to limit the experiments to a reasonable minimum, the strain would have to alternate from various pre-determined levels to zero over several time intervals. Because the stress would not decay to zero during each excitation period, when the strain is forced to go to zero the stress has to become negative (i.e. compressive type). (See Fig. 5.18.)

One obvious consequence of alternating the strain is that for a given total excitation time (e.g. $t_1 + (t_3 - t_2)$), the stress decays less than if the strain had been kept at the same level for the same period. Hence from the design point of view the 'effective' relaxation modulus for intermittent loading conditions is greater than for continuous excitations at equal elapsed times under load. This means that savings in amount of materials used can be made if this factor is taken into account. The requirement of providing tensile and

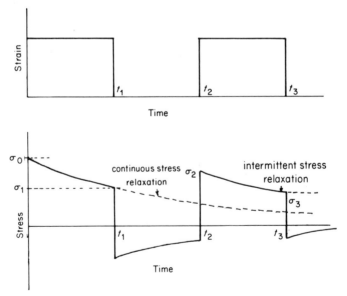

Fig. 5.18.   Variation of stress for alternating strain inputs.

compression loads on specimens in a single evaluation would create additional complexities in the apparatus; for this reason, stress relaxation evaluations are not viable for the purpose of generating data for general design.

(b) Creep Tests
In the light of the above difficulties, it has become common practice to use creep tests under step-input stress conditions as a means of evaluating the deformational behaviour of materials. The equipment is relatively simple and the data are reliable provided due attention is given to the critical factors affecting accuracy. These are described below.

(1) The load must be applied smoothly and over relatively short times in order to avoid shock waves in the initial stages and to minimise the effects of the initial loading history on the subsequent observations (Fig. 5.19). Impact loading would create disturbances in the setting of the extensometer mounted on the specimen and would create high local strains which can produce internal changes, known as 'strain softening' phenomena; and affect the subsequent behaviour of the material. In the case of composites, for instance, the high load peaks could break the fibre/matrix interfacial bond. It is important, however, that the rate of application of the load is not too slow; otherwise, the ramp input in the first stage could produce a 'memory' effect for the later stages of the test. A suitable damper must be fitted, therefore, on the loading device of the machine.

(2) Because of the difficulty of machining representative specimens of sufficiently small cross-sectional area, it is necessary to utilise the mechanical advantage of a lever in order to develop sufficiently high stresses in the specimens. (See typical apparatus in Fig. 5.20.) The top end of the load linking bar terminates in a small gauge chain, which is kept coaxial with the

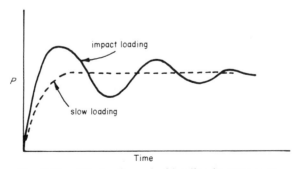

Fig. 5.19.   Effects of speed of loading in creep tests.

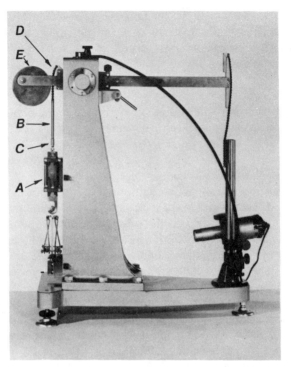

Fig. 5.20.   Lever loading creep machine and extensometer. A, slider mounted in guide block; B, load linking bar; C, universal joint; D, curved guide block; E, adjustable disc for balancing tensioning lever. (After Dunn et al.[14])

axis of the specimen by a curved guide block mounted at the front end of the lever. The other end of the lever has a similar curved guide carrying a chain supporting the loaded pan during a test. The load pan and loading mechanism are below the bench (hence not shown in Fig. 5.20). The heavy disc at the front end of the lever can be adjusted, through an eccentric axis, to balance the lever and to provide some pre-tension on the specimen, and offset the weight of the extensometer, etc.

    (3) Frictions and specimen mis-alignment must be avoided. Hence precision of engineering in constructing the apparatus is absolutely essential.

5.5.2 Modes of Deformation

In order to eliminate the need to evaluate materials under various modes of deformation (e.g. tension, compression or shear, etc.), it is necessary to find out which type of test is likely to provide the most meaningful and accurate data using simple apparatus.

An important point to consider is how to measure accurately the resulting deformation. Under axial loading it would be necessary to use an extensometer fitted on the specimen in order to eliminate errors arising from 'end effects', i.e. restrictions by the clamps for tensile loading and friction between loading plates and end surfaces of the specimens for compressive loading; both of these phenomena create local transverse stresses. Hence readings have to be taken at some distance from the ends of the specimens where pure axial loading takes place. Fixing of strain gauges on specimens, as an alternative to extensometers, and as a means of minimising the length of specimens over which the deformation is to be measured, would be far too expensive and time consuming. Consequently the use of strain gauges is more viable for stress analysis purposes than for materials evaluations.

Experience shows that the modulus in compression is higher than in tension; however, the difference becomes appreciable only above certain strain values, depending on the nature of the material.[12] Consequently by taking measurements in tension alone, one obtains conservative estimates for compression situations, so that any error arising will be on the safe side. With isotropic materials it can be assumed that the variation of Poisson ratio with time and stress level will not produce appreciable errors in the estimates of deformations or deflections. Ultimately, however, the designer has to estimate the extent of error involved and decide what allowances are necessary. The lack of precise knowledge on Poisson ratio effects does not create serious problems because one is aware of the range of values it is likely to assume, i.e. it increases with time, temperature and stress level. This range is 0·35–0·50 for amorphous polymers and 0·4–0·5 for crystalline polymers.

For a circular plate supported at the edges and subjected to a central load the maximum deflection (at the centre of the plate) is proportional to $[(3 + v)(1 - v^2)]/(1 + v)$. This expression assumes values ranging from 2·2 for $v = 0·35$ and 1·75 for $v = 0·5$ and consequently one can make direct estimates of the difference in deflection at the working load.

Problems arise, however, with anisotropic materials, where it is more difficult to make simple estimates like the above. Furthermore Benham and McCammond[21] have found that for materials like polypropylene and at reasonably long elapsed times and high stresses (i.e. in excess of $10 \, MN/m^2$) the contraction ratio (so called because the term Poisson ratio refers normally to elastic materials) can assume values slightly in excess of 0·5. They attribute such an effect to a change in morphology of the material while it is being tested. However, for annealed materials, the increase in contraction ratio above the theoretical maximum of 0·5 was found to be much more modest. Polypropylene, however, is notorious for undergoing secondary crystallisation processes on ageing; consequently the problem outlined here is not one of testing technique but of the state of the material, which must be taken into account (see later). With such limitations in mind, it is possible to conclude that creep tests in tension, carried out over a range of times, stress levels and temperature, and which take into account the effects of the morphological state of the material, are sufficient to obtain the 'essential' data required for a rational engineering design.

5.5.3 Advantages of Creep Tests in Tension Over Equivalent Tests Under Stress Relaxation Conditions

If the points made earlier on the construction of a creep test apparatus are taken into account, the capital cost can be relatively low. Furthermore, extensometers that (a) can be easily clipped onto specimens; (b) are capable of producing amplifications of the displacement produced, either through optical or electrical systems; and (c) can continuously record and store the information are becoming available on the market as standard items. The construction of a stress relaxation apparatus, on the other hand, is much more involved since it requires initial control and subsequent monitoring of the load by means of load-cells, while the strain input has to be determined by means of extensometers. These would have to produce feedback on the mechanism controlling the clamp separation in order to maintain a constant strain. For intermittent step-inputs it is also necessary to provide a mechanism for reversing the load into compression in order to reduce the strain to zero at regular intervals (see Fig. 5.19). With creep tests, producing intermittent step-input loads is simply a matter of introducing an automatic loading and unloading device. In this way the strain on the specimen can be recorded for both creep and recovery stages of the experiment without difficulty.

5.6 PRESENTATION OF DATA FROM CREEP TESTS

5.6.1 Creep Curves Obtained Under Continuous Loading Conditions

The strain monitored as a function of time can be recorded directly for various stress levels, from which the respective values for the creep compliance (or 'creep modulus', taken directly as the reciprocal of the compliance) can be calculated and used directly in standard design formulae.

In addition, from these data one can make cross-plots of stress versus

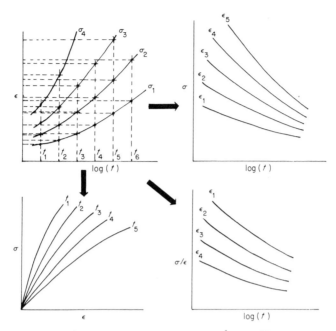

Fig. 5.21. Presentation of creep data.[22]

strain for various elapsed times to produce a family of stress/strain
isochronous curves, and of stress versus time curves for several strain levels in
order to obtain a family of 'isometric' curves (Fig. 5.21).

5.6.2 Recovery from Continuous Loading Creep Conditions
There are many situations in practice where the component is subjected to a
load only for a certain period of time and is then removed. Provided the strain
has not exceeded the 'yield' conditions, the component will recover to its
original geometry and dimensions. A strain which recovers completely is a
good indication that no permanent change has occurred in the morphological structure of the material and that any subsequent loading will produce
the same deformation path as in the previous loading history.

It is necessary, therefore, to know the recovery behaviour of the material
from different loading and creep-time histories. Obviously, if the behaviour
is linear, this information can be obtained directly from the creep data
through the use of the Boltzmann superposition principle. According to the
superposition principle, assigning a zero value to the stress (i.e. removing an
existing stress from the material) is equivalent to adding one stress of equal

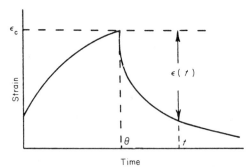

Fig. 5.22. Definition of creep recovery parameters $((\varepsilon_r)_f \equiv \varepsilon(t)/\varepsilon_c;$ $(t_R) \equiv (t - \theta)/\theta)$.

magnitude but of opposite sign (or direction). In other words, removing a tensile stress σ_0 from the component is equivalent to adding a compressive stress $-\sigma_0$, since this will force the strain to follow the opposite direction (which is, in effect, what happens during recovery). In this situation the recovered strain becomes,

$$\varepsilon(t) = -D_0\sigma_0 - \sigma_0 \int_{-\infty}^{\infty} L(\lambda)(1 - e^{-\theta/\lambda})e^{-(t - \theta)/\lambda}\,d\lambda \qquad (5.42)$$

where $\theta =$ creep time and $t =$ time from the application of the stress.

Practical experience shows that the recovered strain of plastics in the non-linear region is higher at first but increases at some later stages at a lower rate than expected from linear theory.[22] Whereas the first observation is to be expected from the stress activation effects illustrated earlier, the latter is not. The reversing of strain from internal stresses in the final stages of the recovery may be hindered by the structural rearrangements involved. This makes it absolutely necessary to consider recovery experiments as an integral part of the evaluation of materials by means of creep tests (Fig. 5.22). To facilitate the presentation of recovery data, one can define new coordinates as follows:

(a)   Fractional recovered strain,

$$(\varepsilon_r)_f \equiv \frac{\text{Strain recovered}}{\text{Creep strain at the time of removal of load}}$$

(b)   Reduced time

$$(t_R) \equiv \frac{\text{Recovery time}}{\text{Duration of creep period}}$$

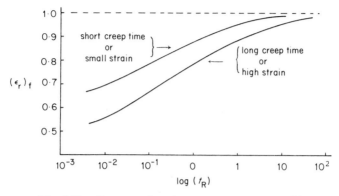

Fig. 5.23.   Recovery data from creep experiments.[23]

In the case of a linear material, one single master curve would have been obtained (Fig. 5.23). A limited number of creep recovery data, however, enable us to obtain a band of values from which approximate interpolations can be made.

5.6.3 Creep and Recovery Under Intermittent Loading Situations

Creep followed by recovery is, perhaps, the most common practical situation. In view of the difficulties involved in the use of the Boltzmann superposition principle, it is necessary to obtain the relevant design data from practical measurements. Obviously, the time to reach full recovery is much longer than the creep time and increases with the level of the creep strain (see Fig. 5.23). Consequently if a load is reapplied on the component, the residual strain from the previous load history will add onto the new strain (Fig. 5.24).

As a result, a plot of the total accumulated strain as a function of the total creep time $t_1 + (t_3 - t_2) + \cdots$ for various ratios of creep to recovery time, shows accumulated strain under intermittent loading to be much lower than that obtained under continuous loading (Fig. 5.25).

If one defines an 'apparent creep modulus' as the ratio of the excitation stress to the strain observed for any given total time under load (i.e. the accumulated strain from the previous loading history plus the strain from the most recent load), one can expect a variation of modulus with time as shown in Fig. 5.25(b). In other words, the decay of the modulus with time for intermittent loading situations is much less than for continuous loadings. The reduction as a function of time being smaller the longer the recovery time relative to the creep time.

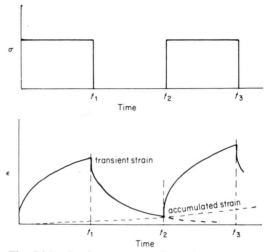

Fig. 5.24. Strain response to intermittent stresses.

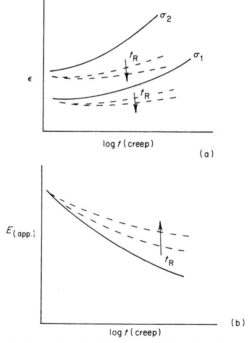

Fig. 5.25. Comparison of strain (a) and modulus (b) from continuous and intermittent creep experiments. —— Continuous loading, --- intermittent loading. Note: the longer the recovery time in relation to the creep period, the flatter the curve.

5.6.4 Effects of Temperature

In view of the sensitivity of the deformational behaviour of plastics to temperature, it is necessary to make measurements over a range of temperatures. Judicious choice of the temperature intervals is essential in order to keep the amount of experimentation within reasonable limits. A knowledge of the main glass transition temperature, as measured by either short term or dynamic tests, sets the absolute maximum temperature limit for the material. Measurements carried out at two, or possibly three, temperatures which cover a wide range are sufficient, since interpolations for intermediate temperatures can be made with some approximation from short term data.

5.6.5 Effects of the Morphological State of the Material and Physical Ageing

Apart from anisotropy resulting from orientation during processing, the morphological state of thermoplastics can vary according to the thermal history of the material, prior to and while under stress in service (ageing). Consequently a thorough evaluation of the deformational behaviour of plastics would have to take into account these effects.

In view of the usual constraints on financial resources and time for the evaluation of materials it is desirable to devise procedures that permit the prediction of long term effects from short term tests.

Despite the complexity of the problem some relatively simple procedures have been suggested by Turner,[24] for the case of low density polyethylene, and Struik[25] for a wide range of glassy polymers, in relation to the effect of physical ageing (densification).

For the case of low density polyethylene, it would seem that physical ageing produces a downward (vertical) shift in the creep curves (see Fig. 5.26), while for glassy polymers there is a horizontal shift along the time axis (see Fig. 5.27).

Consequently from a knowledge of the creep behaviour of the material in a reference state it is possible to make the necessary extrapolations, from tests carried out over short time periods, on samples subjected to a wide range of ageing treatments.

The above procedures, however, have limited applicability insofar as they are valid only for deformations falling within the linear behaviour (i.e. small strains) and, secondly, they cannot be used indiscriminately for densification effects brought about by other means (e.g. by annealing or by altering the cooling rate during processing). In relation to the latter Turner[24] demonstrated that, for the case of low density polyethylene, the

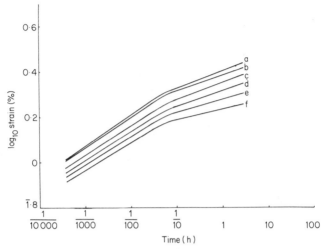

Fig. 5.26.　Effects of physical ageing on the creep behaviour of low density polyethylene. (After Turner.[24])

	Density (g/cm^3)	Storage time (h)
a	0·917 3	$1\frac{1}{2}$
b	0·917 7	$6\frac{1}{2}$
c	0·918 3	25
d	0·918 9	145
e	0·920 2	410
f	0·919 8	3 190

discontinuity point (Fig. 5.26) moves to longer times when the increase in density is achieved by reducing the cooling rate.

Since the latter would require a horizontal shift as well as a vertical shift, it means that the shift procedure used has to take into account also the nature of the fabrication process and cooling conditions (even without going into the complications that would arise from the anisotropy and density fluctuations within the component).

Similar limitations are applicable to glassy polymers. Struik,[25] for instance, suggests that physical ageing does not affect the deformational behaviour within the secondary relaxations regions (Fig. 5.28).

This means that the accuracy of the extrapolations through horizontal shifts will vary considerably according to the position of the curves considered in relation to the secondary relaxation region.

Struik has also observed that the magnitude of the horizontal shift in the

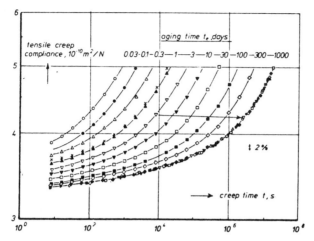

Fig. 5.27.   Small-strain tensile creep curves of rigid PVC quenched from 90 °C (i.e. about 10 °C above T_g) to 40 °C and further kept at 40 ± 0·1 °C for a period of 4 years. The different curves were measured for various values of the time t_e elapsed after the quench. The master curve gives the result of a superposition by shifts which were almost horizontal; the shifting direction is indicated by the arrow. The crosses refer to another sample quenched in the same way, but only measured for creep at a t_e of 1 day. (After Struik.[25])

creep curves brought about by physical ageing decreases with the level of strain. Presumably this results from the hydrostatic components of the applied stress, which counteract the densification brought about by the ageing process, or it simply arises from internal dilatations accompanying the large scale segmental motions.

In the light of the dependency of the shift factor on both ageing

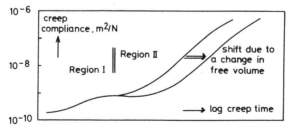

Fig. 5.28.   Effects of physical ageing on the creep compliance of glassy polymers. (After Struik.[25])

temperature and level of strain Struik proposes the use of coefficient μ (the double logarithmic shift rate), defined as

$$\mu = \frac{d(\log a)}{d(\log t_i)} \qquad (5.43)$$

where a = horizontal shift due to physical ageing, and t_i = ageing time prior to testing at the same temperature. The value of μ varies from zero to approximately 1 and goes through a maximum on approaching the T_g of the polymer.

At the same time the value of μ decreases monotonically and tends to zero as the level of the stress or strain approaches the yielding conditions of the polymer (see also Chapt. 6). Furthermore the 'shift rate' is independent of the ageing and testing temperature, provided these are equal.

An important aspect of physical ageing is its reversibility. In other words physical ageing can be erased by any external agency that increases the free volume of the polymer, i.e. by heating it above its glass transition temperature or even by subjecting it to stresses approaching the yield strength of the polymer. This implies also that water absorption after physical ageing under dry conditions would have a similar effect (see Section 5.6.6).

5.6.6 Effects of Additives
When the amount of additives used in a polymer composition is very small (say $< 1\%$) the direct effect on the deformational behaviour of the base polymers can be neglected. These could affect, however, the long term behaviour of the base polymer by altering its ageing characteristics (both physical and chemical ageing).

Large proportions of additives (e.g. fillers, plasticisers and rubbery inclusions), will alter, on the other hand, the overall intermolecular forces, free volume and molecular relaxations, producing substantial changes in the deformational spectrum of the polymer.

(a) Plasticisers
In Chapt. 3 the effects of plasticisers were discussed mainly in relation to volume relaxations. From an engineering point of view it is obviously necessary to know how they affect the deformational behaviour over the whole range or retardation/relaxation times, including stress activation effects.

Depending on the type of interactions with the polymer molecules and the testing temperature in relation to the glass transition temperature, the addition of low levels of plasticisers produces a decrease in the value of the elastic component of the creep compliance (anti-plasticisation) and an increase in the values of $L(\lambda)$ at longer retardation times (i.e. the creep curves become steeper).

With increasing the stress input, the elastic component of the creep function will pass through a minimum and eventually it will assume values greater than that of the base polymer (plasticisation). The effects of low levels of plasticisers on the creep behaviour of PVC, in both linear and non-linear regions, has been demonstrated by the author[26] and is illustrated in Fig. 5.29. The cross-over point (i.e. the transition from antiplasticisation to plasticisation) decreases to shorter elapsed times when (a) the stress input increases, (b) the testing temperature increases, (c) the concentration of plasticiser increases and (d) the conditioning time (physical ageing) decreases.

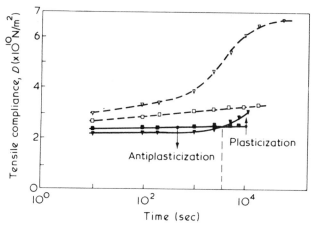

Fig. 5.29.   Effect of plasticiser on creep behaviour of PVC compositions (unaged samples). ■, □, Unplasticised PVC; ▼, ▽, PVC + 10 phr tricresyl phosphate; ▼, ■, nominal stress = 16 MN/m²; ▽, □, nominal stress = 25 MN/m². (After Mascia.[26])

(b) Fillers

The theory of reinforcement in Chapt. 3 was developed on the assumption that filler and matrix exhibit an elastic behaviour, hence it cannot be applied directly to polymers. The theory, can be developed further, however, by assuming that the filler and matrix form a perfect bond (i.e. interfacial phenomena are being neglected). A further assumption can be made that the only contribution to the time dependence of the deformation functions comes from the matrix (i.e. the filler behaves elastically). In this way the same theories can be used by replacing the modulus or compliance terms for the matrix with the respective time-dependent functions. In other words, if we describe the viscoelastic behaviour of the matrix by simple expressions such as precede eqn. (5.36), then the law of mixtures for the upper and lower limits of reinforcement becomes, respectively

$$[E(t)]_c = \phi_f E_f + [(1 - \phi_f)E_\infty]_m + [(1 - \phi_f)\varphi^*(t)]_m \qquad (5.44)$$

and

$$[D(t)]_c = \frac{\phi_f}{E_f} + [(1 - \phi_f)D_0]_m + [(1 - \phi_f)g^*(t)]_m \qquad (5.45)$$

The above equations suggest that the filler produces a vertical displacement in the stress relaxation and creep curves (elastic term effect) as well as a reduction in the steepness of the curves (matrix dilution effect). The same approach can be extended to consider the effects in the non-linear deformation regions if the assumption is made that the functions can still be split into a stress-dependent, time-independent component and a stress-dependent time function as in eqn. (5.36). This means that if the complete stress relaxation or creep curves are known for the base polymer, it should be possible to determine the long-term behaviour of the corresponding composite from short term tests, by establishing the exact relationship between the deformation function of the polymer and that of the composite over the initial part of the curves.

There are indications, however, that interfacial phenomena and other fibre/matrix interaction (e.g. nucleations of the matrix around the fibres) could distort the nature of the deformation function of the composite from that of the matrix in an unpredictable manner. A poor interfacial bond between fibres and matrix could introduce additional relaxations and increase, therefore, the steepness of the creep and stress relaxation curves, contrary to the predictions inferred from the reasoning in the foregoing discussions.

From dynamic tests it has been found, for instance, that the loss factor due to interfacial effects, for the case of glass-reinforced polypropylene homopolymers (a typical poorly bonded system), increases considerably as the temperature approaches the main relaxation regions. On the other hand, the interfacial loss factor remains approximately equal to zero (or it assumes small negative values, possibly as a result of matrix nucleation), for glass reinforced grades of nylon and reactive (grafted) polypropylene (both well-bonded systems).[27]

In view of the time/temperature superposition of the viscoelastic deformation of polymers it is expected that such a performance will be also reflected in creep tests when the interfacial $L(\ln \lambda)$ values are plotted as a function of time. (Note that $L(\lambda)|_\lambda = 1/\omega$ in dynamic tests is, in fact, equal to $(2/\pi)\tan \delta$.)

5.7 EXAMPLES OF THE APPLICATION OF CREEP DATA TO DESIGN PROBLEMS

5.7.1 A Design Involving Continuous Constant Loading

Problem: A book shelf is to be made in polypropylene supported on an aluminium frame and brackets. Extruded sheets (for ease of fabrication) from stocks 0·5 cm thick are to be used. The width of the shelf is to be 20 cm, and it is considered aesthetically undesirable for the deflection between supports to exceed 1 cm. It is estimated that the maximum load will not exceed 1 kg/cm length and will be acting continuously on the shelf for a period of up to 10 years, when the books will be removed and the shelves dismantled for redecoration or other purposes. After this period any unrecovered deformation is not considered to be problematic, since the shelf platforms will be turned upside down. Calculate the minimum distance required for the supporting brackets.

Solution: The maximum deflection for a uniformly loaded beam supported at two ends is given by

$$\Delta_{max} = \frac{5PL^4}{384I} D(t, \sigma) \qquad (5.46)$$

disregarding the deflections resulting from shear stresses.

The problem involves estimating the value of the compliance $D(t, \sigma)$ to use in the above equation so that L can be calculated. The problem, however, is complicated by the fact that the level of stress is itself dependent on the span (L). In fact, from the fundamental relationship $M/I = \sigma/y$, where $M = (PL/2)x - (Px^2/2)$ and $y = d/2$, resolving for $x = L/2$ (i.e. the maximum moment at the mid-point of the shelf), one obtains

$$\sigma_{max} = \frac{3PL^2}{4bd^2} \qquad (5.47)$$

One does not have a mathematically accurate relationship for $D(t, \sigma)$ as a function of the stress into which one could substitute eqn. (5.46) to obtain an expression for Δ_{max}, with L as the sole independent variable. It is, therefore, necessary to find an indirect method of deducing the level of stress for which the compliance is to be calculated from the creep data available.

One approach is to select the maximum level of strain that the material must not exceed. The maximum level of strain can be determined on the basis that the material must not yield locally and produce the classical white marks, or simply on the basis of safety so that the component will not undergo brittle fractures during its estimated life. For polypropylene it is suggested that a maximum strain of 3 % can be used if the material has been processed under 'satisfactory' conditions.[28]

The manufacturer's literature[29] quotes results for a polypropylene homopolymer at 1, 2 and 3 % strain (at 20 °C) up to 10^8 s. A period of 10 years is equivalent to $3 \cdot 15 \times 10^8$ s; consequently extrapolation over this small interval should be quite reliable. The creep modulus obtained from this extrapolation is $2300 \, \text{kg/cm}^2$, i.e. a creep compliance of $4 \cdot 35 \times 10^{-4} \, \text{cm}^2/\text{kg}$ ($\simeq 4 \cdot 25 \times 10^{-9} \, \text{m}^2/\text{N}$). Substituting in eqn. (5.47), one obtains

$$L^4 = \frac{384 I \Delta_{max}}{5 P D(t, \sigma)} = \frac{384 \times 20 \times 0 \cdot 5^3 \times 1}{5 \times 1 \times 4 \cdot 35 \times 10^{-4}} \, \text{cm}^4$$

$$\simeq 44 \times 10^4 \, \text{cm}$$

Therefore

$$\underline{L = 25 \cdot 75 \, \text{cm}}$$

Note that if a value for the modulus obtained by a standard test method were used, i.e. for polypropylene (homopolymers), the compliance would be of the order of $0 \cdot 8 \times 10^{-9} \, \text{m}^2/\text{N}$, and the calculated value of L would be $39 \cdot 7 \, \text{cm}$. However, using eqn. (5.47) one finds that the maximum stress that a load of $1 \, \text{kg/cm}$ will produce over a span of $25 \cdot 75 \, \text{cm}$ is $11 \cdot 2 \, \text{MN/m}^2$. On the basis of the data consulted[30,31] the tensile strain after 10 years would be well in excess of 3 %. To remain within the limits of 3 % strain, the stress cannot be allowed to exceed the value of $8 \cdot 06 \, \text{MN/m}^2$, which means that the maximum load per unit length would have to be $\simeq 0 \cdot 8 \, \text{kg/cm}$. Because this is realistic in terms of loads exerted by books, it can be concluded that the design would be quite appropriate for the application considered.

5.7.2 A Design Involving Intermittent Loads

Problem: A folding chair is to be designed (say, in polypropylene homopolymer) in such a way that in use it acts as a cantilever beam supported at the rear. The load is assumed to act in the centre of the seat, and it is required that the maximum deflection does not exceed 2·5 cm. It can be assumed that at the most there will be a loading cycle of 4 h on, 1 h off and 4 h on again for every day of its working life.[32] Since this is likely to be an overestimate of the loading conditions, it can be assumed that, by fixing a working life of 10 years, the chair would satisfy the expected requirements. The dimensions of the seating area are 40 cm × 40 cm, and the load is likely to be 100 kg at the most. Calculate the required thickness of the seating platform.

Solution:

(i)   The maximum deflection is given by

$$\Delta_{\max} = \frac{P}{6I} (3a^2 L + a^2) D(t, \sigma) \qquad (5.48)$$

where a = distance from load point to supported edge (i.e. 20 cm).

(ii) To estimate $D(t, \sigma)$, one takes into account the reduced time

$$t_R = \frac{16\,h}{8\,h} \frac{(\text{recovery time})}{(\text{total creep time})} = 2$$

for which, on the basis of creep recovery data,[33] the residual strain at the end of each cycle is at the most 10 % of the value assumed at the end of the creep period.

On the question of safety (see later), one can assume that a maximum strain of 3 % for long periods is permissible and that the residual accumulated strain increases in constant steps over the entire working life (10 years) of the chair. In other words, one assumes that the Boltzmann superposition principle can be applied on the basis that an error produced through underestimation can be compensated by the specified over-design conditions. Hence with a total number of cycles equal to 3650, the residual strain at the end of the first cycle can be allowed to be of the order of 0·03/3600, i.e. $8·4 \times 10^{-3}$ %. From extrapolation of the ICI data in ref. 33 a creep modulus of $0·9 \times 10^9$ N/m^2 can be considered a reasonable estimate for a 4 h elapsed time and $8·4 \times 10^{-3}$ % strain level. This creep modulus corresponds to a creep compliance of $1·1 \times 10^{-9}$ m^2/N. Hence solving for I, one obtains a value of 35·5 cm^4, for which the thickness is calculated to be of the order of 2·2 cm. This is, obviously, far too high to be practical. Furthermore the maximum stress (at the outer skin) calculated from

$$\hat{\sigma} = \frac{M}{I} \frac{d}{2} = 40 \times 10^6 \, \text{N/m}^2$$

would produce a strain of 4·4 % after the first loading cycle.

Hence the value of d (cross-section thickness) has to be reduced through appropriate adjustments in the geometry of the seating platform (e.g. by introducing 'strengthening' ribs along the lines of maximum stress while maintaining the same value of I).

5.7.3 A Design Involving a Combination of Stress Relaxation and Creep

Problem: It is proposed to produce connecting components in polypropylene (homopolymer grade) for a squared section steel tubular frame to be used for the construction of display shelves. The connecting components are locked in position by applying a torque to the tube section after they have been slotted in place. The assembly is shown in Fig. 5.30. The load exerting a bending moment on the arms of the connecting pieces and compressive stresses at the edges of the vertical sections can be assumed to be quite small and can be disregarded for the purpose of the analysis.

The problem reduces, therefore, to a design involving stress relaxation of the shear stress τ and creep of the cross-section under the compressive stresses σ_c of the arms of the connecting pieces. Both viscoelastic processes contribute to a reduction in the forces resisting the pulling out of the tubular

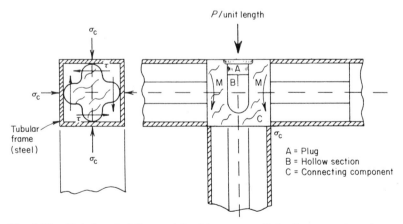

Fig. 5.30. Tubular steel frames joined by means of plastics connector pieces.

frames. Suppose that the compressive stresses have been estimated at approximately $8 \times 10^6 \, \text{N/m}^2$ and the initial shear stress at approximately $11 \times 10^6 \, \text{N/m}^2$. The component can be assumed to cease to function satisfactorily when the compressive strain becomes approximately 3 % and the shear stresses have decayed to a value of $3 \cdot 5 \times 10^6 \, \text{N/m}^2$. Estimate the satisfactory service life of the connecting pieces.

Solution: From the creep data for polypropylene homopolymer in tension, a 3 % strain is reached after $10^8 \, \text{s}$ (just over three years). Since for strains greater than 10 % the modulus in compression is expected to be higher than in tension, the actual useful service life can be assumed to be even longer. Although we do not have stress relaxation data, we can use the isometric creep curves instead. Knowing that stress relaxation curves as a function of time are less steep than isometric creep stress curves, the underestimation will be on the safe side. The isometric curves for creep in tension at 20 °C seem to obey the same time-dependent decay at all level of strains (the only difference being in the value of the time-independent component, which increases, of course, with the level of strain).[29]

Since we have the relationship

$$E(t, \varepsilon) \simeq 2[1 + v(t, \varepsilon)] \times G(t, \varepsilon)$$

where $v(t, \varepsilon)$ increases with time, we expect isometric shear stress to decay faster than an equivalent tensile stress. However, the short term contraction ratio for polypropylene is already in the region of $0 \cdot 4$ and, if anisotropy is absent, it is unlikely that this would exceed a value of $0 \cdot 50$ (or $0 \cdot 52$ when morphological changes take place through ageing). Hence the error involved in disregarding the difference in rate of stress decay between shear and tension is likely to be very small. From the isometric curves in the literature[33] it would seem, therefore, that the decay of a stress from a value of $11 \, \text{MN/m}^2$ to $3 \cdot 5 \, \text{MN/m}^2$ would take place over about 1 year. As it is unlikely that the above performance will be acceptable, more rigid materials, e.g. engineering polymers or glass-reinforced grades, would have to be considered as alternatives to polypropylene.

REFERENCES

1. J. G. WILLIAMS, *Stress Analysis of Polymers*, Longman, London, 1973, p. 147.
2. S. TURNER, *Trans. Plast. Inst.*, **34** (1966) 131.
3. H. LEADERMAN, *Rheology*, Vol. I, F. R. Eirich (ed.), Academic Press, New York, 1958, Chapt. 1.

4. S. TURNER, *Trans. Plast. Inst.*, **34** (1966) 132.
5. W. N. FINDLEY, *SPE Journal*, **16** (1960) 57.
6. G. W. BECKER, *Kolloid-Z.*, **175** (1961) 99.
7. S. TURNER, *Trans. Plast. Inst.*, **34** (1966) 115–35.
8. I. M. WARD and E. T. ONANT, *J. Mech. Phys. Solids*, **11** (1963) 217.
9. S. TURNER, *British Plastics*, **37** (1964) 322–4.
10. L. MASCIA, *Polymer*, **19** (1978) 325.
11. S. TURNER, *Mechanical Testing Plastics*, Iliffe, London, 1973.
12. ICI Ltd (Plastics Division), *Technical Service Note G 123*, p. 10.
13. P. P. BENHAM and F. V. WARNOK, *Mechanics of Solids and Structures*, Pitman, London, pp. 189–90.
14. C. M. R. DUNN, W. H. MILLS and S. TURNER, *British Plastics*, **37** (1964) 386–92.
15. M. J. BONNIN, C. M. R. DUNN and S. TURNER, *Conference on Research on Engineering Properties of Plastics*, Cranfield Institute of Technology, England, Jan. 6–8, 1969.
16. M. R. JONES, *Mechanics of Composite Materials*, McGraw-Hill, New York, 1975, p. 54.
17. S. TURNER, *Trans. Plast. Inst.*, **34** (1966) 133.
18. G. MENGES and R. REISS, *Plastics and Polymers*, **42** (1974) 119–23.
19. P. C. POWELL, *Plastics and Polymers*, **39** (1971) 43.
20. BS 4994, *Specifications for Tanks and Vessels in Reinforced Plastics and Polymers*, 1973.
21. P. P. BENHAM and D. McCAMMOND, *Plastics and Polymers*, **39** (1971) 130–6.
22. ICI Ltd, Plastics Division, *Technical Service Note G 123*.
23. ICI Ltd, Plastics Division, *Technical Service Note G 123*, p. 8.
24. S. TURNER, *British Plastics*, **37** (1964) 503.
25. L. C. E. STRUIK, *Physical Ageing in Amorphous Polymers and Other Materials*, Elsevier, Amsterdam, 1978.
26. L. MASCIA, *Polymer*, **19** (1978) 325–8.
27. L. MASCIA and A. ALLAVENA, *Reinforced Plastics Congress*, Brighton, 1976, p. 179.
28. ICI Ltd, Plastics Division, *Technical Service Note G 117*, p. 33.
29. ICI Ltd, Plastics Division, *Technical Service Note PP 110*, p. 15.
30. ICI Ltd, Plastics Division, *Technical Service Note G 123*, p. 16.
31. ICI Ltd, Plastics Division, *Technical Service Note PP 110*.
32. S. LEVY and J. H. DuBois, *Plastics Product Design Engineering Handbook*, Van Nostrand Reinhold, New York, 1973, p. 249.
33. ICI Ltd, Plastics Division, *Technical Service Note PP 110*, 1973, p. 30.

6

Engineering Interpretations of Mechanical Failures

Within the context of materials evaluation the term 'failure' is used to describe the level of stresses that will cause the material to fracture or to deform irreversibly (yield).

Although failures of structures may not necessarily involve fracture or yielding (see, for instance, examples in Chapt. 5), it is essential to take into account these types of failures in any kind of design exercise.

6.1 FAILURES INVOLVING YIELDING

In the case of materials (like metals) that obey Hooke's law up to the yield point, it is fairly common practice to base design calculations on the maximum stress allowable in order to avoid yielding (say, $\simeq 80\%$ of the value at yield). Since most components in service are subjected to complex stresses, it is important for the designer to know the stress conditions under which yielding takes place.

The criteria proposed in the past predict the value of the principal stresses that cause yielding based on a knowledge of the yield strength of the material, i.e. the value of the stress at yield under monoaxial stress conditions, since this can be easily measured experimentally.

The most widely accepted criteria are those attributed to Tresca and von Mises for isotropic materials and Hill for anisotropic materials.

6.1.1 The Tresca Criterion of Yielding
This criterion postulates that yielding occurs when the maximum shear stress in the material reaches a critical value, which depends on the nature of

the material. In terms of principal stresses this criterion can be written as

$$\tau_{max} = \frac{\sigma_{max} - \sigma_{min}}{2} = \tau_{cr} \qquad (6.1)$$

Under monoaxial stresses $\sigma_{min} = 0$; therefore,

$$\tau_{max} = \frac{\sigma_{max}}{2} = \frac{\hat{\sigma}_Y}{2} = \tau_{cr}$$

where $\hat{\sigma}_Y$ is known as the *yield strength*.
Hence

$$\sigma_{max} - \sigma_{min} = \hat{\sigma}_Y \qquad (6.2)$$

Note that under uniaxial stresses the plane of maximum shear stress is at 45° to the direction of the applied normal stress (Fig. 6.1).

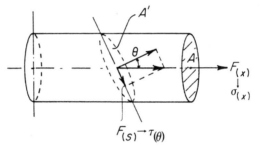

Fig. 6.1.   Resolved shear stresses. $A = A' \cos \theta$; $F_{(S)} = F_{(x)} \sin \theta$; $\tau_{(\theta)}A' = \sigma_{(x)}A \sin \theta$; therefore $\tau_{(\theta)} = \sigma_{(x)} \sin \theta \cos \theta$; $\tau_{(\theta)} = \sigma_{(x)}/2 \sin 2\theta'$; hence, $\tau_{max} = (\sigma_{(x)}/2) \sin (2 \times 45') = \sigma_{(x)}/2$.

The association of yielding with shear deformations can be inferred directly from the observation of thin planar regions of high shear strains (normally known as shear bands or Lüder lines) along the plane of maximum shear stresses. This phenomenon, which was first discovered by Lüder in metals, is also observed with most plastics materials and varies from white lines detectable under normal visible light (as in the case of crystalline polymers) to birefringent microbands visible under polarised light[1] (in the case of amorphous glassy polymers—as shown in Fig. 6.2).

For the case in which the yield strength in tension is the same as in compression (i.e. $\hat{\sigma}_Y|_{tens} = \hat{\sigma}|_{comp} = \hat{\sigma}_Y$), the Tresca criterion can be stated in quite simple terms.

Fig. 6.2.   Shear bands formed under plane strain compression in polystyrene at 60 °C, viewed between cross-polars. (After Bucknall,[1] see also Section 7.7.1.)

Under plane stress conditions ($\sigma_3 = 0$), one can have the following two cases:

(i)   Biaxial tensile stresses, where

$$\sigma_{min} = \sigma_3 = 0$$

$$\tau_{cr} = \frac{\sigma_{max} - 0}{2} = \frac{\hat{\sigma}_Y}{2} \qquad (6.3)$$

(ii)   Biaxial stresses, one compressive and the other tensile, where
$\sigma_{min} = \sigma_{comp}$

$$\tau_{cr} = \frac{\sigma_{tens} - \sigma_{comp}}{2} = \frac{\hat{\sigma}_Y}{2} \qquad (6.4)$$

This criterion becomes somewhat more complex in the case of plastics, where the yield strength in compression is greater than in tension. The Mohr–Coulomb construction can be used, however, to predict the yielding conditions. This criterion reduces to

$$\frac{\sigma_1}{\hat{\sigma}_Y|_{tens}} + \frac{\sigma_2}{\hat{\sigma}_Y|_{comp}} = 1 \qquad (6.5)$$

The geometric construction of the Tresca criterion is shown in Fig. 6.3.

6.1.2 The von Mises Criterion of Yielding
This states that yielding occurs when the maximum distortional energy reaches a critical value. For materials whose yield strength in tension is

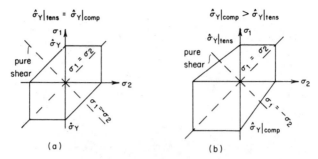

Fig. 6.3.   Geometric representation of the Tresca criterion.

equal to that in compression, one can obtain a relationship between principal stresses and yield strength through the following reasoning:
    The total deformational energy can be divided into two components:

(i)   A volumetric component, associated with hydrostatic stresses.
(ii) A distortional component, associated with shear stresses (Fig. 6.4).

This relationship can be expressed in terms of stresses and strains by using the following reasoning:

$$\sigma_{ii} = 3\sigma_m + \sigma'_{ii} \qquad (6.6)$$

where

$$\sigma_m = \frac{\sigma_1 + \sigma_2 + \sigma_3}{3}$$

σ'_{ii} are the deviatoric normal stresses and σ_m is the mean (or hydrostatic) stress. From the above relationships it follows that, for incompressible materials, $\sigma'_1 + \sigma'_2 + \sigma'_3 = 0$.

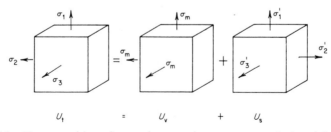

Fig. 6.4.   Decomposition of normal stresses into mean stress (σ_m) and deviatoric stresses (σ'_{ij}).

The deviatoric normal strains can be obtained through the usual relationship between stresses and strains; i.e.

$$\varepsilon'_1 = \frac{\sigma'_1}{E} - \frac{v}{E}(\sigma'_2 + \sigma'_3) \qquad \varepsilon'_2 = \frac{\sigma'_2}{E} - \frac{v}{E}(\sigma'_1 + \sigma'_3); \qquad \text{etc.}$$

Therefore,

$$\varepsilon'_1 + \varepsilon'_2 + \varepsilon'_3 = \varepsilon'_v = \frac{(1-2v)}{E}(\sigma'_1 + \sigma'_2 + \sigma'_3) = 0 \qquad (6.7)$$

This implies that the deviatoric strain components cannot produce any change in volume but only a distortion of the geometry.

Since $U_v = \frac{1}{2}\sigma_m \varepsilon_v$, for materials which are elastic up to the yield point, one obtains

$$U_v = \frac{1}{2}\sigma_m \frac{3\sigma_m}{E}(1-2v)$$

Therefore

$$U_v = \frac{(1-2v)}{6E}(\sigma_1 + \sigma_2 + \sigma_3)^2$$

Also

$$U_t = \frac{1}{2}\sigma_{ii}\varepsilon_{ii} = \frac{1}{2E}[\sigma_1^2 + \sigma_2^2 + \sigma_3^2 - 2v(\sigma_1\sigma_2 + \sigma_2\sigma_3 + \sigma_1\sigma_3)]$$

Hence

$$U_s = U_t - U_v$$

$$= \frac{(1+v)}{6E}[(\sigma_1 - \sigma_2)^2 + (\sigma_2 - \sigma_3)^2 + (\sigma_1 - \sigma_3)^2] \qquad (6.8)$$

Under monoaxial stresses, i.e. $\sigma_2 = \sigma_3 = 0$,

$$U_s = \frac{\hat{\sigma}_Y^2}{2E} \qquad (6.9)$$

Therefore, equating (6.8) to (6.9) and letting $v = \frac{1}{2}$, one obtains

$$(\sigma_1 - \sigma_2)^2 + (\sigma_2 - \sigma_3)^2 + (\sigma_1 - \sigma_3)^2 = 2\hat{\sigma}_Y^2 \qquad (6.10)$$

For conditions of plane stress, i.e. $\sigma_3 = 0$

$$\sigma_1^2 + \sigma_2^2 - \sigma_1\sigma_2 = \hat{\sigma}_Y^2 \qquad (6.11)$$

which is the equation for an ellipse with the major axis forming a 45° angle with the σ_1 and σ_2 axes.

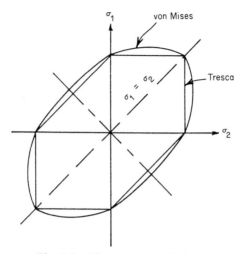

Fig. 6.5.   Plane stress conditions.

The comparison between the Tresca criterion and von Mises criterion is shown in Figs. 6.5 and 6.6, respectively, for the case in which the yield strength in tension is the same as in compression.

The independence of the yielding conditions from the hydrostatic components of the stresses makes it useful to express the von Mises

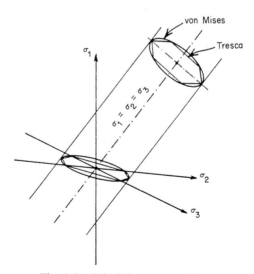

Fig. 6.6.   Triaxial stress conditions.

criterion in terms of octahedral shear stresses. The planes normal to the orthogonal axes, forming equal angle ($54° 44'$) with the principal axes, i.e. those along which $\sigma_1 = \sigma_2 = \sigma_3$, produce an octahedron. The stress acting on each face of the octahedron can be resolved into normal octahedral stress (σ_{oct}) and an octahedral shear stress (τ_{oct}) lying in the octahedral plane. The normal octahedral stress is taken to be equal to the hydrostatic component of the total stress; while the octahedral shear stress is equivalent to the total shear stress, defined as

$$\tau_{oct} = \tfrac{1}{3}[(\sigma_1 - \sigma_2)^2 + (\sigma_2 - \sigma_3)^2 + (\sigma_1 - \sigma_3)^2]^{1/2} \qquad (6.12)$$

Hence the von Mises criterion can be written as

$$3\tau_{oct} = \sqrt{2}\,\hat{\sigma}_Y \qquad (6.13)$$

This in effect shows that the von Mises and Tresca criteria are equivalent insofar as they both attribute the yielding phenomenon to shear stresses and differ only with respect to the magnitude of the shear stress at which yielding occurs. It can be shown, in fact, that for a component subjected to uniaxial tensile stress (σ_x) and pure shear stress (τ_{xy}), e.g. a thin-walled cylinder under axial tension and torsion, the Tresca criterion reduces to[2]

$$\left(\frac{\sigma_x}{\hat{\sigma}_y}\right)^2 + 4\left(\frac{\tau_{xy}}{\hat{\sigma}_y}\right)^2 = 1 \qquad (6.14)$$

and the von Mises criterion becomes

$$\left(\frac{\sigma_x}{\hat{\sigma}_y}\right)^2 + 3\left(\frac{\tau_{xy}}{\hat{\sigma}_y}\right)^2 = 1 \qquad (6.15)$$

The two criteria are exactly the same for conditions of uniaxial stresses and balanced biaxial stresses ($\sigma_1 = \sigma_2$). The greatest difference arises for the state of pure shear ($\sigma_1 = -\sigma_2$) (see Fig. 6.5), but this amounts only to about 15 %.

The advantage of expressing the von Mises yield criterion in terms of octahedral shear stress is that it enables three different stresses to be grouped in one single parameter. If one defines the octahedral linear strain (ε_{oct}) and octahedral shear strain (γ_{oct}) in the same way as for the octahedral stresses, i.e.

$$\varepsilon_{oct} = \frac{\varepsilon_1 + \varepsilon_2 + \varepsilon_3}{3}$$

and

$$\gamma_{oct} = \tfrac{2}{3}[(\varepsilon_1 - \varepsilon_2)^2 + (\varepsilon_2 - \varepsilon_3)^2 + (\varepsilon_1 - \varepsilon_3)^2]^{1/2}$$

then the plots of the octahedral shear stress versus octahedral shear strain values at yield should produce one single master curve irrespective of the state of the stress.

It is often convenient to define 'equivalent stress' and 'equivalent strain' as

$$\bar{\sigma} = \frac{\sqrt{2}}{2} \, [(\sigma_1 - \sigma_2)^2 + (\sigma_2 - \sigma_3)^2 + (\sigma_1 - \sigma_3)^2]^{1/2} \qquad (6.16)$$

and

$$\bar{\varepsilon} = \frac{\sqrt{2}}{3} \, [(\varepsilon_1 - \varepsilon_2)^2 + (\varepsilon_2 - \varepsilon_3)^2 + (\varepsilon_1 - \varepsilon_3)^2]^{1/2} \qquad (6.17)$$

These equations reduce to axial normal components of stress and strain for uniaxial tests, i.e. when $\sigma_2 = \sigma_3 = 0$,

$$\bar{\sigma} = \frac{\sqrt{2}}{2} \, [\sigma_1^2 + \sigma_1^2]^{1/2} = \sigma_1$$

and

$$\bar{\varepsilon} = \frac{\sqrt{2}}{3} \, [2(\varepsilon_1 + v\varepsilon_1)^2]^{1/2}$$

Since at yield only the shear stresses contribute to the deformational process, no change occurs in volume and, therefore, v can be taken to be equal to $\frac{1}{2}$; this will make $\bar{\varepsilon} = \varepsilon_1$.

In the case of plastics one can take into account the difference in yielding conditions between tensile stresses and compressive stresses in terms of the octahedral shear stress or 'equivalent' stress by introducing a dilatability or compressibility coefficient μ, i.e.[3]

$$\tau_{\text{oct}} = \tau_0 \pm \mu \sigma_m \qquad (6.18)$$

where τ_0 is the octahedral shear stress under zero hydrostatic stresses (or in pure shear).

For uniaxial tests, $\sigma_2 = \sigma_3 = 0$, hence

$$\sigma_m = \frac{\hat{\sigma}_Y|_{\text{tens}}}{3}$$

and

$$\sigma_m = -\frac{\hat{\sigma}_Y|_{\text{comp}}}{3}$$

Therefore,

$$\tau_{oct} = \frac{\sqrt{2}}{3} \hat{\sigma}_Y|_{tens} - \mu \frac{\hat{\sigma}_Y|_{tens}}{3}$$

and

$$\tau_{oct} = \frac{\sqrt{2}}{3} \hat{\sigma}_Y|_{comp} + \mu \frac{\hat{\sigma}_Y|_{comp}}{3}$$

$$\frac{\hat{\sigma}_Y|_{tens}}{\hat{\sigma}_Y|_{comp}} = \frac{\sqrt{2} - \mu}{\sqrt{2} + \mu} \tag{6.19}$$

This is equivalent to saying that in the case of plastics not only the distortional energy component but also the volumetric strain energy is involved in the yielding process. Consequently, the coefficient μ determines the extent of involvement of the volumetric energy. This can be quite significant with dilational hydrostatic stresses when yield occurs concomitantly to cavitational processes, such as crazing (to be discussed later).

When the modified yield criterion of polymers is written in terms of yield strength in compression and in tension, it takes the following form[4]

$$3\tau_{oct}(\hat{\sigma}_Y|_{comp} + \hat{\sigma}_Y|_{tens}) = \sqrt{2} \, 3\sigma_m(\hat{\sigma}_Y|_{comp} - \hat{\sigma}_Y|_{tens}) - (2\hat{\sigma}_Y|_{tens}\hat{\sigma}_Y|_{comp})$$

$$\tag{6.20}$$

6.1.3 Yielding of Plastics Within the Context of Product Design

It is clear that the yield criteria for plastics have been adapted from those derived from metals and verified by the same methodology, i.e. under constant straining rate testing conditions. They do not take into account, therefore, the history of the deformation prior to reaching the yielding conditions. The only attempts to take these into account are the adaptations of Eyring's rate theory, which assumes that the large-scale motions of molecular chains at yield obey the same relationship as the displacement of atomic (or small molecular) species in the flow of liquids.

The relationship between strain rate, temperature and yield strength becomes (see Section 7.2.2)

$$\dot{\varepsilon} = A \exp\left[-\frac{\Delta H}{RT}\right] \sinh\left[\frac{v\hat{\sigma}_Y}{RT}\right]$$

where A = constant for the material, ΔH = activation energy (thermal) for the yield process, and v = activation volume for yielding.

The applicability of such a theory has been confirmed for a limited range of temperatures and strain rates, and it is expected to apply also to incompressible materials under multiaxial stress conditions. It is not known, however, whether for plastics the thermal activation energy and the activation volume are affected by the hydrostatic stress component. Nevertheless, the time dependence of the yield strength expressed in terms of strain rate is of much greater value in processing than for designing of products. There is ample experimental evidence, in fact, that, under continuous creep loading conditions, the history of deformations affects the yielding process by producing non-linearity in the stress/time relationship (Figs. 6.7 and 6.8).[5]

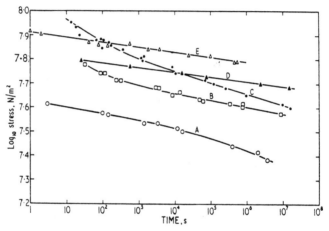

Fig. 6.7.   Ductile failure of non-crystalline plastics in static fatigue at 20 °C. A, Acrylonitrile–butadiene–styrene; B, polyvinyl chloride; C, PMMA (bulk cast); D, polycarbonate; E, polysulphone. (After Gotham.[5])

It is instructive to note that the yield strain is also a function of time; therefore, care must be exercised in comparing isometric stress/time curves with the yield stress curve. Such a relationship is even more complex for the case of intermittent loadings (or fatigue) conditions (see Fig. 6.9).

Little information is available to date on the yield stress/time relationship for intermittent loading evaluations, where the stress is kept constant during the creep period, as a function of the reduced time. No information exists for constant strain (or stress relaxation) evaluation conditions.

Such a paucity of data on yield conditions of plastics imposes severe

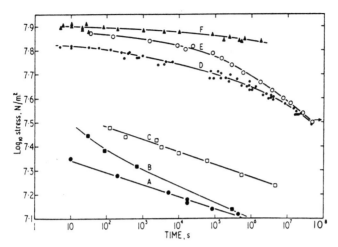

Fig. 6.8.   Ductile failure of crystalline plastics in static fatigue at 20 °C. A, Propylene copolymer; B, poly(4 methylpentene); C, polypropylene; D, acetal copolymer; E, acetal homopolymer; F, Nylon 6,6. (After Gotham.[5])

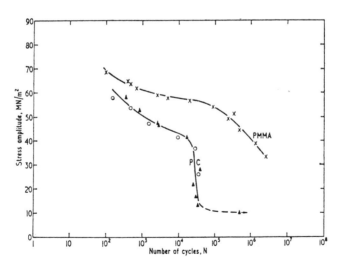

Fig. 6.9.   Fatigure in PMMA and polycarbonate: 20 °C 65 % rh Flexure. $f =$ 0·5 Hz (× ○) 5 Hz (▲). $\sigma_{mean} = 0$. Laboratory atmosphere (no forced cooling). (After Gotham.[6])

constraints on the use of yield stress as a design parameter. This situation is illustrated by the following example.[7]

The cylinder of a hydraulic jack has a bore (internal diameter) of 150 mm and is required to operate at up to $13 \cdot 8 \, \text{MN/m}^2$ internal pressure. Determine the required wall thickness, assuming that Nylon 6,6 is being selected for its excellent resistance to hydraulic oil.

The problem reduces to determining the limiting values for hoop stress (tensile) σ_θ, and for radial stress (compression) σ_r, both of which occur at the inner surface, on the basis of the yield strength of the material.

For materials with a yield strength independent of the loading history, the calculation is quite simple, since $\sigma_r = -p_i = -13 \cdot 8 \, \text{MN/m}^2$; hence the maximum allowable hoop stress can be calculated from the Mohr–Coulomb construction using eqn. (6.5), i.e.

$$\frac{\sigma_\theta}{\hat{\sigma}_Y|_{\text{tens}}} + \frac{\sigma_r}{\hat{\sigma}_Y|_{\text{comp}}} = 1$$

From these values the external diameter can be calculated by solving a simultaneous equation:[7]

$$\sigma_r = A - \frac{B}{r^2}, \qquad \sigma_\theta = A + \frac{B}{r^2}$$

One would solve these for the boundary conditions: $\sigma_r = 0$ at the outer surface when $r =$ external radius (the unknown); $\sigma_r = -13 \cdot 8 \, \text{MN/m}^2$ and $\sigma_\theta =$ value calculated previously when $r =$ internal radius (75 mm).

For the case of Nylon 6,6 the yield strength is time-dependent; consequently, it is necessary to know the likely loading history of the hydraulic jack in service and to have a knowledge of the yield strength as a function of the number of cycles for the appropriate reduced time. The latter should include, of course, data on the time-dependent behaviour of the material in compression. Estimates can be made from values obtained in tension since one can intuitively anticipate that the decrease in yield strength with time is likely to be less in compression than in tension.

If continuous loading is expected, the information required can be obtained directly from Fig. 6.8, which shows that up to about four months the yield strength is only a mild function of time. Extrapolations to longer times, however, could be quite risky, since the yield strength could begin to fall rather rapidly. For the case of acetal polymers, for instance, this occurs in the interval 10^5–10^7 s. It has to be borne in mind, furthermore, that Crawford and Benham[8] have found that for acetal polymers the yield strength under biaxial stresses (compression and tension, as for the case in question) falls more rapidly with time than that under uniaxial stresses. This is a further confirmation of the non-applicability of standard yield criteria in the third quadrant of stresses.

It is worth noting that several authors have expressed creep data under multiaxial stresses in terms of yield criteria[9,10] in a manner analogous to creep phenomena of metals (behaviour at high temperatures and high stresses, where the increase in strain as a function of time takes place via irreversible shear deformations). The fact that plots of equivalent strain as a function of time and cross plots of equivalent stress versus equivalent strain have produced single (master) curves (within the scatter of experimental error), especially when the data have been obtained at high levels of strains,[9] however, is only a manifestation of the predominance of the deviatoric component of the deformation. This is, in fact, responsible for the Poisson ratio's approaching its limiting value of 0·5; that is, the condition in which the only parameter that relates equivalent stress to equivalent strain is the (normal) compliance, i.e.

$$\bar{\varepsilon} = D(t, \sigma) \cdot \bar{\sigma}$$

This cannot be interpreted, however, as a yield process, since (by definition) yielding implies that the deformation has taken place irreversibly (i.e. it will not recover upon removal of the stress); therefore, the observations recorded above serve only to justify the assumption made earlier that the time dependence of the Poisson ratio can be ignored.

6.2 TESTING PROCEDURES FOR THE EVALUATION OF THE YIELD STRENGTH OF PLASTICS

6.2.1 Standard Test Methods

The usual standard test methods (e.g. ISO, ASTM, BS, DIN, etc.) describe procedures for the evaluation of the yield strength in the same manner as for the measurement of the modulus, i.e. at a constant rate of clamp separation. The value of the load at which either a plateau (maximum) or major discontinuity in the curve (see Fig. 6.10) occurs is taken to calculate the yield strength $(\hat{\sigma}_Y)$, using the usual elasticity relationships.

For tests in tension and compression:

$$\hat{\sigma}_Y = \frac{\hat{P}_Y}{A_{(\text{initial})}}$$

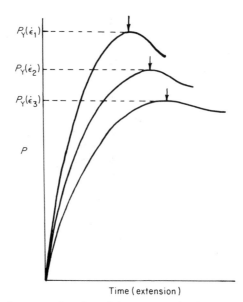

Time (extension)

Fig. 6.10.   Load curves as function of time or extension in constant strain rate tests.

For tests in three-point bending, the yield strength is taken as the value of the maximum skin stress (taking into account 'end effects'):

$$\hat{\sigma}_Y = \frac{3}{2}\frac{\hat{P}_Y L}{bd^2}\left\{1 + 6\left(\frac{\Delta}{L}\right)^2 - 4\left(\frac{d}{L}\right)\left(\frac{\Delta}{L}\right)\right\} \tag{6.21}$$

where $\Delta =$ deflection, $\hat{P}_Y =$ maximum (or discontinuity) load, $L =$ length, $d =$ thickness, and $b =$ width.

Obviously, the value of the yield strength in the three-point bending test will be intermediate between those in tension and in compression, but this is rarely a true yield stress value because of the violation of the assumption of small deflections implicit in the above formula.

6.2.2 Rationalisation of Testing Procedures for the Measurement of the Yield Strength of Plastics

The limitations of constant strain rate data for designing purposes have been discussed at great length in earlier sections. What is needed, therefore, is to obtain data which can be used to describe the relationship between yield strength and history of the deformation.

If one accepts that the errors arising from the application of the standard Tresca or von Mises criteria can possibly be ignored and that the discrepancy between the values in tension and compression produce conservative estimates when evaluations are carried out with a tensile mode of deformation, it is possible to reach the following conclusions:

(1)   Creep tests in tension are sufficient to obtain all the data required for design purposes.

(2)   It is sufficient to record the elapsed time required to cause yielding at several stress levels, under both continuous and intermittent loading conditions.

(3)   The data can be presented as an extension of the isometric stress/ time curves where the yield stress curve represents the limiting condition of the applicability of the data for design purposes (see Fig. 6.11).[11] Note, however, that a higher isometric stress curve would intersect the yield stress curve, which indicates that the strain at yield is not constant.

(4)   To eliminate the uncertainties regarding the decision as to what constitutes a 'yield failure', one could use creep rate as the parameter. In other words, one could define 'yielding' as the condition in which the creep rate reaches a critical value which is deemed to be unsafe for the performance of the component in actual service. This approach not only would redefine the concept of 'yield strength' but also would provide a mechanism for deriving a time/temperature superposition principle.

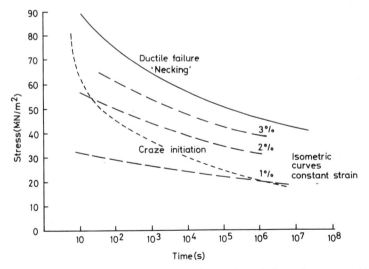

Fig. 6.11.   Isometric stress/time curves for polymethyl methacrylate. (After Ogorkiewicz.[11])

By using the Eyring rate equation as a means of relating the 'yield stress' to temperature, i.e.

$$\varepsilon|_{cr} = A \exp\left[-\frac{\Delta H}{RT}\right] \sinh \frac{v\hat{\sigma}_Y}{RT} = \text{constant}$$

it would be inferred that the dependence of the yield strength on time is the result of the kinetic nature of the activation volume. There is no experimental verification of this hypothesis to date; so that accelerated tests at high temperatures as a means of obtaining yield stress data at longer times are not yet possible.

Tests at high temperatures are useful, however, since they may reveal important features in the yield stress/time relationship (e.g. abrupt changes or discontinuities in the curves) which may serve as a warning that similar phenomena may occur at lower temperatures over longer periods of time.

6.3   FAILURES INVOLVING BRITTLE FRACTURES

For materials like thermoset plastics (highly cross-linked)—diamond, etc.—there is no mechanism for the structural rearrangements required to produce the large scale movements involved in yielding; hence they will always fail in a brittle fashion, i.e. fracture will involve breaking of chemical bonds in the material without any permanent deformations in the fractured area. Strictly speaking, the fracture of vulcanised rubbers above their glass transition temperature may be considered to be a brittle fracture.

Other materials, like metals, thermoplastics, etc., can exhibit a brittle fracture when the stresses are predominantly of the tensile type but will fail by a yielding mechanism under shear and compressive stresses.

6.3.1 Stress Criteria for Brittle Fractures

Very little attention has been given to studies of brittle fractures of plastics with the purpose of developing criteria similar to those for yielding. There are some indications, however, that when tensile stresses are the prevalent causes of fracture, the criterion resembles one of maximum principal stress.[12]

Intuitively, however, it can be expected that brittle fractures are likely to follow a 'maximum principal strain' criterion. Therefore:

(1)   For the maximum principal stress criterion

$$\hat{\sigma}_1 = \hat{\sigma}_F$$

where

$$\sigma_1 > \sigma_2 > \sigma_3 \qquad (6.22)$$

(2)   For the maximum principal strain criterion

$$\hat{\varepsilon} = \hat{\varepsilon}_F$$

where

$$\varepsilon_1 > \varepsilon_2 > \varepsilon_3$$

hence

$$\hat{\varepsilon}_1 = \frac{\sigma_1}{E} - \frac{\nu}{E}(\sigma_2 + \sigma_3) = \hat{\varepsilon}_F = \frac{\hat{\sigma}_F}{E} \qquad (6.23)$$

where $\hat{\varepsilon}_F$ and $\hat{\sigma}_F$ are the strain and stress to fracture in monoaxial tensile tests.

Under plane stress conditions the maximum principal strain criterion reduces to

$$\sigma_1 - \nu\sigma_2 = \hat{\sigma}_F$$

The comparison between the 'maximum principal stress' and 'maximum principal strain' criteria is shown in Fig. 6.12 for conditions of plane stress and tensile stresses bias.

6.3.2 The 'Fracture Mechanics' Approach to Brittle Fractures

The failure criteria discussed so far are based on continuum mechanics; i.e. no consideration is given to the level of stress that may exist on the atomic,

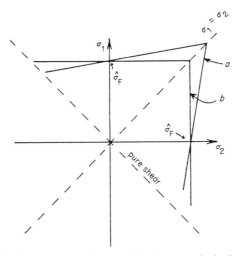

Fig. 6.12.   Brittle fracture criteria. $a =$ Maximum principal strain criterion, $b =$ maximum principal stress criterion.

molecular or microscopic structural entities of the material. Following the discovery of Griffith in 1920[13] that brittle materials (e.g. glasses) exhibit a strength which is only a fraction of the value calculated from the magnitude of interatomic forces and that this results from stress intensifications around discontinuities, such as minute cracks, scientists have focused their attention on the relationship between size of cracks and strength.

This has given rise to a new branch of mechanics known as 'linear-elastic-fracture mechanics', which considers the fracture condition in relation to the geometry of an artificial crack introduced in the specimen tested. The rationale of this approach is that the artificial crack will promote fracture under well-controlled conditions and, therefore, the events associated with the fracture process can be easily followed and quantified. Although it may be argued that the concepts of fracture mechanics may not be valid when the discontinuities are not in the form of microcracks in the material (as, indeed, may well be the case with plastics), the utility of such concepts in comparing materials cannot be overlooked, especially if it is counter-argued that any discontinuity in the structure of the material can be treated as an 'equivalent crack' in terms of fracture mechanics.

Furthermore, a brittle fracture (by definition) is one that takes place through propagation of a fine crack. Hence even though such a crack may not exist originally in the material, it has to form at some stage during the stress history of the material (crack initiation or nucleation). In the case of a thermoset plastics material, there are microscopic discontinuities at the matrix/filler interface and molecular discontinuities with respect to the cross-linking density.

(a) *The Concept of Critical Stress Intensity Factor as a Property of the Material*
The concepts of fracture mechanics have been developed from studies of large plates containing a central crack of dimensions much smaller than the plane dimensions of the plate, i.e. $a/w \to 0$, where $a =$ length of the crack, and $w =$ width of the plate. If the plate is subjected to a tensile stress σ_0 (Fig. 6.13), the stress distribution at the tip of the crack (i.e. for $r \to 0$) can be described by the relationship

$$\sigma_{ij} = \sigma_0 \left(\frac{a}{2r} \right)^{1/2} f_{ij}(\theta) \qquad (6.24)$$

which can be rewritten as

$$\sigma_{ij} = \frac{K}{\sqrt{2\pi r}} f_{ij}(\theta)$$

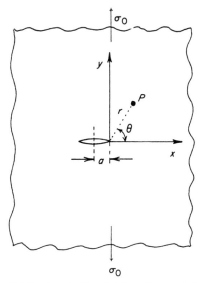

Fig. 6.13.   Stress distribution in the vicinity of a crack in an infinite plate.

where

$$K = \sigma_0 \sqrt{\pi a}$$

K is known as the 'stress intensity factor', which is a parameter that determines the height of the stress distribution curve ahead of the crack tip.

For conditions of *plane stress* the function $f_{ij}(\theta)$ varies according to the stress considered, so that

$$\sigma_{yy} = \frac{K}{\sqrt{2\pi r}} \cos \frac{\theta}{2} \left[1 + \sin \frac{\theta}{2} \sin \frac{3\theta}{2} \right]$$

$$\sigma_{xx} = \frac{K}{\sqrt{2\pi r}} \cos \frac{\theta}{2} \left[1 - \sin \frac{\theta}{2} \sin \frac{3\theta}{2} \right]$$

$$\tau_{xy} = \frac{K}{\sqrt{2\pi r}} \cos \frac{\theta}{2} \sin \frac{\theta}{2} \cos \frac{3\theta}{2}$$

When $\theta = 0$, i.e. along the major axis of the crack, the above equations reduce to

$$\sigma_{yy} = \frac{K}{\sqrt{2\pi r}}; \qquad \sigma_{xx} = \frac{K}{\sqrt{2\pi r}}; \qquad \tau_{xy} = 0$$

Therefore, since $\sigma_{yy} = \sigma_{xx}$, the stress intensity factor K describes the level of stresses along the major axis of the plate.

For thick plates, i.e. under conditions of plane strain, the stress distribution in the direction of the thickness can be obtained by placing $\varepsilon_{zz} = 0$, i.e.

$$\varepsilon_{zz} = \frac{\sigma_{zz}}{E} - \frac{v}{E}\left(\sigma_{xx} + \sigma_{yy}\right) = 0$$

therefore

$$\sigma_{zz} = v(\sigma_{xx} + \sigma_{yy})$$

Consequently, for $\theta = 0$, we obtain

$$\sigma_{zz} = \frac{2vK}{\sqrt{2\pi r}} \quad \text{and} \quad \tau_{xz} = \tau_{yz} = 0 \tag{6.25}$$

The stress distribution ahead of a crack tip for plane strain condition is shown in Fig. 6.14. Note that σ_{xx} at the actual crack tip must be zero since a stress cannot act on a free surface, and that if $\sigma_{zz} \neq 0$, then $\sigma_{yy} > \sigma_{xx}$.

For a given crack length ($2a$ if at the centre of a plate, or simply a if at the edge of a plate), any increase in the value of the 'gross' stress, σ_0, produces a concomitant rise in the level of stresses (σ_{xx}, σ_{yy} and σ_{zz}) in the vicinity of the crack.

Since these stresses are not constant but depend on the distance

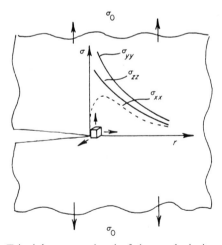

Fig. 6.14.   Triaxial stresses ahead of the crack tip in a thick plate.

considered from the crack tip (Fig. 6.14), it is convenient to express the fracture conditions in terms of a critical value of the stress intensity factor, K_c, i.e.

$$K_c = \hat{\sigma}_0 \sqrt{\pi a}$$

(b) Effects of the Crack Length in Relation to the Width of the Plate (a/w) and Type of Loading

The treatment in the previous section has been derived for infinite plates, i.e. $a/w \to 0$, and implies that K is a single-valued parameter only very near the crack tip, i.e. for $r \to 0$. In fact, the stress at the crack tip is normally expressed in the form of a series, i.e. for plane stress

$$\sigma_{yy} = \sigma_{xx} = \frac{K}{\sqrt{2\pi r}} + \cdots \text{(higher terms)}$$

where K is the first term and, therefore, predominates only when r is very small. Furthermore, since r is never infinity in practice, the stresses σ_{xx} and σ_{yy} will never decay to zero. This means that the constraints imposed by the edges in a finite specimen will actually make the value of the stress intensity factor K bigger than is predicted from the finite plate assumption; K is a function, therefore, of the ratio a/w.

We can rewrite the definition of K in the form

$$K = Y\sigma_0 \sqrt{a} \tag{6.26}$$

where $Y = f(a/w)$; so that $Y = \sqrt{\pi}$ for $a/w \to 0$. Y is called the 'geometrical factor' and depends not only on the a/w ratio (Fig. 6.15) but also on the way

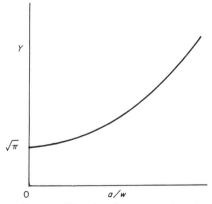

Fig. 6.15. The compliance calibration factor as a function of the a/w ratio.

the load is applied in relation to the crack and geometry of the specimen. The values of the Y function can be obtained either experimentally (see later) or by means of stress analysis.[14]

Typical functions for various types of loadings are shown below.

(1) *Single edge notch (SEN) specimens* (see also ref. 15)

   (i)   Tensile loads:

$$Y = 1{\cdot}99 - 0{\cdot}41(a/w) + 18{\cdot}70(a/w)^2 - 38{\cdot}48(a/w)^3 + 53{\cdot}85(a/w)^4$$

   (ii)   Three-point bending loads $(L/W = 4)$:

$$Y = 1{\cdot}93 - 3{\cdot}07(a/w) + 14{\cdot}53(a/w)^2 + 25{\cdot}11(a/w)^3 + 25{\cdot}80(a/w)^4$$

(2) *Double edge notch (DEN) specimens*

These are tested in tension and the stress intensity factor is normally expressed as

$$K = \sigma_0 \left[w \left(\tan \frac{\pi a}{w} \right) + 0{\cdot}1 \sin \frac{2\pi a}{w} \right]^{1/2}$$

(c) *Effects of Mode of Fracture*

So far we have assumed that fracture occurs by a crack opening mode, which is by far the most common way for fracture to occur. There are cases, however, where fracture can take place as a result of shear stresses. The nature of the functions that relate the local shear stresses to the 'gross' shear stresses are such that they become equal to 1 for $\theta = 0$.

In general, therefore, we can have three possible modes of fracture as shown in Fig. 6.16.

Hence for infinite plates and under plane stress conditions, we have the following relationships:

$$\sigma_{xx} = \sigma_{yy} = \frac{K_{\mathrm{I}}}{\sqrt{2\pi r}} \qquad \text{where} \quad K_{\mathrm{I}} = \sigma_0 \sqrt{\pi a} \qquad (6.27)$$

$$\tau_{xy} = \frac{K_{\mathrm{II}}}{\sqrt{2\pi r}} \qquad \text{where} \quad K_{\mathrm{II}} = [\tau_\infty]_{\mathrm{F}} \sqrt{\pi a} \qquad (6.28)$$

and

$$\tau_{yz} = \frac{K_{\mathrm{III}}}{\sqrt{2\pi r}} \qquad \text{where} \quad K_{\mathrm{III}} = [\tau_\infty]_{\mathrm{P}} \sqrt{\pi a} \qquad (6.29)$$

The critical value of K for each mode of fracture will be different in each

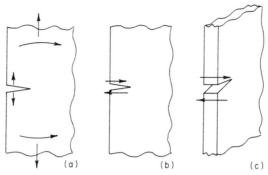

Fig. 6.16.   The three basic fracture modes. (a) Mode I (K_I), crack opening mode. (b) Mode II (K_{II}), forward shear mode. (c) Mode III (K_{III}), antiplane (parallel) shear mode.

case, so that we need to know all three K_c values to characterise fully the fracture characteristics of the material.

Very little attention has been given, however, to fracture under Mode II and III loading conditions, and consequently the 'geometrical' factors are not known. However, Stanford and Stonesifer[16] carried out tests on unidirectional composite materials under combined 'crack opening/ forward shear' modes and concluded that (for these materials) the following relationship applies:

$$C_1 K_I^2 + C_2 K_{II}^2 = \text{constant}$$

where C_1 and C_2 are functions of the elastic moduli in the two principal directions, such that

$$\sqrt{C_1/C_2} = \sqrt{E_\parallel / E_\perp}$$

While this can be considered a criterion for brittle fractures, it is not known to what extent it can be generalised and whether it would be at all applicable to isotropic materials. In the latter case we would have

$$K_I^2 + K_{II}^2 = \text{constant}$$

which has been suggested to apply to the case of metals.[17]

6.4   PRE-YIELDING PHENOMENA AS PRECURSORS TO BRITTLE FRACTURE OF THERMOPLASTICS

Quite often microcavitation phenomena can be observed for both metals and thermoplastics at levels of stresses that are below those required for large scale yielding. For the case of plastics these can be detected even by the naked eye and are known as stress-whitening and crazing phenomena. In other words, the microcavitations can reach dimensions greater than the wavelength of visible light and create internal scatter of the incident light.

6.4.1 Structure and Mechanism of Formation of Microcavitations and Crazing

The exact origin of these phenomena has not yet been established, but it is suspected that they result from triaxial tensile stresses set up at the boundaries of molecular and morphological discontinuities within the structure of the material. Impurities, such as residual monomer, and additives (particularly the incompatible types) can also give rise to or, at least, assist the nucleation process for their formation. Such events occur even more in the case of polymer blends and crystalline block copolymers. The structure of microcavitations can vary considerably depending on the nature of the heterogeneous entities that have caused their formation and on the manner in which they spread out subsequent to their nucleation.

In any case, however, the voids appear to be connected by some fibrillar elements which are believed to consist of molecularly oriented matter. Morphologically homogeneous materials, i.e. those in which there is no fluctuation in density within the structure, e.g. glassy amorphous polymers, produce the characteristic craze structure; while heterogeneous systems undergo stress whitening, the structure of which is much more complex and varied. Typical examples of microcavitation structures are shown in Figs. 6.17(a) and (b). Note that crazes generally consist of an open network of polymer fibrils between 10 and 40 nm in diameter, interspersed by voids of about 10–20 nm.

A coarser structure is sometimes observed in the central region of the craze.[20] Noting that, contrary to yield phenomena, craze zones are formed at right angles to the direction of the maximum principal stress, Sternstein and Myers[21] have proposed the following mechanism: the triaxial tensile stresses set up around regions of molecular discontinuities decrease the glass transition temperature and bring small domains of the material into its rubbery state (Fig. 6.18). The hydrostatic stress component required to nucleate a craze, therefore, is a function of the dilatability coefficient μ, the coefficient of thermal expansion α (volumetric), and the difference between the glass transition temperature and the ambient temperature; i.e.

$$\sigma_m = \beta(T_g - T) \qquad \text{where } \beta = \left(\frac{\alpha_r - \alpha_g}{\mu_r - \mu_g}\right) \qquad (6.30)$$

Fig. 6.17(a).   Electron micrograph of stress-whitened PVC (Darvic 118), showing voids elongated in the direction of the applied tensile stress. (After Vincent.[18])

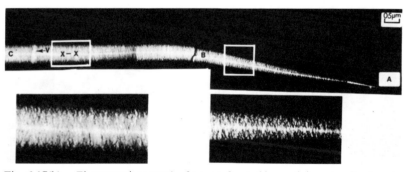

Fig. 6.17(b).   Electron micrograph of a craze formed by applying a tensile stress to a cast polystyrene film. (After Beaham.[19])

The subscripts r and g refer to the rubbery state and glassy state respectively.

As the stress increases, the rubbery domains will eventually form small holes ($\simeq 2$ μm), while the walls will extend in the direction of the maximum principal stress, forming the characteristic craze structure (Fig. 6.18).

The fibrils stabilise the molecular orientation through a process of

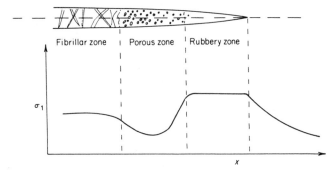

Fig. 6.18.   Stress distribution in the various stages of the formation of a craze.

'devitrification' since the triaxial stresses, which have nucleated the formation of the pores, are no longer present during the extension of the connecting walls; i.e. these extend mainly under the influence of the axial stress. Consequently, they are brought back in the glassy state.

This redistribution of stresses causes the rubbery zone to extend laterally while the inner fibrillar zone increases in magnitude accordingly. Figure 6.18 reveals two important characteristics of crazes in respect to the mechanical behaviour of materials:

(1)   The stress intensification at the tip of the craze similar to that produced by a crack.

(2)   The craze can carry loads (unlike cracks) through the interconnecting fibrils.

It can be easily appreciated that both microvoiding and crazes can produce favourable conditions for the occurrence of brittle fractures. Whether the material reaches the value of the yield strength, consequent to the formation of microvoids, or breaks in a brittle manner, depends on the 'critical stress intensity factor'. The level of stress to reach the K_c value of the material depends, in turn, on the dimensions which the craze assumes during the loading history of the material. (See later discussions.)

6.4.2 Mechanical Criteria of Crazing

Sternstein and Myers[21] have put forward a criterion for crazing under plane stress conditions. This was derived on the basis of the observation that crazing occurs only if there is a 'tensile bias' of principal stresses. In other words, crazing does not occur under conditions of pure shear (where $\sigma_1 = -\sigma_2$) and whenever the compressive stresses are greater than the tensile stresses. The following empirical equation seems to produce a good fit for the loci of the stress coordinates under which crazing occurs:

$$|\sigma_1 - \sigma_2| = A_{(T)} + \frac{B_{(T)}}{I_1} \qquad (6.31)$$

where $|\sigma_1 - \sigma_2| = $ normal stress bias (tensile); $I_1 = $ first invariant of the stress tensor, i.e. $I_1 = \sigma_1 + \sigma_2 + \sigma_3$; $A_{(T)}$ and $B_{(T)} = $ materials constants depending on temperature.

It is instructive to note that the plane of the craze is always normal to the maximum tensile stress but has no preference under balanced biaxial stresses (i.e. for $\sigma_1 = \sigma_2$), when crazing becomes purely the result of dilatations; i.e. it arises solely from the first invariant I_1.

Sternstein and Myers[21] arrived at the geometric representation of eqn. (6.31) through the following reasoning: a plot of the normal stress bias versus the reciprocal of the first invariant consists of straight lines that intersect at the same point the hyperbola curve obtained for conditions of $\sigma_2 = 0$. This indicates simply that under uniaxial tension the crazing stress is constant (Fig. 6.19).

Note that curve 4 represents the conditions of dilatational crazing, i.e. for $\sigma_1 - \sigma_2 = 0$. Therefore

$$\frac{1}{I_1} = \text{constant}$$

When these curves are replotted on principal stresses axes, all the points that fall below the hyperbola of Fig. 6.19 (i.e. when $\sigma_1 > 0$ and $\sigma_2 > 0$) will

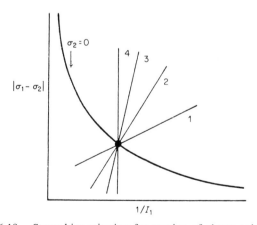

Fig. 6.19.   Stress bias criterion for crazing of glassy polymers.

lie in the first quadrant, while those above the hyperbola ($\sigma_1 > 0$ and $\sigma_2 < 0$) will lie in the second quadrant. (See Fig. 6.20.) All the curves pass through a common point at $\sigma_2 = 0$ but will never intersect the pure shear axis (i.e. for $\sigma_1 = -\sigma_2$). The validity of this criterion has been confirmed also under triaxial stress conditions.

If a hydrostatic pressure component, $-P$, is added to the biaxial stress field (σ_1, σ_2) the superposition results in a triaxial stress field with principal stresses ($\sigma_1 - P$, $\sigma_2 - P$, $-P$). This has the effect of reducing the first

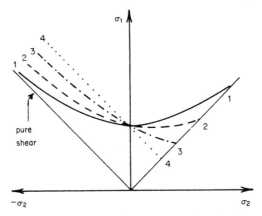

Fig. 6.20.   Geometrical representation of the stress bias criterion on biaxial stress quadrants.

invariant I_1 term, and, therefore, the loci of the points defining the crazing envelope in Fig. 6.20 are displaced upwards. The opposite is true when a hydrostatic dilatational stress is superimposed on the biaxial stress field. A similar effect is produced when the temperature is increased. The dilatation resulting from the increased temperature can be considered, in fact, to be equivalent to that resulting from the application of isotropic (hydrostatic) tensile stresses.

Because the stress bias criterion is purely empirical and lacks physical meaning Oxborough and Bowden[22] have suggested a slight modification which results in a 'criterion of maximum tensile principal strain'. Rewriting the stress bias criterion as

$$E\varepsilon_1 = \sigma_1 - v(\sigma_2 + \sigma_3) = A(t, T) - \frac{B(t, T)}{I_1}$$

one obtains a more general expression where the left-hand term can be replaced by either

$$E(t, \sigma, T)(\varepsilon_1)_{t,T} \quad \text{or} \quad \frac{(\varepsilon_1)_{t,T}}{D(t, \sigma, T)}$$

Hence

$$(\varepsilon_1)_{t,T} = D(t, \sigma, T) \left[A(t, T) - \frac{B(t, T)}{I_1} \right]$$

$$(\varepsilon_1)_{t,T} \simeq \frac{1}{E(t, \sigma, T)} \left[A(t, T) - \frac{B(t, T)}{I_1} \right] \tag{6.32}$$

where $(\varepsilon_1)_{t,T} = $ maximum principal strain for crazing (time- and temperature-dependent); $E(t, \sigma) = $ relaxation modulus of the material; $D(t, \sigma) = $ creep compliance of the material; and $A(t, T)$ and $B(t, T) = $ material parameters that are responsible for the deviation of the time and temperature dependence of the craze strain from the creep strain (Fig. 6.11).

The main difference between 'stress bias' and 'maximum tensile principal strain' criteria occurs in the areas approaching the pure shear axis. It would seem, however, that the latter is more accurate insofar as it can explain the shear microbands observed at around 45° to the direction of the principal stress in uniaxial compression tests.[23] These microbands consist,

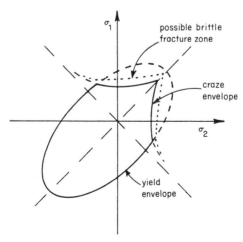

Fig. 6.21.   Failure envelope for glassy polymers under plane stress conditions.

in fact, of microvoids or craze matter and can be considered to be the result of tensile components around the heterogeneous entities present in the material.

Futhermore, the modified criterion incorporates time-dependent terms, i.e. the compliance or modulus of the material and the parameters A and B, but it is not known how these interact to produce a time-dependent critical 'crazing' strain. In any case the importance of craze criteria is that they may be used to determine the maximum allowable stress in a design in order to reduce the likelihood of brittle fractures. The global envelope for the loci of points of the safe stress in a design is shown in Fig. 6.21, which also indicates that the criterion for brittle fractures follows very closely that for the craze envelope.

6.5 ENERGY CONSIDERATIONS IN FRACTURES: THE CONCEPT OF TOUGHNESS

Toughness is a term normally used to denote the amount of total energy absorbed by the material to induce fracture; i.e.

$$U_f = V \int_0^{\varepsilon_{ij}|_f} \sigma_{ij} \, d\varepsilon_{ij} \qquad (6.33)$$

where V is the total volume of the specimen and the integral term is the 'total' strain energy density of the specimen at the point of fracture. For a material that is linear elastic up to the fracture strain, this becomes simply

$$U_f = \frac{V \hat{\sigma}_f \hat{\varepsilon}_f}{2} = \frac{V \hat{\sigma}_f^2}{2E}$$

From eqn. (6.33), above, it can be easily inferred that for a material that undergoes yielding the toughness is substantially higher than for the same material when it fails in a brittle fashion.

To be able to predict the toughness of plastics, therefore, it is essential to have a criterion for the strain at break so that the integration of eqn. (6.33) can be performed. Furthermore, if the material yields only in a restricted region of the specimen, as in the case of notched specimens, there is a distribution of stress and strain, and the recorded energy at fracture decreases as the volume of the material that undergoes yielding prior to fracture becomes increasingly small.

The conditions that induce such a transition from a maximum (i.e.

ductile failure) to a minimum (i.e. brittle failure) are depicted in Fig. 6.22. Temperature is taken as the main parameter since it influences the extensibility of the molecular chains. The other parameters, e.g. sharpness of the notch (r), the strain rate ($\dot{\varepsilon}$) and the hydrostatic (tensile) stress components, can be considered to displace the temperature range over which the transition takes place in the direction of the arrows. Note that

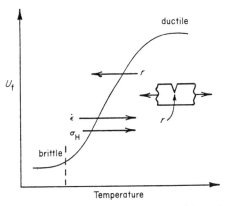

Fig. 6.22.   Factors affecting the ductile/brittle transition of thermoplastics.

such a transition will occur also for the case of thermoset plastics where there is no yielding involved in the fracture, and it arises mainly from increased chain flexibility as the material passes through the glass/rubber transition. The difference in level of energy, however, is much smaller than for thermoplastics.

Notches and hydrostatic tensile stress components hinder the uncoiling of the molecular chains and, therefore, reduce the extension at break. The straining rate, on the other hand, affects the characteristic time of the chain rotations in relation to the time taken by the stress to reach the value of the yield stress. Higher strain rates are more likely to produce brittle fractures because they offer a greater opportunity for internal 'defects' (e.g. microvoids or crazes) to grow to critical dimensions for rapid crack propagation, thereby preventing the material from reaching yielding conditions, or the molecules from reaching their potential extensibility if fracture is preceded by yielding.

Empirically one can, therefore, evaluate the toughness of materials by studying the effects of all the parameters mentioned earlier, but it is difficult to use the data in design calculations and, furthermore, experience shows that the values obtained are dependent on the geometry of the specimen used.[24] This has stimulated researchers to take a more rational approach to the evaluation of the toughness of materials through the fracture mechanics analysis.

6.5.1 The Griffith Interpretation

Griffith was the first to invoke energy considerations (in a rational manner) in fracture processes. Realising that the inability of glass to reach its theoretical strength was due to the presence of internal cracks which acted as stress raisers, he attempted to estimate the value of the 'observed' strength on the basis of the Inglis stress analysis for elliptical holes in an infinite plate, i.e.

$$\sigma_{max} = \sigma_0 \left(1 + 2 \sqrt{\frac{a}{r}} \right)$$

where σ_{max} = the value of the stress at the edge, σ_0 = nominal stress calculated from the force divided by the perpendicular area, a = half-crack length, and r = radius at the tip of the crack.

On the basis of this equation, as the tip radius approaches zero (the case of natural cracks), the maximum stress tends to infinity, which would make the nominal stress at break extremely small. This weakness in the theoretical prediction of the strength of materials based solely on stress considerations led Griffith to stipulate an energy balance as a criterion of fracture; i.e. the loss of strain energy in the plate must be equal to the gain in surface energy as a result of the formation of the crack. From stress analysis one obtains an equation for the loss of strain energy of an elastic plate resulting from the formation of a very small central crack; i.e.

$$U_S = -\frac{\pi a^2 \sigma B}{E} \qquad \text{for conditions of } a/w \to 0$$

where a = half-crack length, w = width of the plate, B = thickness of the plate, E = Young's modulus, and σ = the gross stress acting in the direction perpendicular to the crack plane. Note that the minus sign denotes a negative increment, i.e. a release of energy, in the formation of the crack. The energy acquired by the surface of the crack, on the other hand, is given by

$$U_C = 4\gamma a B$$

where γ = surface energy of the material.

The total energy change in the process of formation of the crack is therefore

$$\Delta U = U_S + U_c = -\frac{\pi a^2 \sigma^2 B}{E} + 4\gamma aB$$

The total energy change is independent of the crack length, since the above applies for conditions $a/w \to 0$; hence

$$\frac{\delta(\Delta U)}{\delta a} = 0$$

and, therefore

$$\frac{\delta(\Delta U)}{\delta a} = 4\gamma - \frac{2\pi a \sigma^2}{E} = 0$$

From this one obtains the familiar equation of Griffith for brittle fractures of elastic materials:

$$\hat{\sigma} = \sqrt{\frac{2\gamma E}{\pi a}} \qquad (6.34)$$

This equation relates the strength ($\hat{\sigma}$ at fracture) to the size of the internal crack a and to the surface energy and modulus of the material. According to eqn. (6.34), therefore, the strength of brittle materials cannot be regarded as an intrinsic property of the material.

Note that the Griffith equation is derived from stress analysis considerations; however, one can arrive at an expression very similar to the above with much simpler reasoning. Assume that when a crack is formed there will be an area around the crack tip where the strain energy is zero, since the stress on the two surfaces of the crack is zero (Fig. 6.23).

If a crack extends by an infinitesimal amount da, there will be an increase in the region of the plate where the strain energy is zero; i.e. the annulus is of thickness da, so that the balance strain energy and the increase in surface energy becomes

$$U\,dV = 4\gamma B\,da$$

If it is assumed that the annulus is circular (for simplicity), the volume, dV, is

$$dV = \pi[(a + da)^2 - a^2]B = \pi[-2a\,da + (da)^2]B$$

Neglecting the $(da)^2$ term, we obtain

$$dV = 2\pi Ba\,da$$

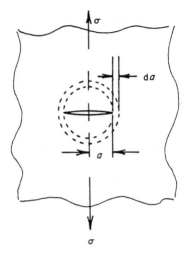

Fig. 6.23.   Stress-free circular zone around the surfaces of a crack.

Since U is the strain energy density of the plate (i.e. strain energy per unit volume) for an elastic material, we can write:

$$U = \tfrac{1}{2}\sigma\varepsilon = \frac{1}{2}\frac{\sigma^2}{E}$$

and, therefore,

$$\frac{\sigma^2}{E}\,\pi Ba\,\mathrm{d}a = 4\gamma B\,\mathrm{d}a$$

which gives

$$\sigma = \sqrt{\frac{4\gamma E}{\pi a}} \tag{6.35}$$

This only differs from the equation of Griffith derived from stress analysis by a factor of 2 but does not alter the nature of the relationship between σ and a.

6.5.2 The Irwin Concept of 'Critical Strain Energy Release Rate'

Obviously the concept of surface energy as a criterion for brittle fractures cannot be used when crack propagation involves the formation of a yield zone ahead of the crack tip (see discussions in Section 6.6.1). Orowan[25] suggested, in fact, that a term γ_p (i.e. the energy required to produce the plastic deformations over the fracture surfaces) should be added to the surface energy term. Because of the difficulties in estimating the value of γ_p and other energy terms (e.g. cavitational work to produce microvoids, etc., may be involved in the fracture process), Irwin[26] proposed that the concept of 'critical strain energy release rate', G_C, should be used instead. In other words the extension of a crack involves the release of strain energy and catastrophic fracture takes place when this reaches a critical value.

Consequently Griffith's equation should be written as

$$\sigma = \sqrt{\frac{G_C E}{\pi a}}$$

Substituting into this the expression for the critical stress intensity factor for an infinite plate, i.e. $K_C = \sigma\sqrt{\pi a}$, one obtains a relationship between K_C and G_C, i.e.

$$K_C^2 = EG_C \qquad \text{(plane stress conditions)}$$

and

$$K_C^2 = E'G_C \qquad \text{(plane strain conditions)}$$

where $E' = E/(1 - v)^2$, known as the 'effective' modulus of the material under plane-strain stresses. Note, in fact, that under plane strain (i.e. $\varepsilon_3 = 0$) one obtains $\sigma_3 = v(\sigma_1 + \sigma_2)$. When this is substituted in the equations for the strains ε_1 and ε_2 respectively, we obtain

$$\varepsilon_1 = \frac{1}{E} \left[(1 - v)^2 \sigma_1 - v(1 + v)\sigma_2 \right]$$

$$\varepsilon_2 = \frac{1}{E} \left[(1 - v)^2 \sigma_2 - v(1 + v)\sigma_1 \right]$$

Consequently, when each of the two stresses σ_1 and σ_2 are, in turn, equal to zero, we can write the relationships between stress and strain for conditions of plane strain as

$$\frac{\sigma_1}{\varepsilon_1} = \frac{\sigma_2}{\varepsilon_2} = \frac{E}{(1 - v)^2} = E' \qquad \text{(effective modulus)}$$

6.5.3 Relationship between 'Critical Stress Intensity Factor' and 'Critical Strain Energy Release Rate'

(a) *Measurement of the 'Critical Strain Energy Release Rate'*
Irwin's greatest contribution in the area of fracture mechanics is, possibly, his approach to the measurement of the critical strain energy release rate.

Measurements can be made on specimens of any size or shape. The only important parameter to control is the fracture mode, provided that an appropriate calibration of the compliance of the specimen as a function of the crack length is known.

The compliance of the specimen is defined simply as

$$C = \frac{\Delta}{P} \tag{6.36}$$

where Δ = extension or deflection, and P = load (see Fig. 6.24).

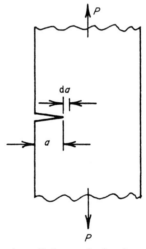

Fig. 6.24.   Load applied to a single-edge-notch specimen.

Irwin stipulates two extreme conditions for fracture, one occurring under fixed extension or deflection and the other under constant load. The rate of change in energy per infinitesimal increase in crack length must be the same, irrespective of the manner in which fracture of the specimen is induced. This quantity is the strain energy release rate G, which reaches its limiting (critical) value when catastrophic fracture occurs. The invariance of the strain energy release rate for the two extreme fracture conditions is illustrated in Fig. 6.25.

Therefore

$$\frac{\delta U}{\delta A} = G = \tfrac{1}{2}\Delta \delta P = \tfrac{1}{2}P \delta \Delta$$

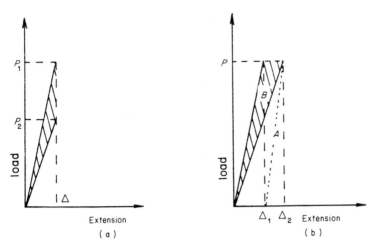

Fig. 6.25.   Irwin's analysis of fractures at constant extension (left) and constant load (right). Left: Area $= \delta U/\delta A = \frac{1}{2}\Delta(P_1 - P_2)$, or $\delta U/\delta A = \frac{1}{2}\Delta\delta P$, where $A =$ surface area of the extended crack. Right: Area $= A - B$. When $(\Delta_2 - \Delta_1) \to 0$, $A \simeq 2B$. Area $B = \delta U/\delta A = \frac{1}{2}P(\Delta_2 - \Delta_1) = \frac{1}{2}P\delta\Delta$.

Catastrophic fracture occurs, therefore, when

$$-\frac{\delta U}{\delta A}\bigg|_{\delta a \to 0} = G_C \quad \text{or} \quad -G_C = \lim_{\delta a \to 0} \frac{\delta U}{\delta A}$$

where $\delta a =$ crack extension length (i.e. $\delta A = B\delta a$), and $B =$ plate thickness.

The energy stored in the plate (or specimen) prior to crack extension is simply

$$U = \frac{1}{2}P\Delta = \frac{1}{2}P^2 C$$

The rate of loss of potential energy in the plate as a result of crack extension is

$$G = -\frac{\delta U}{\delta A} = CP\frac{\delta P}{\delta A} + \frac{1}{2}P^2\frac{\delta C}{\delta A} \tag{6.37}$$

Therefore:

(1)   For crack extension at constant load

$$G = -\frac{\delta U}{\delta A} = \frac{1}{2}P^2\frac{\delta C}{\delta A} = \frac{P^2}{2B}\frac{\delta C}{\delta a} \tag{6.38}$$

(2)   For crack propagation at constant extension we have

$$CP = \Delta = \text{constant}$$

Therefore

$$\frac{\delta(CP)}{\delta A} = P\,\frac{\delta C}{\delta A} + C\,\frac{\delta P}{\delta A} = 0$$

i.e.

$$\frac{\delta P}{\delta A} = -\frac{P}{C}\,\frac{\delta C}{\delta A}$$

Substituting in (6.37) above, we obtain

$$G = -\frac{\delta U}{\delta A} = -P^2\,\frac{\delta C}{\delta A} + \frac{1}{2}\,P^2\,\frac{\delta C}{\delta A} = -\frac{1}{2}\,P^2\,\frac{\delta C}{\delta A}$$

i.e.

$$G = -\frac{\delta U}{\delta A} = -\frac{P^2}{2B}\,\frac{\delta C}{\delta a}$$

Hence the only difference between the rates of change in potential energy for infinitesimal crack extensions in the two different cases is with respect to the sign, which is only a manifestation of the difference in direction of the energy change involved; i.e. at constant extension there is an obvious loss of potential energy, while for a crack to extend under constant load conditions it is necessary that the plate gains stored energy. (Compare the two cases in Fig. 6.25 and see later discussions.)

At the point of catastrophic fracture the rate of energy change during crack extension reaches its limiting (critical) value so that we can write

$$G_c = -\frac{1}{2}\,\frac{P_f^2}{B}\,\frac{\delta C}{\delta a} \qquad (6.39)$$

where $P_f = $ load at fracture.

From the above discussions it can be inferred that the critical strain energy release rate can be calculated from a calibration curve of the compliance as a function of the crack length for a particular type of material and specimen used and by recording the load at which fracture takes place. Note that, if rectangular specimens are used, the curve of the compliance as a function of the crack length has a changing slope (Fig. 6.26).

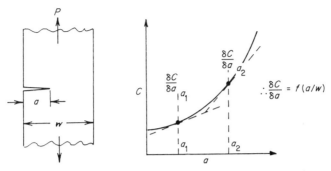

Fig. 6.26. Change of compliance with crack length for a rectangular specimen.

This means primarily that there can be considerable error in tracing the tangents, especially when the crack lengths are quite large. Secondly, unless crack initiation is very rapid, there could be further error arising from slow crack growth while the load is increased. To overcome this inconvenience Mostovoy et al.,[27] have developed a technique which utilises a double cantilever beam specimen, the contour of which is such that the rate of change of the compliance $(\delta C/\delta a)$ is independent of the crack length. These authors have derived the following expression for the rate of change of the compliance:

$$\frac{\delta C}{\delta a} = \frac{6}{EB}\left[\frac{4a^2}{h^3} + \frac{1+v}{h}\right] = \text{constant}$$

where E = Young's modulus of the material, v = Poisson ratio of the material, a = crack length, h = specimen half-height at the crack tip, and B = thickness.

This equation can be rewritten as

$$\frac{4a^2}{h^3} + \frac{(1+v)}{h} = \frac{\delta C}{\delta a}\frac{EB}{6} = m$$

hence $mh^3 - (1+v)h^2 - 4a^2 = 0$. By choosing a convenient value for m and determining h as a function of a, the contour for the linear compliance specimen can be obtained (Fig. 6.27).

The value of m is chosen in such a way that the specimen has sufficient rigidity to resist lateral twisting during the application of the load. Often, side grooves along the axis of the specimen are introduced to ease the crack propagation and to minimise the lateral twisting mentioned above. When specimens of sufficient thickness cannot be prepared, stiffening plates above and below the groove can be used.

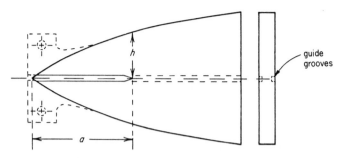

Fig. 6.27. Linear compliance specimen.

(*b*) *The Determination of Y Calibration Curves from Compliance Measurements*

Starting with the definition of K_C, we can obtain an expression for Y as a function of $\delta C/\delta a$ through the relationship between K_C and G_C, i.e. for rectangular specimens:

$$K_C = Y\sigma\sqrt{a} = \frac{YP}{Bw}\sqrt{a}$$

Since

$$K_C^2 = E'G_C = \frac{E'P^2}{2B}\frac{\delta C}{\delta a}$$

we obtain

$$\frac{Y^2P^2}{B^2w^2}a = \frac{E'P^2}{2B}\frac{\delta C}{\delta a}$$

and therefore,

$$Y = \sqrt{\frac{Bw^2E'}{2a}\frac{\delta C}{\delta a}}$$

Alternatively we can express the Y factor as a function of $\delta C/\delta(a/w)$, which is more widely used since it groups together the two important geometrical parameters of the specimens used. We can actually write the definition of K_C for a rectangular specimen

$$K_C = \frac{YP}{Bw^{1/2}}\sqrt{\frac{a}{w}} \qquad (6.40)$$

If we let $Y' = Y\sqrt{a/w}$, we can write

$$K_C = \frac{Y'P}{Bw^{1/2}}$$

and therefore,

$$K_C = \frac{Y'}{\sqrt{a/w}}\, \sigma\sqrt{a} = Y'\sigma\sqrt{w}$$

Thus,

$$K_C^2 = Y'^2\sigma^2 w = E'G_C = \frac{E'P^2}{2B}\frac{\delta C}{\delta a}$$

or

$$\frac{Y'^2}{Bw} = \frac{E'}{2}\frac{\delta C}{\delta a}$$

Therefore

$$Y'^2 = \frac{E'Bw}{2}\frac{\delta C}{\delta a}$$

By keeping w constant we can rewrite the above as

$$Y' = \sqrt{\frac{Bw^2 E'}{2}\frac{\delta C}{\delta(a/w)}}$$

Note that, since

$$Y = f\left(\sqrt{E'\frac{\delta C}{\delta a}}\right) \quad \text{or} \quad Y' = f'\left(\sqrt{E'\frac{\delta C}{\delta(a/w)}}\right)$$

the product $E'\delta C$ is constant; therefore, Y and Y' are factors that depend only on the geometry of the specimen and *not* on the nature of the material.

For rectangular specimens, however, Y and Y' are a function respectively of a and a/w. Hence by experimentally measuring the compliance as a function of a or a/w and plotting the calculated values of Y and Y' against each value of a/w (Fig. 6.28) we can obtain the polynomials for the geometrical factors, which are expressed as shown in Fig. 6.28.

$$Y = C_0 - C_1(a/w) + C_2(a/w)^2 - C_3(a/w)^3 + C_4(a/w)^4$$

and

$$Y' = C_1(a/w)^{1/2} - C_2(a/w)^{3/2} + C_3(a/w)^{5/2} - C_4(a/w)^{7/2}$$

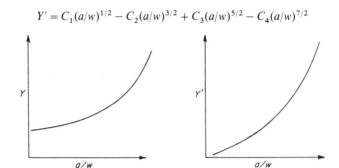

Fig. 6.28. Polynomial expressions and graphical presentation of the calibration factors Y and Y' in function of the a/w ratio.

6.6 FRACTURE MECHANICS APPLIED TO THERMOPLASTICS

The LEFM concepts described earlier raise three fundamental questions in the evaluation of the toughness of thermoplastics:

(1)   Do the events, such as yielding or crazing, at the tip of a spreading crack disturb the singularity of the stress distribution around the crack tip from which the definition of K_C is derived?

(2)   Is the analysis applicable to cases where fracture is initiated from 'blunt' notches rather than from 'sharp' natural cracks?

(3)   Does the non-linear viscoelastic nature of the material invalidate the underlying theory based on linear-elastic behaviour?

These questions and other more general points are discussed in the following sections.

6.6.1 Fractures Through a Confined Yield Zone

According to the equations of the stress distribution around the crack tip (Section 6.3.1(a)), the stress σ_y would go to infinity (or at least it would rise very rapidly to very large values) at the tip of the crack. Obviously if a material is capable of yielding, the stresses cannot rise above the values determined by the yield criteria. Consequently to maintain the constancy of strain energy, there has to be a readjustment of the stress distribution as shown in Fig. 6.29.

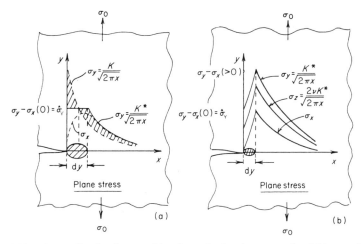

Fig. 6.29.   Stress distribution resulting from the development of a yield zone ahead of a crack.

Note that in the case of plane stress the maximum value of σ_y is $\hat{\sigma}_Y$ (according to the Tresca criterion $\sigma_y - \sigma_x(0) = \hat{\sigma}_Y$), while under plane strain conditions this will rise from the value of $\hat{\sigma}_Y$ at the crack tip to much higher values at the far edge of the yield zone, since at this point the minimum stress (σ_x) is greater than 0. Accordingly the size of the yield zone is much smaller than in the case of plane stress because of the singularity of the stress distribution.

When the yield zone is formed, there will be a shift in the stress distribution curve so that the work done by the stress considered remains constant, i.e.

$$\int \frac{K}{\sqrt{2\pi x}} \, \mathrm{d}x = \text{constant}$$

The stress intensity factor ahead of the yield zone assumes a different value (K^*). However, if the yield zone remains very small compared to the width of the specimen (say, about $1/50$), the nature of the function of the stress distribution does not alter.

Consequently, when the fracture conditions are reached, the crack spreads through the middle of the yield zone and the analysis for the calculation of K_C is still applicable, provided that a modified value for the crack length, called the 'effective crack length', is used. In the analysis it is normally assumed that effective crack length during fracture extends through the midpoint of the yield zone; hence the new value of the crack length can be taken as

$$a_{\text{eff}} = a_{\text{true}} + \frac{dy}{2} \qquad (6.41)$$

The dimension of the yield zone can be estimated by assuming that the cross-section of the yield zone is circular. Under plane stress conditions, in fact, at $x = dy/2 = r_y$ the area under the stress distribution curve is given by

$$A = \int_0^{r_y} \frac{K}{\sqrt{2\pi r}} \, dr = \sqrt{\frac{2}{\pi}} \, K r_y^{1/2}$$

Also at $x = r_y$ we have $\sigma_y = \hat{\sigma}_Y = K/\sqrt{2\pi r_y}$, i.e.

$$r_y = \frac{K^2}{2\pi \hat{\sigma}_Y^2}$$

thus,

$$A = 2\hat{\sigma}_Y r_y$$

Since the area below the line $\sigma_y = \hat{\sigma}_Y$ and up to the point $x = r_y$ is obviously $\hat{\sigma}_Y r_y$ and the shaded area above this line (Fig. 6.29(a)) is $2\hat{\sigma}_Y r_y - \hat{\sigma}_Y r_y = \hat{\sigma}_Y r_y$, the curve of the stress distribution in absence of yielding will intersect the curve $\sigma_y = $ constant at the midpoint of the yield zone (i.e. at $x = r_y$). Therefore, we can take the effective crack length to be $a_{\text{eff}} = a + r_y$, which provides a justification for the assumption in eqn. (6.41).

The calculation of the size of the yield zone for plane strain conditions is more difficult, but is often assumed to be approximately $\frac{1}{3}$ of the value obtained under plane stress.[17]

Further complications arise when the yield zone is neither regular (in geometry) nor circular. For the case of plastics, the yield zone is often elliptical and the length of the yield zone can be estimated by approximating it to a rectangular geometry. The solution obtained by Dugdale[28] for such a case is

$$\frac{dy}{2} = \frac{\pi}{8} \left(\frac{K}{\hat{\sigma}_Y} \right)^2 \qquad (6.42)$$

A more accurate estimate of both geometry and size of the yield zone can be obtained by applying the appropriate criterion and substituting for $\hat{\sigma}_Y$ in the equations for the stress distribution in Section 6.3.1. Under plane strain conditions, for instance, we obtain $\sigma_z = v(\sigma_x + \sigma_y)$, which, substituted in the von Mises criterion, gives

$$\hat{\sigma}_Y^2 = (\sigma_x^2 + \sigma_y^2)[1 - v(1 - v)] - \sigma_x \sigma_y[1 + 2v(1 - v)]$$

Further substitution in the equations in Section 6.3.1(a) gives the loci of the points of the yield zone at $x = r_y$, i.e.[29]

$$\hat{\sigma}_Y^2 = \frac{K^2}{2\pi r_y} \cos^2 \frac{\theta}{2} \left\{ 4[1 - v(1 - v)]3\cos^2 \frac{\theta}{2} \right\}$$

For the extreme conditions we can place $v = \frac{1}{2}$ for plane strain and $v = 0$ for plane stress, since in the first case it is assumed that there is no dilatation (i.e. yielding occurs completely by shear deformations), and in the second the constraint imposed by the material surrounding the yield zone prevents it from undergoing a Poisson contraction along the width of the plate (i.e. $|\varepsilon_x|_Y = 0$). This gives rise to two distinct geometries of the yield zone as shown in Fig. 6.30. The respective distances from the point of each zone are:

$$r_y = \frac{1}{2\pi} \left(\frac{K}{\hat{\sigma}_Y} \right)^2 \qquad \text{for plane stress} \qquad (6.43)$$

and

$$r_y = \frac{1}{6\pi} \left(\frac{K}{\hat{\sigma}_Y} \right)^2 \qquad \text{for plane strain} \qquad (6.44)$$

i.e. they are equivalent to the values assumed for a circular cross-section but do not correspond to the midpoint of the yield zone.

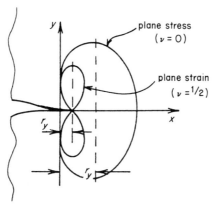

Fig. 6.30.   Shape of the yield zone ahead of a crack tip.

In general, on the outer surface of the specimen the conditions are typically plane stress, while in the middle regions the material is under plane strain. When the specimen is extremely thin the plane stress regions are much larger than those in plane strain. The reverse is true, on the other hand, for specimens that are very thick.

Consequently it is to be expected that, since the net volume of the yield zone varies considerably with the thickness of the specimen, the value of the critical stress intensity factor will vary accordingly (Fig. 6.31).

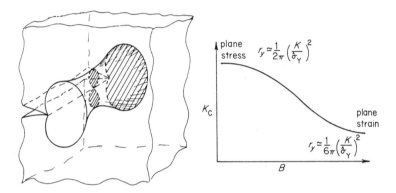

Fig. 6.31.   Transition from plane stress to plane strain across the thickness (B) of a specimen.

Note that when a craze zone develops ahead of the crack tip, either this can be considered as a rectangular strip of the yield zone, and its size calculated using the Dugdale equation, where the yield stress is replaced by the crazing stress ($\hat{\sigma}_C$), or the true stress distribution is taken into account as in Fig. 6.18 and the stress intensity factor is modified accordingly,[29] i.e.

$$K = \sigma_0\sqrt{\pi a} - \frac{2}{\pi}\hat{\sigma}_C\sqrt{2\pi r_C} \qquad \text{(plane stress)} \qquad (6.45)$$

where r_C = half length of the craze zone.

6.6.2 Fractures Through Diffused Yield Zones

(a) *Blunt Notches as Originators of Brittle Fractures in Thermoplastics*
All the arguments so far have been based on events occurring ahead of 'sharp' natural cracks. There are many situations where fracture can take place in the regions of stress concentrations, such as blunt notches, holes,

etc., without the development of 'sharp' natural cracks. With materials capable of yielding, in fact, there can be considerable opening (or displacement) at the notch before fracture occurs. Contrary to the case where fracture propagates through natural cracks, with blunt notches the tip radius is an important consideration. In this particular case we consider fractures which occur through a yield zone which is still very small in relation to the width of the specimen (or component) and a yield zone which has the shape of a narrow band which extends along the direction of the notch (Fig. 6.32).

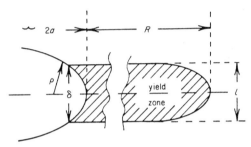

Fig. 6.32.   Concept of crack opening displacement (COD).[30]

The displacement at the opening of the notch is

$$\delta = \varepsilon l = \varepsilon 2\rho$$

where ε = strain at the crack tip, ρ = notch tip radius, l = height of the yield zone, and δ = crack opening displacement (COD).

The COD for a notch of length $2a$ in a thin plate (plane stress) of infinite width is given by[31]

$$\delta = \frac{8\hat{\sigma}_Y a}{\pi E} \ln \left(\sec \frac{\pi \sigma}{2\hat{\sigma}_Y} \right) \qquad (6.46)$$

At low stress levels this may be simplified to give

$$\delta = \frac{\pi \sigma^2 a}{E \hat{\sigma}_Y} = \frac{K^2}{E \hat{\sigma}_Y} = \frac{G}{\hat{\sigma}_Y}$$

i.e.

$$G = \hat{\sigma}_Y \delta \qquad (6.47)$$

Substituting for δ in eqn. (6.47), we obtain

$$G = 2\hat{\sigma}_Y \rho \varepsilon$$

At fracture then we would have

$$G_C = \hat{\sigma}_Y \delta_C \qquad (6.48)$$

or

$$G_C = 2\hat{\sigma}_Y \rho \hat{\varepsilon}_f \qquad (6.49)$$

Hence the critical strain energy release rate can be obtained from measurement of the COD (i.e. the crack face separation) up to the point of fracture.

Equation (6.49) is not really satisfactory insofar as for $\rho \equiv 0$ (i.e. natural crack), G_C becomes zero. For this reason eqn. (6.48) is preferred to eqn. (6.49). However, there is a definite dependence of the fracture energy parameter G_C on the tip radius; i.e. it becomes increasingly large as ρ increases. Hence it is better to differentiate the terms used, i.e.

$$G_C \quad \text{for } \rho \to 0 \qquad \text{and} \qquad G_B \quad \text{for } \rho > 0$$

Plati and Williams,[32] in fact, suggest the use of the Inglis equation for the stress concentration around elliptical holes, i.e.

$$\sigma_C = \sigma \left[1 + 2 \sqrt{\frac{a}{\rho}} \right]$$

which, for conditions of $\rho \ll a$ may be written as

$$\sigma_C = \frac{2\sigma\sqrt{a}}{\sqrt{\rho}} \qquad (6.50)$$

From the definition of K_C and its relationship to the critical strain energy release rate (in this case G_B), we obtain

$$G_B = \frac{Y^2 \sigma^2 a}{E} \quad \text{(plane stress)} \qquad (6.51)$$

Substituting eqn. (6.50) in eqn. (6.51):

$$G_B = \frac{Y^2 \sigma_C^2}{4E} \rho \qquad (6.52)$$

This is also unsatisfactory since it implies that $G_B = 0$ when $\rho = 0$. However, this controversy can be removed by stipulating that at failure the stress σ_C reaches the yield strength value ($\hat{\sigma}_Y$) at a distance r_y from the notch tip. Equation (6.50) then becomes:[29]

$$\hat{\sigma}_Y = \pi\sigma\sqrt{a}\,\frac{(r_y + \rho)}{(2r_y + \rho)^{1/3}} \qquad (6.53)$$

so that for sharp cracks (i.e. $\rho = 0$) we have

$$\hat{\sigma}_Y = \frac{\pi\sigma\sqrt{a}}{2\sqrt{2r_y}}$$

which is the same as the Dugdale equation for the relationship between size of the yield zone and yield strength.

When $r_y = 0$ this obviously reduces to the equation for the stress concentration (eqn. (6.50)). It would seem that by rearranging eqn. (6.53) we obtain the following expression for G_B:[33]

$$G_B = G_C\,\frac{(1 + \rho/2r_y)^2}{(1 + \rho/r_y)^2}$$

and, therefore, a plot of G_B (calculated from the Irwin's expression in Section 6.5.3) versus ρ will yield a curve which gives G_C at the intercept and a positive slope depending on the value of r_y. Note that in the above reasoning it is implied that r_y (at fracture) remains constant and is independent of the notch tip radius ρ. This is true only during crack propagation but not in the initial stage of the crack extension.

6.6.3 Fractures Preceded by Overall Yielding of the Section of the Specimen (or Component): The J Integral Approach

Both the COD and the blunt notch strain energy release rate (G_B) measurements attempt to estimate the toughness of material where the yield zone spreads further than might be expected from LEFM considerations. In both situations, however, the singularity of the stress distribution (i.e. $\sigma_{ij} = (K/2\pi r)f_{ij}(\theta)$) is assumed.

If we now consider the extreme case where the yield zone extends through the entire section of the specimen prior to crack extension, the above assumption is clearly invalidated. The stress distribution at the yield point, in fact, is constant along the x axis (Fig. 6.33); and, for the case of an ideal elastic/plastic material, i.e. in absence of strain softening or strain hardening phenomena, this remains the same up to the point of fracture.

For a double notch specimen in plane stress, for instance, using the slip-field theory, the following relationship applies:

$$\sigma_y - 2\hat{\tau}_Y\theta = \text{constant}$$

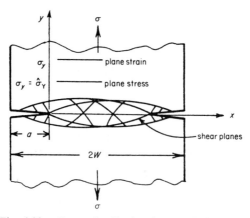

Fig. 6.33.   Stress distribution in spread yield zones.

Therefore along the line of the crack, i.e. when $\theta = 0$:

$$\sigma_y = 2\hat{\tau}_Y \qquad \text{(Tresca criterion)}$$

or

$$\sigma_y = \sqrt{3}\hat{\tau}_Y \qquad \text{(von Mises criterion)}$$

The work done to extend the yield zone along the direction of the load $\hat{P}_Y$ by an amount u is simply

$$U = \hat{P}_Y u$$

assuming that $\hat{P}_Y$ is independent of u (i.e. an ideal elastic/plastic material).

The loss of potential energy in the plate when the crack extends by an infinitesimal amount da, is simply

$$\left.\frac{\delta U}{\delta A}\right|_{\delta a \to 0}$$

This corresponds to the so-called J integral (i.e. the equivalent of the strain energy release rate in the LEFM analysis), and therefore we can write[34]

$$J = \frac{\delta U}{\delta A} = -\frac{u}{B}\frac{\delta \hat{P}_Y}{\delta a} \qquad \text{(for conditions } u = \text{constant)} \qquad (6.54)$$

and

$$J = \frac{\delta U}{B\delta a} = \frac{\hat{P}_Y}{B}\frac{\delta u}{\delta a} \qquad \text{(for } \hat{P}_Y = \text{constant)} \qquad (6.55)$$

As for the strain energy release rate for elastic fracture we can stipulate a critical value for the J integral (J_C) as the condition for catastrophic fracture.

For the analysis to be valid it is essential that the J integral reduce to the strain energy release rate (G) for elastic fractures. This, in fact, is the case insofar as for elastic deformations the potential energy stored in the specimen or component is:

$$U = \tfrac{1}{2}Pu = \tfrac{1}{2}CP^2 \qquad \text{(since } u = CP\text{)}$$

Therefore

$$J = \frac{\delta U}{\delta A} = \frac{P^2}{2B} \frac{\delta C}{\delta a} \qquad \text{(for conditions } P = \text{constant)} \qquad (6.56)$$

For those cases where there are both elastic and plastic deformations in crack extension we have the situation depicted in Fig. 6.34. The area under the curve represents the total loss of potential energy involved in the extension of the crack by an infinitesimal amount da, i.e.

$$\text{Area} = JB\,da$$

Substituting further into eqn. (6.56), i.e. for $U = CP$, we obtain

$$J = \frac{1}{2} \frac{\delta C}{\delta a} \frac{u^2}{C^2 B}$$

Hence, for a linear elastic behaviour the dependence of J on the displacement (u) is parabolic (for a given crack length), while for fully

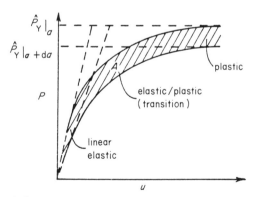

Fig. 6.34. Load/displacement curves for crack extensions involving elastic and plastic deformations.

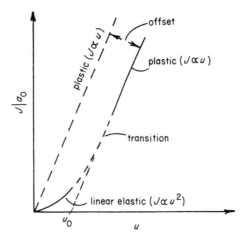

Fig. 6.35. Dependence of J at a given crack length (a_0) on displacement for elastic/plastic crack extension (After Knott.[31])

plastic deformations the relationship is linear, i.e. $J \propto u$. Consequently the overall dependence of J (at a given crack length) on displacement is as shown in Fig. 6.35.

Therefore there is a region which cannot be treated in simple analytical terms (i.e. the transition from linear elastic to plastic behaviour). However, if the dependence of J on u in the linear elastic region is very mild (i.e. the curve is very shallow) we can express the relationship between J and u in the form of a linear equation

$$-J = u_0 + \frac{u}{B} \frac{\delta \hat{P}_Y}{\delta a}$$

where the term u_0 is known as the 'offset displacement'.

A practical evaluation of J_C, therefore, involves first producing a curve of the yield load ($\hat{P}_Y$) as a function of the crack length, from which we obtain the required value of $\delta \hat{P}_Y / \delta a$ and then measuring the displacement u during fracture.

However, the concept of the J integral for the measurement of the fracture toughness of materials is still in its infancy and much more work needs to be done fully to explore its potential, particularly with respect to the effect of specimen geometry. It would seem that the approach has paramount importance in the case of thermoplastics since it offers a methodology for the evaluation of the toughness over a wide range of temperatures (i.e. from the glassy state to the onset of the rubbery state). Through interpolation of the data in the transition region between the 'elastic' and the 'plastic' analysis, we can obtain a full curve of the behaviour as a function of the temperature.

6.6.4 Implications of the Viscoelastic Behaviour of Plastics in the Use of Linear Elastic Fracture Mechanics Theory

(a) *Viscoelasticity Effects in Crack Initiation*

Both LEFM and the J integral concept for ductile fractures stipulate that the potential energy is consumed exclusively for the propagation of the crack, i.e. the potential energy in the two halves of the specimen reduces to zero upon completion of the fracture process. With plastics materials there is an additional energy term to be taken into account, i.e. the term associated with viscoelastic deformations. This energy remains stored in the specimen after fracture and reduces to zero only after complete recovery of the strain and/or relaxation of internal stresses. The overall energy balance can be written, therefore, as[35]

$$-\frac{\delta U}{B\delta a} = J + \beta \delta W / \delta V$$

where $\delta W / \delta V$ is the work/unit volume done by the specimen against the surroundings after the propagation of the crack, while J is the energy actually involved in the formation of the crack (see Fig. 6.36). β is a geometric factor that takes into account the size of the specimen.

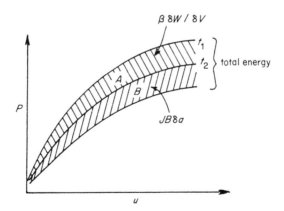

Fig. 6.36. Crack extension at constant displacement (i.e. stress relaxation conditions) for viscoelastic materials.

In other words the total energy input is the sum of the energy 'losses' through viscoelastic deformations and those associated with the fracture process.

The viscoelastic term, however, can be taken into account in the determination of the calibration curve of the compliance as a function of the crack length. The compliance would have to be measured as a function of the elapsed time and of the load applied, in view of the non-linearity of the viscoelastic behaviour of plastics. In other words the increase in compliance with time and level of the applied load is a manifestation of the energy expenditure in the viscoelastic deformations.

This means that we could use LEFM solutions for the fracture conditions written in the usual viscoelastic nomenclature. For example,

$$G_C = -\frac{P^2}{2B}\frac{\delta C}{\delta a}(t, P)$$

where the value of $\delta C/\delta a$ taken corresponds to that at the particular elapsed time and load level at which fracture occurs. In other words the time/load dependence of G_C is the result of the viscoelastic nature of the material and it is constant under isometric/isochronous conditions.

For fractures in the absence of a yield zone, the K_C value is also constant for specified time and strain (or stress) level conditions, i.e. the relationship $K_C^2 = E'(t, \sigma) \times G_C$ would apply. In this case we could use extrapolation procedures for the long term evaluation of $\delta C/\delta a$, using the stress/strain/ time relationships for the particular material considered (if these are known).

For thermoplastics, on the other hand, the compliance of the specimen is also a function of the size of the yield zone (i.e. $r_y = (\pi/8)(K_C/\hat{\sigma}_Y)^2$), which is also dependent on the elapsed time in view of the time dependence of the yield stress.

In this situation we could use the K_C approach for the determination of the toughness using the standard relationship written in the form of

$$K_C = Y\sigma\sqrt{(a + r_y)}$$

where

$$r_y = \frac{\pi}{8}\left(\frac{K_C}{\hat{\sigma}_Y(t, \varepsilon)}\right)^2$$

However, the non-linearity of the deformational behaviour makes it difficult to relate K_C to G_C because of the variations in levels of stresses in strains in the specimen.

At a given elapsed time t_1, for instance, the strain in the specimen (i.e. at a large distance from the crack tip) would be ε_1, while at the crack tip it would be ε_2 (where $\varepsilon_2 > \varepsilon_1$). Hence the relationship between K_C and G_C becomes

$$K^2_{C(t_1,\varepsilon_2)} = E'_{(t_1,\varepsilon_1)} G_C$$

where

$$K_{C(t_1,\varepsilon_2)} = Y\sigma \sqrt{a + \left[\frac{K_C}{\hat{\sigma}_Y|_{(t_1,\varepsilon_2)}}\right]^2 \frac{\pi}{8}}$$

Similar discussions apply to J_C, where $\delta P_Y/\delta a$ has to be considered as a viscoelastic parameter.

(b) Viscoelastic and Kinetic Effects in Crack Propagation
So far it has been assumed that fracture occurs through a catastrophic increase in crack length (for metals this has been estimated to be about $\frac{1}{3}$ of the speed of sound), while crack growth prior to reaching critical conditions is believed to be negligible. This is also true for metals, except for dynamic stress conditions, at low temperatures and in the presence of corrosive environments. For plastics, on the other hand, there is ample evidence to suggest that fracture can take place by slow crack growth and that the value of the K_C obtained is a function of the speed at which the crack is driven.[36] The double cantilever tapered specimen (Fig. 6.37) is particularly useful for the study of the effect of K (the input stress intensity factor) on crack speed, since in this case K is independent of crack length and, therefore, the slow

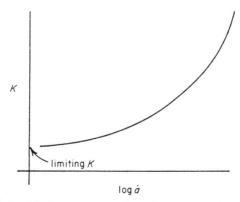

Fig. 6.37.   Relationship between K_C and crack speed for viscoelastic materials.

crack growth prior to the onset of rapid crack propagation will not affect the results. (Note that $G_C = (P^2/2B)(\delta C/\delta a) = $ constant, i.e. $\delta C/\delta a = $ constant, hence $K_C^2 = (E'P^2/2B) \times $ constant.) Therefore, the crack speed will remain constant throughout the entire crack propagation process. With rectangular specimens, on the other hand, the increase in $\delta C/\delta a$ with crack length accelerates the rate of crack growth. It seems, however, that the crack growth process is due to both viscoelastic deformations and the kinetic nature of the events along the path of the crack.

There are, in fact, microcavitational phenomena to be considered in addition to deformational phenomena (such as yielding). In any case, and irrespective of whether plane stress or plane strain conditions prevail, the stress along the path of the crack (σ_x) is zero at the very tip of the crack. Consequently, dilatational stresses are highest at some distance from the crack tip where microcavitations, such as voids and crazes, will develop and always be larger than elsewhere.

It is expected, therefore, that the growth of such microcravitations will eventually cause rupture of the surrounding material and increase the crack length. The rate at which these phenomena take place and produce a time-dependent growth of the crack is expected to depend on the viscoelastic nature of the material surrounding the microvoids (even the deformations of yielded material have a viscoelastic character) and on statistical aspects of their rupture.

Although little is known about the net effect of these events on the growth rate of the crack, it would seem that the relationship between K_C and crack speed follows the general behaviour described in Fig. 6.37 (see also ref. 37). Therefore, there could be a limiting value of K for which the crack is stable. If this is the case, then this value would constitute a safe level of stress intensity for design.

6.7 FACTORS AFFECTING THE FRACTURE TOUGHNESS OF PLASTICS

In developing, or selecting, plastics compositions it is helpful to have a knowledge of the effects of structural parameters, temperature, and environmental agents on the properties considered.

On the basis of the mechanics and mechanisms described earlier it is possible to make some inferences about the effects of the above-mentioned parameters on the fracture energy (G_C).

6.7.1 Effects of Structural Parameters on Fracture Toughness

In discussing structural parameters we must distinguish: (a) intrinsic characteristics of the base polymer and (b) the interactions with the additives or other components of the composition considered. Of the intrinsic structural parameters of the base polymers, molecular weight is particularly relevant, as are the distribution of the molecular weight and polydispersity (e.g. chain branches) and morphological parameters such as degree of crystallinity, spherulite size and orientation parameters. Factors like chain flexibility and intermolecular forces have to be considered in connection with time and temperature parameters insofar as these affect the relaxation characteristics of the polymer chains. For the case of thermosets in the absence of fillers it is obviously the cross-linking density and the distribution of cross-links within the overall tri-dimensional network. For thermosets one can exclude the possibility of micro-cavitational processes. Consequently the energy requirements in the creation of two fractured surfaces depend on the number of chemical bonds that have to be broken, while the kinetics of the crack growth is a function of the rate of stress transfer on individual bonds. In other words it is suggested that uneven distributions of cross-links in the network would cause faster rates of crack propagation in view of the lower efficiency of the load transfer within the molecular structure. That is, the covalent bonds will carry much higher loads than in the case where the cross-links are uniformly distributed.

It can be expected that similar arguments apply in the case of thermoplastics. Longer molecular chains and a low degree of polydispersity provide a more efficient mechanism for the transfer of stresses along the covalent bonds. This would raise the level of external stresses (and, therefore, the necessary strain energy) for the occurrence of local events at the crack tip, e.g. crazing or yielding. At the same time, the crack propagation rate will be reduced by the greater extensibility of the material surrounding the microvoids which lowers the rate at which these coalesce and extend the crack.

Increasing the degree of crystallinity and size of spherulites, on the other hand, lowers the critical strain energy release rate and increases the rate of crack propagation,[36] possibly because of the lower incidence of the microvoids which would grow faster and coalesce into extended cracks.

In general any structural feature that promotes a high incidence of microvoids or crazes (e.g. heterogeneous morphological structures such as blends, crystalline block copolymers, etc.), will enhance the toughness. At the same time it is necessary that the extension of the material (like fibrils) surrounding the microvoids does not occur under the constraints of triaxial tensile stresses as these would reduce the elongation at break. For instance, this could happen when fillers or reinforcing fibres are strongly bound onto the polymer matrix. When poor adhesion is exhibited at the fibre/matrix interface, there will be a substantial enhancement of toughness since (a) the stored energy at the crack tip can be dissipated in breaking the interfacial bond and forming microvoids and (b) the matrix can extend under plane stress (rather than plane strain) conditions. Plasticisers tend to increase the toughness of plastics by increasing the extensibility of the polymer as the result of the reduction in the potential energy barriers for molecular rotations. On the other hand, low levels of plasticisers may exhibit an anomalous behaviour through antiplasticisation and/or stress corrosion (see later).

6.7.2 Effects of Temperature and Time of Duration of the Load

In addition to the effect on the energy dissipated by viscoelastic deformations, which is reflected in the value of the modulus, the loading history and the ambient temperature can also directly affect the value of G_C and the crack growth rate.

For a given history of deformation the effect of a rise in temperature depends on the balance between the reduction in plastic energy/unit volume resulting from the reduction in yield strength (i.e. $Wp \simeq \frac{1}{2}(\hat{\sigma}_Y^2/E)$; note that the decrease of Wp for plane stress conditions is expected to result from the steeper decrease in the $\hat{\sigma}_Y^2$ term than the E term) and from the possible increase in the volume of the yield zone ahead of the crack tip. The latter is expected to result from an increase in critical crack opening displacement, i.e. greater extensibility of the material within the yield zone. This has been confirmed by Plati and Williams,[38] who have found an increase in G_C for many polymers in the temperature regions of secondary relaxations. It would seem, on the other hand, that the value of K_C decreases with an increase in temperature.[36] This is probably due to the reduction in modulus.

It would also seem that experimental results of the energy to fracture through LEFM studies are in agreement with the empirical results illustrated as a generalised scheme in Fig. 6.22.[39] In other words, the temperature range over which the value of the critical strain energy release rate exhibits a significant increase is related to the size of the tip radius of the notch, which in turn determines the value of the critical crack opening displacement δ_C.

Again, the reduction in yield stress with increasing temperature is more than compensated by the higher extensibility of the polymer in the yield zone ($G_C = \hat{\sigma}_Y \delta_C$), while the higher crack tip radius at a given temperature spreads the size of the yield zone and involves a greater volume of material in plastic work.

6.7.3 Effects of Environmental Agents

Most plastics materials suffer from some form of embrittlement or accelerated crazing in the presence of certain liquids.

Although the reasons or mechanisms for the occurrence of this phenomenon are not known, it would seem that they are of a physical rather than a chemical nature, in contrast to stress corrosion of metals.

In terms of viscoelastic fracture mechanics, the effect of the environment is to lower the K or G curve as a function of the time to fracture or crack growth rate as shown in Figs. 6.38 and 6.39.

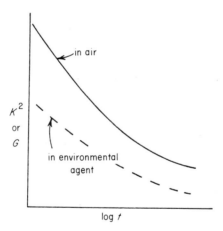

Fig. 6.38.   Effect of environment on the K or G value in relation to the fracture time.

In both cases the effect of the environment is to accelerate the rate of crack growth, possibly by supplying additional energy (as interfacial energy of wetting) at the crack tip. Similarly, the environmental agent lowers the curve of the crazing stress as a function of time since the liquid agent can diffuse in the microcavities and accelerate the rate of formation of the craze.

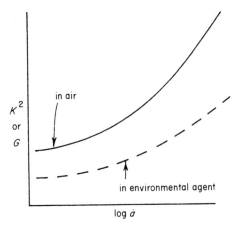

Fig. 6.39.   Effect of environment on the K or G value in relation to the rate of crack growth.

6.8   EMPIRICAL EVALUATIONS OF TOUGHNESS

There are many empirical tests that have been designed to estimate the toughness of materials in an attempt to obtain data which can be directly converted into life expectancy or confidence in the performance of an article. In general these tests are carried out on specimens that closely resemble a particular component. The more widely used tests for this purpose are briefly discussed in the following sections.

6.8.1 Creep Rupture Tests

Creep rupture tests are creep tests carried out up to the point of failure. Normally either a tensile specimen or a tube or pipe inflated at constant pressure is used. The time to failure and observations about the nature of the failure (i.e. whether brittle or ductile) are recorded. In general, the curve of applied stress (as in the case of tensile specimens), or hoop stress, i.e. the maximum stress for pipes as a function of time to failure, has the shape indicated in Fig. 6.40.

The transition from ductile to brittle fracture (point A on the curve) is considered to be the most important parameter in relation to the toughness of the material. In design one should therefore avoid the onset of brittle fractures by choosing a working stress that is well below that which would produce brittle fractures for a given estimated lifetime of the component. The occurrence of crazing is also recorded since this is believed to be the

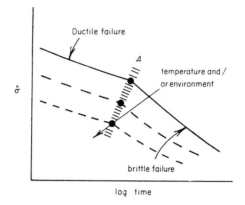

Fig. 6.40.   Ductile/brittle transition of thermoplastics in creep rupture tests.

precursor to the onset of brittle fractures, although in some cases, e.g. for polycarbonates and rubber toughened plastics, the crazes can be quite stable and the material fails in a ductile fashion.

6.8.2 Fatigue Tests

Fatigue is a term that was introduced in the early part of the 19th century to denote the failure of metal components at stress levels that were well below the tensile strength of the material. It was recognised that these failures were occurring when the stresses imposed were not steady in magnitude but varied in a cyclic manner. However, as already discussed, such failures can also occur when the imposed stress is constant. Hence the modern tendency is to refer to creep ruptures as static fatigue and to cyclic ruptures as dynamic fatigue. The discussions that follow, however, refer to the latter type of failure. For equipment construction convenience either the actual stress imposed or the strain (or deflection) is normally allowed to vary sinusoidally, with the upper and lower limits kept constant throughout the duration of the test. The mean stress and the upper and lower limits can obviously be made to vary from one test to another in an endless number of combinations.

When both upper and lower limits are either positive or negative, the type of stress cycle is termed fluctuating. On the other hand, when one stress limit is positive and the other is negative, it is known as a reversed cycle. There are two particular cases of the reversed and fluctuating cycles that arise frequently in engineering evaluations.[40] The first is a symmetrical reverse cycle, in which the mean stress is zero and the upper and lower limits are equal and respectively positive and negative. The second is a fluctuating cycle in which the mean stress is half the maximum, with the minimum at zero. The important stress parameters in (cyclic or dynamic) fatigue tests are, therefore,

(i)   Mean stress:

$$\bar{\sigma} = \frac{\sigma_{max} + \sigma_{min}}{2}$$

(ii)   Stress range:

$$2\sigma_a = \sigma_{max} - \sigma_{min}$$

(iii) Stress ratio:

$$R = \frac{\sigma_{min}}{\sigma_{max}}$$

In addition to the above variations, fatigue tests are sometimes carried out in bending or by cyclic torsion, or even in combination with any of the above. Furthermore the frequency of the oscillations can be varied over a very wide range.

Fatigue data are usually presented as plots of the applied stress(es) as a function of the number of cycles (N) to fracture either on a linear–log or a log–log scale, as in Fig. 6.41, at a given frequency.

It is obvious, however, that it would be practically impossible to obtain experimental data for all possible stress ranges, mean stress, stress modes and frequencies. Consequently, empirical rules have been developed to represent the variation of mean and range of stress relative to the static strength values (yield strength $\hat{\sigma}_Y$ or brittle strength $\hat{\sigma}_F$ determined from constant strain rate tests) and the fatigue curve for reversed cycles (i.e. zero

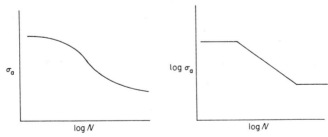

Fig. 6.41.   Typical methods of presenting fatigue curves.

mean stress). This combination is a very convenient method for obtaining experimental data in view of the simplicity of the equipment used, i.e. a rotating cylindrical beam.

The conditions for fatigue tests are normally chosen from a diagram constructed from a plot of the semi-range of the stress σ_a against the mean stress $\bar{\sigma}$ (Fig. 6.42).

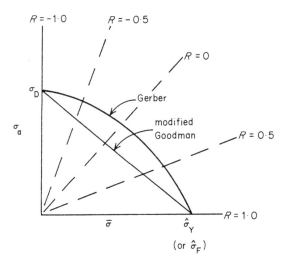

Fig. 6.42.   Diagram of the mean stress versus stress range for fatigue tests.

The limiting conditions are, therefore,

(i) $\bar{\sigma} = 0$, $\sigma_a = \sigma_D$ (fatigue limit for reversed stress)
(ii) $\sigma_a = 0$; $\bar{\sigma} = \hat{\sigma}_Y$ (or $\hat{\sigma}_F$).

(1) A straight line that joins the limiting coordinates represents the empirical rule known as the 'modified Goodman' relationship, which is given by the algebraic expression

$$\sigma_a = \sigma_D \left(1 - \frac{\bar{\sigma}}{\hat{\sigma}_Y} \right)$$

where σ_D = the chosen value for the semi-range at which the material will fail after N cycles.

To present the actual data from the σ_a–N plots, the value of σ_D is chosen as that at which the curve flattens out, i.e. the fatigue (or endurance) limit of the material, since this is considered to be the safe value for designing purposes.

(2) Another relationship is obtained by joining the limiting coordinates by a parabola known as the *Gerber* parabola, which is given by the algebraic expression

$$\sigma_a = \sigma_D \left[1 - \left(\frac{\bar{\sigma}}{\hat{\sigma}_Y} \right)^2 \right]$$

The rationale for these empirical rules is that one expects a reduction in the allowable range of stress (for a particular fatigue limit σ_D) as the mean (tensile stress) in a cycle is increased.

There is little data available on plastics to support either of these general rules. However, from the experience gained on metals[41] it may be expected that those materials (or test specimens) that fail in a brittle fashion may follow the 'modified Goodman' relationship, while those that fail in a ductile fashion may fall closer to the parabolic curve.

The problems arising for plastics when using the same testing conditions as for metals have been summarised by Gotham.[42] Three aspects are considered to be of particular importance:

(1) *Frequency.* This affects the damping coefficient, i.e. the energy losses through viscoelastic deformations, and consequently there can be 'thermal softening effects' resulting from the temperature rise. Hence, to avoid these difficulties, it is recommended that the frequency be kept to very low values, i.e. 0·1–10 Hz.

(2) *Level of the mean and semi-range stress.* Because of their visco-elastic nature the presence of a non-zero mean stress will give rise to accumulative residual strain with time as the number of cycles increases.

(3) *Limits for the excitation stress or strain.* Because of the non-linearity of the viscoelastic behaviour of plastics, it is expected that the results will depend on whether a constant stress or constant strain is used as the upper and lower limits of the test.

Furthermore, because of the complexity of the relationship between stress and strain, it is preferable to use square wave excitations rather than the usual sinusoidal types. In this way the data can be converted into stress to failure versus total time under stress and compared with static fatigue or creep rupture results. The experiments carried out in the laboratories of ICI[42] seem to indicate that the data from continuous load experiments (creep rupture) and those under intermittent (low-frequency) loadings (fatigue) are superposable. The only difference is with respect to the time for the occurrence of the ductile/brittle transition, which is much shorter for the cyclic fatigue tests. This phenomenon is probably associated with localised increases in temperature in the microvoid regions that precede the nucleation and growth of cracks. In other words, it is unlikely that temperature increases can be eliminated in the case of plastics in view of their very low thermal diffusivity, except at extremely low frequencies.[43]

In the light of this it would seem that rather than trying to prevent increases in temperatures by cooling the specimen during testing, it is better to record the true temperature of the specimen and to consider the 'thermal softening' behaviour as a characteristic (or even a property) of plastics in dynamic fatigue. Hence, by using a zero mean stress, one can use the accumulated strain and heat content as the additional parameters that determine fatigue failures. It is quite possible that one could use as an extreme condition of testing that which gives no (apparent) increase in heat content but allows a residual strain to accumulate during the test. However, there are still no rules available that can be used as a guide for the selection of testing conditions in a manner similar to the Goodman and Gerber diagrams for fatigue life in the absence of thermal softening.

6.8.3 Tear Tests

Tear tests were developed for flexible materials to obtain an indication of the toughness of films and sheeting which may develop cracks during handling. As in the case of strength measurements the standard test methods are carried out in a universal mechanical testing machine at a constant rate of clamp separation, and the force required to maintain the grip separation rate during the propagation of a tear (or notch) is recorded (e.g. ASTM D1004, ASTM 1938, BS 2782—Method 308 A, etc.). Often the tear strength is expressed as the force/unit thickness of the film or sheet. There is an additional test for plastics films known as the Elmendorf test (e.g. ASTM D1922, BS 2782—Method 308 B, Din 53363), in which the tear is propagated by the oscillation of a heavy pendulum and the energy consumed is recorded (see the discussions on impact strength in the next section).

6.8.4 Impact Strength Measurements

Impact tests have been developed for the purpose of assessing the fracture resistance of plastics for situations where a component may be subjected to rapid blows. In the interests of simplicity the tests are devised in such a

manner that the *energy* required to break the specimen is recorded rather than the *force* to break.

Impact tests are divided into two main classes:

(a) Pendulum Tests

In these tests a blow is delivered on a clamped or supported specimen through the release of the arm of a 'pendulum' or 'hammer'. The energy absorbed by the specimen during fracture is read directly by means of a pointer which rotates with the pendulum after the specimen has been struck. This, in fact, records the height to which the mass of the pendulum rises after the impact with and subsequent fracture of the specimen (Fig. 6.43). A scale is directly calibrated in terms of energy absorbed which is manifested as a reduction in the height (and hence the potential energy) of the pendulum, i.e.

$$\Delta U = mg(h_0 - h_f)$$

These tests can be carried out in tension (e.g. ASTM D 1822) or in bending. In the latter case the load acts on the specimen either as a cantilever (Izod tests, e.g. ASTM D256 (A), BS 2782—Method 306 A, etc.)

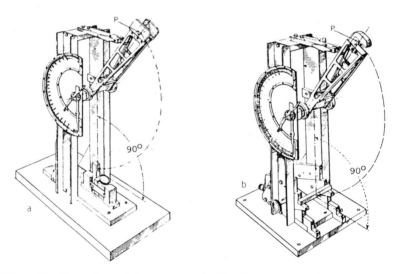

Fig. 6.43. Typical pendulum apparatus for impact strength measurement. P = Point of impact. (a) Cantilever beam (Izod-type) impact machine. (b) Simple beam (Charpy-type) impact machine. (After ASTM.[44])

or at the centre as a freely supported beam (Charpy tests, e.g. ASTM D256 (B), BS 2782—Methods 306 D and 306 E, ISO/R 179, etc.). (See Fig. 6.44.) Normally the Charpy and Izod tests are carried out on specimens containing a central V-notch which acts as a stress raiser and consequently localises the fracture and limits the spread of the yield zone over a narrow region (as already discussed). Tensile impact tests are carried out, on the other hand, on unnotched specimens.

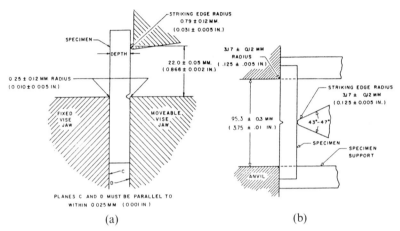

(a)   (b)

Fig. 6.44.   Typical supporting, clamping and striking position of specimen in pendulum impact tests. (a) Relationship of vise, specimen, and striking edge to each other for Izod test methods A and C. (b) Relationship of anvil, specimen, and striking edge to each other for Charpy test method. (After ASTM.[44])

(b) *Falling Weight Tests*
In these tests a sheet or disc is rested on a cylindrical support. Films are usually held in a taut position by means of suction pads around the periphery of the support. A striker with a hemispherical tip is dropped from a given height on the sheet of film at the midpoint (Fig. 6.45).

The evaluation of the impact strength of the material is carried out either by the 'staircase method' or on a 'probability of failure' basis at several levels of input energy, known as the 'Probit' method. The staircase method consists of starting with a certain energy level ($U = mgh$) and examining whether or not this is sufficient to cause fracture of the sheet.

If failure has occurred, the energy is decreased by one 'step' (through a reduction of the mass carried by the striker) for the next specimen. Alternatively the energy is increased by one step if the specimen survives the

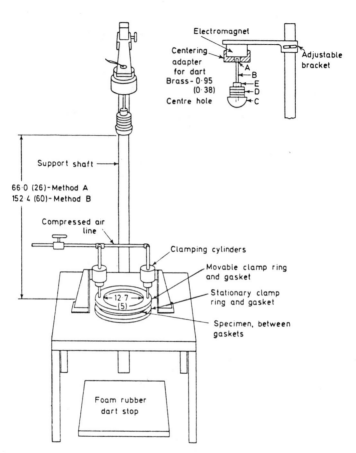

Fig. 6.45. Typical apparatus for the measurement of the impact strength by the falling weight method—free falling dart impact test for polyethylene film. A, steel shaft tip 0·64 (0·25) o.d. by 1·27 (0·50) long. B, dart shaft, method A—aluminium 0·65 (0·25) o.d. by 12·70 (5) long, $\frac{1}{4}$–20 thd (N.C.) 1·27 (0·50) long on bottom, No. 5–40 thd, (N.F.) for steel tip; method B—same as in method A except dart shaft is 11·43 (4·50) long. C, hemispherical head, method A—3·81 (1·500 ± 0·005) in diameter; method B—5·08 (2·000 ± 0·005) in diameter. D, removable weights. E, collar and screw. (Dimensions in centimetres with inch equivalents in parentheses.) (After ASTM.[45])

blow by the falling dart. The impact strength is taken as the mean of the input energies for about 20 trials within the range of possible failures.

With the 'Probit' method, on the other hand, a number of specimens (usually 10) are subjected to impact at each of a series of input energies, and the fraction of specimens that have fractured at each level is recorded. These values are plotted against the input energy level, and the value that corresponds to 50 % failure is recorded as the impact strength of the sheet. Obviously it is also useful to record the quartile values (i.e. energy at 25 % and 75 % failures, respectively) to obtain an indication of the spread or range of the incidence of failures. If a sufficiently large number of samples is tested and the number of levels of input energies is realistic (at least 5), the failure distribution curve should be Gaussian. Consequently the plot of % failure versus energy level should produce a sigmoidal curve on a linear scale, while a 'normal probability' scale (i.e. the % specimens fractured is plotted on a logarithmic scale) should give a straight line. The steepness or slope of the curve would, in this case, give an idea of the spread, while the median of the fracture energy (the so-called F_{50} value) gives a measure of the impact strength.

Such tests are also used sometimes to evaluate actual components, such as pipes or bottles. For the case of bottles or similar articles it is often considered more realistic, in terms of service performance, to drop the actual component (in the case of bottles these could be filled with water or the liquid contained in actual service) and to change the height from which they are dropped as a means of varying the level of energy input. The staircase method seems preferable for this purpose, perhaps because it uses fewer samples to carry out a full evaluation. The grand mean of the heights of 'failures' and survivals gives a measure of the impact strength, while the standard deviation is used to measure the spread of failures.

6.9   RATIONALISATION OF STANDARD TEST PROCEDURES

As can be expected from previous discussions, evaluation procedures by the empirical methods described in Section 6.8 produce data that vary according to the dimensions and geometry of the specimens used. As a result they are mainly useful for quality control purposes rather than as predictors of service performance. Alternatively, evaluations based on fundamental principles, when these are well-established, may be too time consuming and may require costly equipment which sometimes is not even available commercially. Consequently a reasonable compromise can often be achieved through elaboration of the standard test procedures.

In this section particular attention is drawn to fatigue (both static and dynamic) evaluations and impact strength measurements. Two approaches have been shown to be useful for the purpose of rationalising the standard testing procedures, one using the stress concentration approach and the other based on fracture mechanics.

6.9.1 The 'Stress Concentration Factor' Approach

Since a sudden change in the geometry or dimensions of a component produces sharp increases in the stress distribution around the discontinuity, engineers have tried to obtain a more realistic assessment of the service performance of materials by using specimens that contain well-defined stress raising features, e.g. circular holes, notches with controlled tip radius, etc. As already mentioned in Section 6.3.1, the stress distribution around such discontinuities can be related to their geometry. For a material that is linear elastic and fails in a brittle fashion, the same stress distribution will be maintained up to the point of fracture. If the maximum principal stress (or strain) criteria apply to such conditions, the nominal (or gross) stress for fracture can be estimated from a knowledge of the strength of the material, $\hat{\sigma}_F$, i.e. the value obtained on specimens without notches. To make such an estimate, it is usual to define the stress concentration factor K_t as

$$K_t = \frac{\sigma_{max}}{\sigma_{(nominal)}}$$

the value of which can be calculated (for simple geometries and within certain boundaries) or can be evaluated experimentally.

It is interesting to note that when such an approach is used for the case of polymethyl methacrylate (i.e. one of the thermoplastics materials that comes closest to the conditions specified above), a master curve is obtained from the fatigue data when the stress amplitude is plotted as the maximum stress around the notch (i.e. $\sigma_{max} = K_t \sigma_{(nominal)}$),[46] while the curve obtained for unnotched specimens (i.e. $K_t = 1$) lies below the master curve for the notched specimens (Fig. 6.46).

While the discrepancy mars the validity of the maximum principal stress criterion, it is a relief to find at least that the master curve lies on the safer side from the design point of view. To what extent such a discrepancy is due to thermal softening effects (which are expected to increase the compliance of the specimen, therefore producing a lower failure stress if the maximum strain criterion applies) is a matter that has yet to be resolved. On the other hand, it could simply signify that a more fundamental approach, such as fracture mechanics, is necessary to predict the occurrence of brittle failures.

A similar approach has been tried for the measurement of the impact strength of thermoplastics.[24] Unification of the results has been obtained from plots of the energy to fracture as a function of the stress concentration factor only for the case of polymethyl methacrylate. It is obvious that the

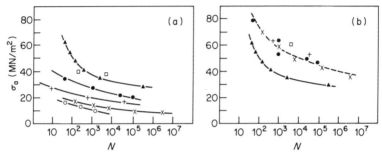

Fig. 6.46.   Fatigue data for cast sheets of polymethyl methacrylate. (a) Effect of notch tip radius on fatigue of PMMA: 20 °C 65 % rh. (b) Data of (a) adjusted to take account of stress concentration factor. ▲ Unnotched. □ Notched, $r = 2$ mm. ● Notched, $r = 1$ mm. + Notched, $r = 500\,\mu$m. × Notched, $r = 250\,\mu$m. ○ Notched, $r \approx 10\,\mu$m.   Uniaxial   tension/compression. $f = 0\cdot5$ Hz. $\sigma_{\text{mean}} = 0$.   Laboratory atmosphere (no forced cooling). (After Gotham.[46])

most disconcerting aspect of this approach is the estimation of the stress concentration factor K_t. When a yield zone develops ahead of the notch, the estimated values of K_t from elastic solutions are much higher than the true values since in this case the stress within the yield zone will be substantially lower than that predictable from the use of the Inglis equation in Section 6.5.1.

For metals attempts have been made to take this fact into account by means of a plastic-strain concentration factor (i.e. $K_\varepsilon = \varepsilon_{\text{max}}/\varepsilon_{\text{(nominal)}}$) and a plastic-stress concentration factor (i.e. $K_\sigma = \sigma_{\text{max}}/\sigma_{\text{min}} \simeq 1 + (K_t - 1)(E_s/E)$, where E_s is the secant modulus at the yield point). No such approach, however, has been reported to date for plastics.

Despite the many uncertainties concerning the relationships between toughness and the stress concentration factor, the evaluations carried out on specimens tested at various levels of severity of stress concentrations are very useful to designers. Metallurgists, for instance, frequently use the concept of notch-sensitivity factor q in fatigue evaluations as a means of rating materials. The definition of q is given by

$$q = \frac{K_f - 1}{K_t - 1}$$

where K_f is the so-called fatigue-notch factor or fatigue-strength reduction factor. In other words for materials that show no reduction in fatigue strength in the presence of notches ($K_f = 1$) the value of q is zero; i.e. the material is *not* notch-sensitive.

At the other extreme when the notch produces the maximum (theoretical) effect, i.e. with brittle fractures ($K_f = K_t$), the value of q is 1. Thus although q is not constant, but depends on the actual geometry of the notch which produces the calculated value of K_t, at least it gives an indication of its deviation from the two extremes of notch sensitivity. For the case of plastics, the stress concentration approach has received the greatest attention with respect to evaluation of the impact strength by the pendulum tests. Plots of the impact strength as a function of the notch tip radius at a given notch depth produce curves, the steepness or slope of which can be used to obtain an indication of the notch sensitivity of materials. Such a plot for five different polymers is shown in Fig. 6.47. It is instructive to note that two materials such as polymethyl methacrylate and ABS, which are normally considered as extremes in respect to toughness, display a very low notch sensitivity in these empirical evaluations.

To minimise the number of tests in the study of the notch sensitivity, and in view of the fact that only qualitative judgements can be made from such results, the suggestion is often made that tests should only be carried out at two levels of severity of notch for the assessment of the notch sensitivity. Furthermore, tests should be carried out over a wide range of temperatures in order to obtain an indication of the transition between ductile and brittle fractures since the impact strength varies very rapidly (i.e. over a fairly narrow temperature range) when such a change of behaviour occurs.

From these graphs it may also be possible to obtain an estimate of the dependence of the ductile/brittle transition as a function of the notch tip radius, which is probably the most significant parameter to consider in the design of components for impact resistance.

Another practical method for the assessment of the notch sensitivity of materials is the preparation of discs or sheets having an embossed surface on one side and a smooth surface on the other. The embossing has a stress-raising effect, and consequently in a falling weight impact test different values will be obtained according to which side of the sheet is struck. When

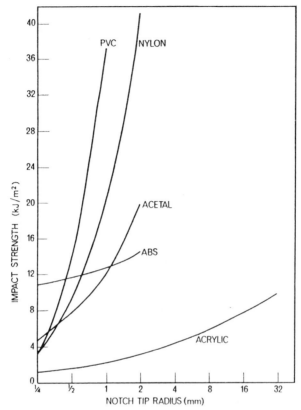

Fig. 6.47.   Impact strength as a function of notch tip radius for samples of five different polymers. (After Vincent.[24])

the embossed side is in tension (i.e. as in the case where the opposite side is being struck by the falling dart) there will be a considerable reduction in the level of energy to cause fracture of the sheet.

6.9.2 The 'Fracture Mechanics' Approach
Having established that the intrinsic parameters that characterise the toughness of materials are those derived from the fracture mechanics analysis, it is instructive to inquire whether and to what extent it may be possible to use standard tests to measure these fundamental parameters.

(a) Fatigue Evaluations
Since, for the case of plastics (thermoplastics in particular), fracture occurs

by slow crack propagation, it would seem that fatigue evaluations should be carried out with the aim of establishing the law that governs the rate of crack growth ($\dot{a}$). This, in turn, can be used to predict the time to fracture. In other words, the crack growth rate measurements will be used as accelerated tests for the prediction of long term fracture behaviour.

What is required, therefore, is a knowledge of the following functions:

$$\dot{a} = f(K) \quad \text{for static fatigue}$$

and

$$\dot{a} = g(\Delta K) \quad \text{for dynamic fatigue}$$

From these functions the time to fracture would then be obtained through integration, i.e.

$$t_f = \int_{a_1}^{a_c} \frac{1}{\dot{a}(K)} \, da$$

where a_1 is the initial length of the crack and a_c is the crack length for which catastrophic fracture occurs at the nominal K input.

Alternatively, this can be directly expressed in terms of K values by substituting in the equation above, i.e.

$$t_f = \frac{2}{\sigma^2 Y^2} \int_{K}^{K_c} \frac{K}{\dot{a}(K)} \, dK$$

where K_C is the critical stress intensity factor measured under constant straining (or stressing) rate conditions in a conventional tensile testing machine and at fairly high rates of clamp separation.

If fatigue tests are carried out at low frequencies and with square wave stress inputs, then the nature of the functions given above should be the same if the relationship is linear. Although there is no indication from the literature as to whether such a relationship is, in fact, linear, this has often been assumed.[47] In the light of the above uncertainties a sound investigation would include studies of the crack growth rate as a function of ΔK at various levels of K_{max}.

Needless to say, the effect of the environment, e.g. temperature, liquids, etc., should also be taken into account in such evaluations. Since in fatigue evaluations the emphasis is on crack growth rate rather than on catastrophic fractures, it is important to remember that neither the arguments on stress intensity factors for blunt notches nor the J integral considerations are particularly applicable. First, crack growth occurs by spreading of natural (i.e. very sharp) cracks; and second, it is unlikely that slow crack propagation is going to take place through general yield of the section ahead of the crack tip.

(b) *Tear Strength Measurements*

As far as tear strength measurements are concerned, the standard test procedures can be rationalised through fracture mechanics by taking into account the rate of change in compliance for the particular fracture mode considered (i.e. crack opening for the ASTM D1004 test and antiplane shear mode for ASTM D1938). Since Y calibration curves for the particular loading conditions and specimen geometry for the standard tests of tear strength are not available, it is perhaps better to use the Irwin approach.

(c) *Impact Strength Measurements*

In standard tests the impact strength is normally quoted as the energy absorbed by the pendulum per 'unit area of the cracked cross-section (or energy/unit thickness when the crack-ligament is specified and the thickness allowed to vary within a certain range, hence the reason why one often finds the Izod impact strength quoted in 'energy per unit notch' length). For metals (steels in particular) it has been observed even before the advent of fracture mechanics[48] that such a quantity is not constant but depends on the geometry of the specimen and in particular on the crack-ligament area or the a/w ratio.

For Charpy tests (or three-point bending in general) such a quantity, i.e. U/A, decreases with increasing a/w,[49] while quite recently the author has observed a minimum in the specific fracture energy curve for the case of cast unsaturated polyesters using Izod tests.[50]

Turner *et al.*[51] have reported from considerations of linear elastic fracture mechanics that the specific fracture energy can vary within a range of factors of $\frac{1}{2}$ to 2 depending on whether gross yielding (i.e. high ductility) fractures or pure brittle fracture occurs, without any yield zone developing at the crack tip.

It is for this reason that G_C, and not the specific fracture energy, should be considered as the true parameter that characterises the fracture toughness of materials. The specific fracture energy corresponds, in fact, to the mean strain energy release rate over the fracture path $(w–a)$. It is possible, however, to convert the specific fracture energy, directly obtained from conventional impact tests, into the appropriate value for the critical strain energy release rate by means of a correction factor Φ,[52] i.e.

$$U/A = G_C\Phi \tag{6.57}$$

The energy absorbed by the specimen during fracture corresponds to the work performed on it by the impact striker, i.e.

$$U = \frac{P^2 C}{2} \qquad (6.58)$$

This can be related to G_C by making the appropriate substitutions. From the definition of G_C one obtains

$$G_C = \frac{P^2}{2B} \frac{dC}{da}$$

which can also be written as

$$G_C = \frac{P^2}{2BW} \frac{dC}{d(a/w)}$$

Substituting for P^2 in eqn. (6.58) above, one obtains

$$U = G_C BW \frac{C}{dC/d(a/w)} \qquad (6.59)$$

If one lets $\Phi = C/[dC/d(a/w)]$, eqn. (6.59) becomes equivalent to eqn. (6.57). The correction factor Φ can be obtained either experimentally or from the geometrical factor Y.

Integrating Irwin's equations, one obtains an expression for the compliance, which, after substitution for G_C (i.e. $G_C = K_C^2/E' = Y^2\sigma^2 a/E'$) yields

$$C = \frac{2\sigma^2 B}{E'P^2} \int Y^2 a\,da + C_0$$

where C_0 is the specimen compliance at *zero* crack length.

By letting $\alpha = (\sigma/P)BW$ (i.e. a factor that relates the stress to the load and geometry of the specimen, conveniently chosen so that it becomes equal to 1 for rectangular specimens in tension), the above equation can be written in general terms as[52]

$$C = \frac{2\alpha^2}{E'BW^2} \int Y^2 a\,da + C_0$$

or

$$C = \frac{2\alpha^2}{E'B} \int Y^2 \frac{a}{w}\,d(a/w) + C_0$$

For three common types of pendulum impact tests one obtains the following relationships for the compliance and correction factor Φ respectively:

(1) Tensile impact specimens (notched)

$$C_0 = \frac{\Delta}{P} = \frac{\varepsilon L}{\sigma B W} = \frac{L}{E' B W}$$

where L = gauge length

$$\alpha^2 = 1$$

$$C = \frac{2}{E'B} \left[\int Y^2 \frac{a}{w} \, \mathrm{d}(a/w) + \frac{L}{2W} \right]$$

and therefore

$$\Phi = \frac{\int Y^2 \frac{a}{w} \, \mathrm{d}(a/w) + \frac{L}{2W}}{Y^2 \frac{a}{w}}$$

(2) Three-point bending specimens (Charpy tests)

$$C_0 = \frac{\Delta}{P} = \frac{S^3}{4E'BW^3}$$

where S is the span

$$\alpha^2 = \frac{\sigma^2 B^2 W^2}{P^2} = \frac{9}{4} \left(\frac{S}{W} \right)^2 \qquad \left(\text{since } \sigma = \frac{3}{2} \frac{PS}{W^2 B} \right)$$

$$\alpha = \frac{3}{2} \left(\frac{S}{W} \right)$$

$$C = \frac{S^2}{2WE'B} \left[9 \int Y^2 \frac{a}{w} \, \mathrm{d}(a/w) + \frac{S}{2W} \right]$$

and therefore

$$\Phi = \frac{\int Y^2 \frac{a}{w} \, \mathrm{d}(a/w) + \frac{S}{18W}}{Y^2 \frac{a}{w}}$$

which, for very short notches (i.e. placing $Y^2 = \pi$), becomes

$$\Phi \simeq \frac{1}{2}\left(\frac{a}{w}\right) + \frac{1}{18\pi}\left(\frac{S}{W}\right)\left(\frac{W}{a}\right)$$

(3) Cantilever beam specimens (Izod tests)

$$C_0 = \frac{4l^3}{E'W^3 B}$$

where l is the distance from the clamped end to the point where the mass strikes the specimen. Substituting for $\sigma = 6Pl/BW^2$ in the expression for α and then in the compliance equation we obtain:

$$C = \frac{4l^2}{W^2 E' B}\left[18\int Y^2 \frac{a}{w}\,\mathrm{d}(a/w) + \frac{l}{w}\right]$$

and therefore,

$$\Phi = \frac{\displaystyle\int Y^2 \frac{a}{w}\,\mathrm{d}(a/w) + \frac{l}{18W}}{Y^2 \dfrac{a}{w}}$$

Note that this expression is the same as for the Charpy test except that the span S has been replaced by the loading distance l.

However, this is at variance with the expression derived by Plati and Williams.[52] The discrepancy results from the fact that the above authors have used expressions for the compliance and α that correspond to those of a specimen in tension.

That the correction factor Φ should follow the same relationship for both Izod and Charpy tests can also be inferred from the experimental results of Vincent and members of the British Standard Institution Sub-Committee PLC/36/2.[24] Specimens of several thermoplastics materials gave the same results when tested by the Izod and Charpy tests respectively. Despite this, Plati and Williams[38] suggest that the Φ values should be higher for the Izod test in view of the rotation of the specimen during fracture.

Once a calibration curve for a given type of test has been obtained, either experimentally or from the theoretical analysis discussed earlier, then it is simply a question of measuring the energy to fracture as a function of the crack length (or a/w) which produces a straight line when plotted versus $BW\Phi$. The slope of this curve corresponds to the G_C value according to the equation $U_f = G_C BW\Phi$. In pendulum tests, however, such a plot produces a straight line which does not pass through the origin but intercepts the axis of the fracture energy. This corresponds to frictional energy losses and

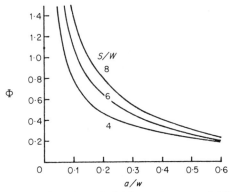

Fig. 6.48.   Factor Φ as a function of a/w for various span/width ratios for three-point bending tests. (After Plati and Williams.[53])

(mostly) to the kinetic energy imparted to the specimen when it is being thrown away by the pendulum after fracture. One can easily allow for such energy by testing specimens which have already been fractured. Such specimens would have a zero compliance which makes the factor Φ also equal to zero; hence, the value of the energy recorded must correspond to the value assumed at the intercept.

Typical curves for the correction factor Φ as a function of the non-dimensional crack length (a/w) for three-point bending tests and plots of the measured fracture energy versus $BW\Phi$ are shown in Figs. 6.48 and 6.49 respectively.

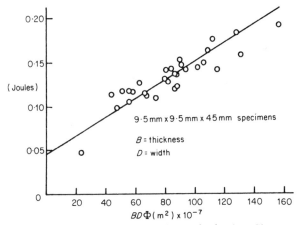

Fig. 6.49.   Typical results for cast PMMA sheets obtained on Charpy tests. (After Plati and Williams.[53])

Note, however, that a thorough analysis of the impact strength would also involve measurements of the fracture energy for various notch tip radii over a wide range of temperatures. The analysis would also include the effect of thickness as a means of assessing the relative changes in G_C and G_B values within the range of conditions varying from plane stress to plane strain. Enormous differences in G_C values have been reported by Plati and Williams,[38] largely depending on the constraints that the material imposes on the development of the yield zone in the transition from plane stress to plane strain conditions.

Table 6.1 shows some interesting differences of the effect of notch tip radius on the strain energy release rate as a function of temperature.

Table 6.1: Typical Values of G_C of Common Plastics Materials Measured at Room Temperature by the Charpy Impact Strength Method[38]

Material	G_C values (kJ/m^2)	
	Plane strain	Plane stress
Polymethyl methacrylate (cast)	1·06	1·28
Polystyrene (unmodified)	0·35	0·90
Rigid PVC	1·23	1·44
Polycarbonate	3·5	5·02
Nylon 6,6 (dry)	0·25	4·15
Polyethylene (medium density)	1·3	11·90
Polyethylene (low density)	5·0	35·00
High impact polystyrene	1·0	15·00

Compare, for instance, the results for polymethyl methacrylate (typical brittle glassy polymer) with polyvinyl chloride (typical tough glassy polymer) in Figs. 6.50 and 6.51.

Note that when gross yielding occurs at the crack tip (as for the case of very ductile materials such as ABS, low density polyethylene and high impact polystyrene) the specific fracture energy is constant (i.e. independent of the non-dimensional crack length). This indicates that the yield load changes proportionally to the ligament area. Expressing this in terms of J_C we can write

$$\frac{u_f}{A} = J_C$$

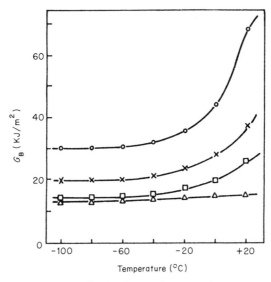

Fig. 6.50.   Blunt notch data for PVC. $\triangle$, sharp notches; $\square$, $\rho = 0{\cdot}25$ mm; $\times$, $\rho = 0{\cdot}5$ mm; $\bigcirc$, $\rho = 1{\cdot}0$ mm. (After Plati and Williams.[38])

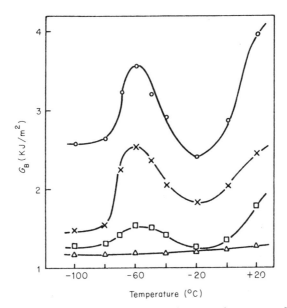

Fig. 6.51.   Blunt notch data for PMMA. $\triangle$, sharp notches; $\square$, $\rho = 0{\cdot}25$ mm; $\times$, $\rho = 0{\cdot}5$ mm; $\bigcirc$, $\rho = 1{\cdot}0$ mm. (After Plati and Williams.[38])

where $A = B(w - a)$, so that

$$U_f = J_C BW \left(1 - \frac{a}{w} \right)$$

The correction factor Φ (or Φ_Y to differentiate it from the elastic solution) varies in a linear fashion from one for unnotched specimens to zero for a completely cracked specimen. The difference between correction factors for the elastic and plastic solutions as a function of the dimensionless crack length is shown in schematic fashion in Fig. 6.52.

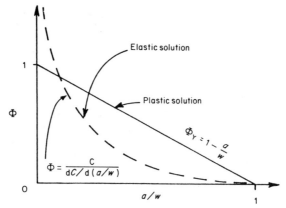

Fig. 6.52. Variation of Φ as a function of a/w for elastic and plastic crack propagation.

Note that in bending, the average plastic displacement u is half of that occurring in tension, so that the actual value of J_C is obtained from the expression

$$J_C = \frac{2U_f}{BW \left(1 - \frac{a}{w} \right)}$$

One great attraction of impact tests for the measurement of the critical strain energy release rate of plastics is that the time element is very short and the presence of notches prevents the development of high loads within the specimen. This means that the viscoelastic effects mentioned earlier can be neglected and that the linear-elastic approach is applicable.

One can also use this approach for the measurement of the impact strength by the falling weight method. What is required is an appropriate experimental evaluation for the Φ factor as a function of a/r, where a is the length of a small central crack, and r is the radius of the circular support.

Note that although an approximate linear relationship has been reported between the fracture energy and the thickness of the sheet,[54] this has to be regarded as fortuitous. Alternatively, since the material tested was quite ductile, the energy measured is that required to deform plastically the fractured region, which is proportional to the J_C value of the material. Experience shows that this relationship (as can be inferred from previous discussions) is not universally applicable.

One of the problems arising in the use of a fracture mechanics approach to falling weight impact tests is that, since the energy applied normally exceeds the actual critical strain energy release rate of the material, crack branching is likely to result. It would seem, however, that if the total crack length of all branches is taken into account, the fracture mechanics analysis is still applicable.[55]

6.10. UTILISATION OF FRACTURE DATA IN DESIGN

6.10.1 The Fracture Mechanics Approach

Traditionally the concept of a critical stress intensity factor has been used to establish the maximum size of internal defects (usually cracks) allowable in a component for a particular working stress. In the case of metals (in the absence of stress corrosion environments) this is relatively simple since the cracks are stable at stress intensification levels below the K_C value. Geometrical features other than defect size must often be taken into account because of the difference in loading conditions from those used in the experimental evaluation of K_C.

For instance, in the design of thin wall pipes containing cracks in the longitudinal direction, the maximum allowable internal pressure can be estimated by expressing the K_I value as a function of the hoop stress. As suggested in Section 6.3.1, the magnitude of the longitudinal stress does not affect the K_I value since it does not contribute to crack opening. The equation usually used is:[56]

$$K_I = \sigma_H\sqrt{\pi a}\left(1 + 1\cdot6\,\frac{a^2}{Rt}\right)$$

where it can be seen that the normal plane stress definition of K_I must be modified by a factor which takes into account the radius of curvature of the pipe (R) and wall thickness (t). This extra factor arises from the elastic bulging at the crack tip which enhances the local tensile stresses.

In the case of plastics one would certainly have to consider the reduction in nominal K_I value that leads to brittle fractures with time arising from both the viscoelastic nature of the material (i.e. $K_C^2(t, \varepsilon) \simeq E(t, \varepsilon)G_C$), and the slow crack growth which increases the true K_I value until it reaches the critical conditions for fracture. In other words, one would have to specify the service life expectancy of the pipe at the required wall thickness in order to predict the maximum allowable internal pressure. Conversely, one may have to estimate the minimum allowable wall thickness from the required working pressure.

Although the above information may be available, or accessible, by experimentation, it could be argued that one would not know which nominal K_I value to choose in the analysis since the internal defect might neither exist nor be detectable.

This is, in fact, where fracture mechanics fails as a predictor of performance and becomes useful only as a means of selecting the best material or design from possible alternatives.

It is obvious that for the above case the best performance is likely to be achieved with a material having the highest K_{IC} value and showing the least reduction in K_I value (nominal) to failure as a function of time. If it is anticipated that the thickness of the pipe is likely to be greater than what is expected to produce plane stress conditions, then the material which shows the least change in K_C value in the transition from plane stress to plane strain is also likely to give the best performance. In the absence of an analytical tool to estimate the life expectancy of the product, one would have to resort to empiricism and evaluation of prototypes to acquire a certain confidence in the likely performance in service. The maximum working stress (e.g. internal pressure) or geometrical parameter (e.g. thickness, diameter of pipe, etc.) would be determined according to the degree of risk involved in the eventuality of a premature failure.

6.10.2 A Practical Approach to Prevention of Brittle Fractures

In the light of the limitations of fracture mechanics in design, the approach which is frequently used in practice is to specify the maximum allowable strain. By choosing a maximum strain value which is well below the level at which yielding or microcavitational phenomena (such as crazing, fibre/matrix debonding in reinforced plastics, etc.) occur, one would guard against the possibility of brittle fracture. In addition to this, the limiting strain would provide a basis for the estimation of the value of the modulus to be used in the pseudo-elastic analysis of the deformations.

Some guidance in the selection of the maximum allowable strain, however, is necessary and can only be estimated from a careful analysis of fracture mechanics data.

Substituting, for instance, the stress for $\varepsilon E(t, \varepsilon)$ in the definition of K_1 one obtains:

$$\frac{K_1}{E(t, \varepsilon)} = Y\varepsilon\sqrt{a}$$

If one were to select the critical conditions (i.e. the K_{1C} and the equivalent a_{crit}) and underestimate the value for the modulus, one would have a basis for the calculation of the maximum working strain. Any reduction in K_{1C} value due to the presence of environmental stress cracking agents can also be taken into account in the estimation of the maximum strain value. Suggestions such as those given in the manufacturers' literature[57] are only guesses that are not really reliable in view of the lack of available and sufficient case histories.

6.11   EVALUATION OF COMPLEX MECHANICAL PROPERTIES

Under this heading are included those properties of the material that cannot be easily expressed in terms of single parameters (or functions) and/or those properties which are related to other fundamental properties. These can be conveniently subdivided into hardness and frictional properties. A combination of these determines the wear resistance of the material. Since interface contact between two bodies results in transfer or removal of material it is opportune to treat these phenomena as failure processes.

6.11.1 Definition and Measurement of Hardness

The hardness of a material denotes its resistance to the penetration of an 'infinitely rigid' body. In other words, any deformation is considered to take place entirely in the material that suffers the penetration. The action of a rigid indentor on the surface of a material produces a complex distribution of stresses, depending on the geometry of the contact area and the degree of penetration of the indentor. The distribution of stresses, in turn, determines the deformation nature of the material in these areas, i.e. elastic, plastic, viscoelastic, etc. Because of these complexities it is almost impossible to obtain from tests an invariant property of the material which we may call hardness; therefore, the values obtained give an indication only of the behaviour of the material. In view of the simplicity of the apparatus required for such measurements, hardness tests find widespread uses for quality control purposes.

For the case of 'rigid' plastics the two most widely used tests are the Brinell and Rockwell tests, both of which make use of spherical indentors to cause a penetration in the material. Tests for softer plastics and rubbers, on the other hand, use a needle type of indentor with either a spherical or truncated conical tip; this type of test is often known as the Shore hardness test. The Brinell hardness number (BHN) is expressed as the load per surface area of indentation. From the geometrical considerations in Fig. 6.53[58] an expression for the BHN can be derived:

$$BHN = \frac{P}{(\pi/2)D^2(1 - \cos \theta)} \tag{6.60}$$

or

$$BHN = \frac{P}{(\pi D/2)(D - \sqrt{D^2 - d^2})}$$

From eqn. (6.60) above it can be seen that reproducibility will be achieved if the angle θ remains constant; hence when selecting a suitable load and sphere, the ratio P/D^2 must be kept constant.

The Rockwell test differs from the Brinell test only with respect to some procedural details such as the application of a pre-load before the major

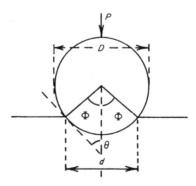

Fig. 6.53.   Indentation created by a loaded sphere onto the surface of a material.

load is applied. Furthermore, the Rockwell hardness is expressed in terms of depth of penetration which is recorded automatically and expressed as an arbitrary number on a scale from zero to 100. A special scale is selected according to the degree of penetration. For plastics the most widely used scales are R, L, M, E and K (e.g. ASTM D785). For the Shore hardness test the type of indentor is normally identified with the letters A, B, C and D, and the test consists of measuring the force required to produce a given penetration, which is recorded on a scale also in arbitrary units.

In all cases the major problem encountered in the testing of plastics is the viscoelastic nature of the material, which requires special attention to keeping a careful control of the time factors involved and the retarded recovery of the indentation after penetration (i.e. Rockwell test).

6.11.2 The Coefficient of Friction
The coefficient of friction between the surfaces of the two materials 'rubbing' or moving relative to one another is defined as the ratio of the shear stress to the normal stress (that opposes the movement) acting on the surface, i.e.

$$\mu_{ij} = \frac{\tau_{ij}}{-\sigma_{ii}}$$

The normal stress is obviously of the compression type. Since there can be only one normal (principal) stress acting on a surface, we can write

$$\mu_{12} = \frac{\tau_{12}}{-\sigma_{11}} \qquad \mu_{13} = \frac{\tau_{13}}{-\sigma_{11}}$$

For isotropic materials $\mu_{12} = \mu_{13}$.

Normally one distinguishes two types of coefficients of friction: a 'static' and a 'dynamic' coefficient. The first refers to the resistance met by the two surfaces in initiating the motion of the two surfaces, while the second refers to the resistance encountered by the two surfaces while they move (or slide) with respect to each other.

In a practical evaluation of the coefficient of friction, one uses the overall surface area of contact (Fig. 6.54) to calculate the shear and normal stress acting at the interface; consequently, one can write the expression for μ simply as $\mu = F_s/P$. The value obtained, however, is not constant but varies with the level of the applied load and the speed with which the two surfaces move relative to one another. The latter is associated with the viscoelastic nature of the material since, in effect, the coefficient of friction is related to the modulus (or compliance) of the material. However, since there is also

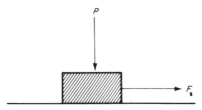

Fig. 6.54.   Forces considered in measurements of the coefficient of friction.

some transfer of material across the interface, it is expected that strength and configuration aspects of the polymer at the interface will play an important role. Consequently, whereas a low yield stress would favour the transfer of material from one surface to another and possibly reduce the coefficient of friction, a high compliance would increase the depth of the deformation created by the normal stress, which would increase accordingly the coefficient of friction.

As a consequence, the coefficient of friction goes through a maximum with an increase in the temperature and/or the sliding velocity of the surfaces. This phenomenon has often been related to the damping coefficient of the material,[59] which would be expected from viscoelastic considerations alone; i.e. the energy required to maintain a steady velocity of the surfaces in contact is related to the amount of energy dissipated by viscoelastic deformation of the surfaces. However, since this consideration does not take into account the material transfer and configurational aspects, as already mentioned, it may be more applicable to the dynamic coefficient than to the static coefficient. This would also explain why the static coefficient is greater than the dynamic coefficient. In other words the transfer of material and the acquisition of new configurations at the interface take place in the initial stage of the sliding process; thereafter the viscoelastic parameters become the predominant factors.

Regarding the configurational aspects of the polymer at the interface, it is noteworthy that linear molecular chains give rise to lower coefficients of friction than those containing side groups and branches. This is apparently related to the fact that the material at the interface is in the oriented state; and, consequently, the more irregular chains are less stable (i.e. less oriented) than linear chains. From this it is expected, therefore, that polypropylene would exhibit a higher coefficient of friction than high density polyethylene, while polytetrafluoroethylene has a lower coefficient of friction than low density polyethylene. This comparison is based on the observation that the polymers which have been compared display approximately the same yield strength and viscoelastic behaviour but differ substantially with respect to side groups and branches.

It is instructive to note, however, that on the basis of the non-linear viscoelastic behaviour of polymers, the coefficient of friction is expected to decrease with an increase in the level of normal stresses (or load) acting at the interface; a behaviour which is accentuated by the increase in true surface area of contact. Furthermore, the slip–stick effect which is often observed in the measurements of the coefficient of friction is associated with the difference between the velocity imposed on the two surfaces in contact by the apparatus and the 'natural' velocity acquired by the sliding members after the static friction has been overcome. The recovery of the deformation behind the sliding member will also contribute to this phenomenon, since it will tend to exert a force which opposes the direction of the movement of the sliding member.

6.11.3 Abrasion and Wear Properties

Abrasion and wear of sliding members will result after prolonged contact. The amount of material which is being transferred to the interface will increase and eventually will form 'debris' which increases local stress levels and produces an even larger transfer of material.

For this reason the presence of 'hard' inorganic particles in a plastics composition would increase the amount of wear, while the incorporation of very soft polymer particles, which display a low coefficient of friction (e.g. PTFE), into hard plastics (e.g. thermosets) will increase appreciably their resistance to wear.

Since wear is the net result of the transfer of material at the interface, it is expected that wear will be inversely related to the yield strength or shear strength of the material. Because the strength of polymers can decrease dramatically with temperature, effective removal of heat will reduce the amount of wear. For this reason it is not surprising that the amount of wear of plastics is less when rubbed against metals than against plastics. In selecting the best material, or combinations of materials, design engineers often make use of the so-called P–V values (i.e. the product of the normal stress and the velocity). The amount of wear is believed to be related to this product, which for a given material or combination of materials is determined on the basis of the maximum allowable amount of wear.

REFERENCES

1. C. B. BUCKNALL, *Toughened Plastics*, Applied Science, London, 1977, p. 142.
2. G. E. DIETER, *Mechanical Metallurgy* (2nd edn.), McGraw-Hill, New York, 1976, p. 83.

270   THERMOPLASTICS: MATERIALS ENGINEERING

3. S. STERNSTEIN and L. ONGCHIN, *Amer. Chem. Soc. Polym. Repr.*, **10** (1969) 1117.
4. R. RAGHAVA, R. M. CADDEL and G. S. YEH, *J. Mat. Sci.*, **8** (1973) 225.
5. K. V. GOTHAM, *Plastics and Polymers*, **40** (1972) 59–64.
6. K. V. GOTHAM, *Plastics and Polymers*, **41** (1973) 273–80.
7. P. P. BENHAM and F. V. WARNOCK, *Mechanics of Solids and Structures*, Pitman, London, 1978, p. 364.
8. R. J. CRAWFORD and P. P. BENHAM, *Plastics and Polymers*, **43** (1975) 140–2.
9. S. OKUDA and T. TAKAYUKI, *Mechanical Properties of Materials*, Vol. 3, pp. 581–9.
10. P. D. EWING, S. TURNER and J. G. WILLIAMS, *The Journal of Strain Analysis*, **7**(1) (1972) 9–22.
11. R. M. OGORKIEWICZ (ed.), *Thermoplastics: Properties and Design*, Wiley, New York, 1974, p. 58.
12. P. P. BENHAM and F. V. WARNOCK, *Mechanics of Solids and Structures*, Pitman, London, 1978, p. 337.
13. A. A. GRIFFITH, *Phil. Trans. R. Soc.*, *London*, **221**(A) (1920) 163–98.
14. E. F. WALKER and M. J. MAY, BISPA MG/E/307/67.
15. B. GROSS, J. E. STRAWLEY and W. F. BROWN, *Tech. Note D2395*, NASA, August 1964.
16. R. J. STANFORD and F. R. STONESIFER, *J. Comp. Mat.*, **5** (1971) 241–5.
17. J. T. BARNBY, *Non-destructive Testing*, **6** (1972) 32–7.
18. P. I. VINCENT, *Second International Conference, Yield, Deformation and Fracture of Polymers*, 5/1, Cambridge, England, March 26–29, 1973.
19. P. BEAHAM, A. THOMAS and M. BEVIS, *J. Mat. Sci.*, **11** (1976) 1207.
20. P. BEAHAM, M. BEVIS and D. HULL, *J. Mat. Sci.*, **8** (1973) 162.
21. S. S. STERNSTEIN and F. A. MYERS, *Solid State of Polymers*, P. H. Geil, E. Baer and Y. Wada (eds.), Marcel Dekker, New York, 1974, p. 540.
22. R. J. OXBOROUGH and P. B. BOWDEN, *Phil. Mag.*, **28** (1973) 547.
23. E. J. KRAMER, *Journal of Macromolecular Science*, *Phys. B*, **10** (1974) 191.
24. P. I. VINCENT, *Impact Tests and Service Performance of Thermoplastics*, The Plastics and Rubber Institute, London, 1971, p. 18.
25. E. OROWAN, *Fatigue and Fracture in Metals, Symposium at the Massachusetts Institute of Technology*, Wiley, New York, 1950.
26. G. R. IRWIN, Fracture, in *Encyclopaedia of Physics*, Vol. 6, Springer, Heidelberg, 1958.
27. S. MOSTOVOY, P. CROSLEY and E. RIPLING, *Journal of Materials*, **2** (1967) 661.
28. D. S. DUGDALE, *J. Mech. Phys. Solids*, **8** (1960) 100.
29. J. G. WILLIAMS, *Stress Analysis of Polymers*, Longmans, London, 1972.
30. G. E. DIETER, *Mechanical Metallurgy* (2nd edn.), McGraw-Hill, New York, 1976, p. 274.
31. J. F. KNOTT, *Fundamentals of Fracture Mechanics*, Butterworths, London, 1973, p. 69.
32. E. PLATI and J. G. WILLIAMS, *Polym. Eng. Sci.*, **15** (1975) 470–7.
33. E. PLATI and J. G. WILLIAMS, *Polym. Eng. Sci.*, **15** (1975) 476.
34. J. F. KNOTT, *Fundamentals of Fracture Mechanics*, Butterworths, London, 1973, pp. 165–75.
35. L. MASCIA and M. BAKAR, *Polym. Eng. Sci.*, **21** (1981) 577.

36. N. H. LINKINS, L. E. CULVER and G. P. MARSHALL, *Second International Conference, Yield, Deformation and Fracture of Polymers*, 12/1, Cambridge, England, March 26–29, 1973.
37. J. G. WILLIAMS, *Int. J. Fract. Mech.*, **8** (1972) 441–6.
38. E. PLATI and J. G. WILLIAMS, *Polymer*, **16** (1975) 915–20.
39. E. PLATI and J. G. WILLIAMS, *Polymer*, **16** (1975) 917.
40. P. P. BENHAM and F. V. WARNOCK, *Mechanics of Solids and Structures*, Pitman, London, 1978, p. 512.
41. G. E. DIETER, *Mechanical Metallurgy* (2nd edn.), McGraw-Hill, New York, 1976, p. 437.
42. K. V. GOTHAM, in *Thermoplastics: Properties and Design*, K. M. Ogorkiewicz (ed.), Wiley, New York, 1974, pp. 59–62.
43. L. C. CESSNA, J. A. LEVENS and J. B. THOMSON, *Polym. Eng. Sci.*, **9** (1969) 339–49.
44. ASTM, *Plastics methods of testing*, Part 27, D256, pp. 84–7.
45. ASTM, *Standard methods for testing*, Part 26, D1709, p. 36.
46. K. V. GOTHAM, *Plastics and Polymers*, **41** (1973) 278, Figs. 12 and 13.
47. H. R. BROWN, J. S. HARRIS and I. M. WOOD, *Second International Conference, Yield Deformation and Fracture of Polymers*, 13/1, Cambridge, England, March 26–29, 1973.
48. J. G. DOCHERTY, *Engineering*, **172** (1932) 135 and 645; **175** (1935) 139 and 211.
49. R. W. DAVIDGE and G. TAPPIN, *J. Mat. Sci.*, **3** (1968) 165.
50. Author's unpublished work with assistance from A. Braham and F. Algui, Institute Algérien du Pétrole, Boumerdes, Algeria, 1978.
51. C. E. TURNER, L. E. CULVER, J. C. RADON and P. KEMISH, *Practical Applications of Fracture Mechanics to Vessel Technology*, Institute of Mech. Eng., London, 1971, p. 38.
52. E. PLATI and J. G. WILLIAMS, *Polym. Eng. Sci.*, **15** (1975) 470.
53. E. PLATI and J. G. WILLIAMS, *J. Mat. Sci.*, **8** (1973) 949–56.
54. A. C. MORRIS, *Plastics and Polymers*, **36** (1968) 433.
55. W. DÖLL, *Second International Conference, Yield, Deformation and Fracture of Polymers*, 8/1, Cambridge, England, March 26–29, 1973.
56. G. W. PARRY and J. MILLS, *Strain Analysis*, **3** (1968) 159.
57. ICI Ltd, Plastics Division, *Technical Service Note G 123*, p. 16.
58. G. E. DIETER, *Mechanical Metallurgy* (2nd edn.), McGraw-Hill, New York, 1976, p. 391.
59. L. E. NIELSON, *Mechanical Properties of Polymers*, Van Nostrand Reinhold, New York, 1962, p. 224.

7

Deformation Behaviour of Thermoplastics in Relation to Processing

In ordinary terminology the word processability denotes the *ease* with which a given material can be *converted* into a *product* (normally the end product). The concern of a processing technologist is to quantify this statement by establishing the relationships between the italicised key words above. The term 'ease' includes parameters that denote the necessary efforts, e.g. materials, labour and energy; while in the term 'end product' are implicit quality and market satisfaction, e.g. sales appeal and serviceability. Finally the processing or conversion operations involved represent the constraints on the feasibility of achieving product quality at low cost.

Logically, to formalise the approach, one would have to start from the constraint. This means that processing methods would have to be grouped according to the manner in which the material acquires the desired shape and is perceived by the human eye when converted into a product. In other words, one would have to consider:

(1) The deformability characteristics of the material and the associated morphological changes brought about by the combined effects of thermal and stress history during processing.
(2) The receptivity of any auxiliary materials deposited on the surface of the component in order to alter its response to light and other environmental agents.

Note that compounding is not considered in these discussions since, although it may involve shaping operations, e.g. formation of a cylindrical or cubical granule, it does not convert the material into an end product.

As a first approximation we can divide the conversion operations into:

(1) Primary processes—those which impose the desired shape on the material.
(2) Finishing (or secondary) processes—those which involve surface treatment.

7.1 CLASSIFICATION OF PRIMARY PROCESSES

A cursory look at plastics processes suggests that a first, if only crude, classification can be made as follows:

Machining processes—shaping by local removal of surplus material from a given stock.

Forming processes—shaping by deforming a standard stock (simple geometry) so that is acquires the desired configuration and dimensions (more-complex geometry).

Fusion processes—shaping by joining small components to the required geometry and dimensions of the end product.

Transfer (or flow) processes—shaping by transporting a 'melt' or liquid system into a cavity or through a confined geometry, where it acquires and consolidates its final shape and dimensions.

In order to put into perspective the problem of processability of plastics, it is necessary to analyse their deformability and morphological characteristics (as already mentioned) over the range of temperatures and stress conditions likely to be experienced during processing. Three aspects are particularly relevant.

(1) Relationships between stresses and strains and identification of the appropriate material parameters that can be used to assess the processing characteristics.
(2) Ductility or maximum possible deformations achievable.
(3) State of the material after it has been submitted to different types of deformations, i.e. dimensional stability and physical properties that govern their service performance.

The assumption will be made that in the case of thermoplastics any incidental chemical changes (i.e. degradation reactions) have a negligible effect on the behaviour described. On the other hand, the influence of the components of a particular material (i.e. type of polymer and additives) will be taken into account.

7.2 INTERPRETATIONS OF THE SHAPING CHARACTERISTICS OF THERMOPLASTICS IN TERMS OF DEFORMATION BEHAVIOUR

The long chain molecules of polymers in thermoplastics compositions confer on them some unique characteristics. These are outlined in the following sections.

7.2.1 Retention of the Behaviour of Solids up to High Temperatures and the Acquisition of the Characteristics of Liquids

According to the theories of viscoelasticity (see Chapt. 5), the deformational behaviour of a polymer can be described by means of mechanical models and defined in terms of parameters such as compliance (when the excitation stresses are constant) and relaxation modulus (at constant strain conditions) as a function of the ratio t/λ, where t is the duration of the excitations or relaxations observed and λ is a property of the material known as the retardation or relaxation time.

For a more complete description one would have to use the concept of distribution of retardation or relaxation times and would have to take into account the non-linearity of the above relationships. For the sake of simplicity it is sufficient, for the time being, to note that single models can, qualitatively, describe the solid characteristics of polymers, so that a plot of compliance versus $\ln t/\lambda$ yields the master curve traced in Fig. 7.1.

As already explained in Chapt. 2 the behaviour of linear polymers is

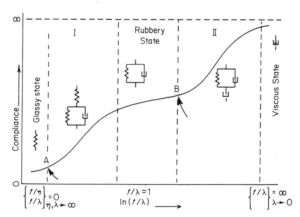

Fig. 7.1.   The solid-like behaviour of polymers described by means of mechanical analogues.

characterised by three states, i.e. the glassy state, the rubbery state and the viscous state, in which the compliance or relaxation modulus is only a mild function of the time, and by two transitional states, I and II respectively in Fig. 7.1, where the deformability parameters are strongly dependent on the time scale of the action of a particular stress on an element of the material.

The initial stages of these two transitional states have been marked with an arrow in Fig. 7.1 at points A and B. These represent, respectively, the onset of two distinctly different mechanisms responsible for the time-dependent characteristics of the polymer. Point A signifies the onset of sequential rotations that take place in the deformational process, while at point B the process of self-recoiling of the molecules begins to take place, resulting in true flow, i.e. irrecoverable deformations.

Note the presence of a Voigt element in transition Zone I and the addition of a viscous element in Zone II. The changeover from one transitional state to the other occurs via the intermediate rubbery state, which can be approximately represented by a Voigt type of element. In other words, the elastic element, representing the constraint on molecular rotations, disappears in view of the low intermolecular forces and rotational energy barriers. However, since λ acquires values of the same order of magnitude as t, the compliance will change only slightly as a function of $\ln t/\lambda$.

It is important to note that the curve in Fig. 7.1 could have been generated through phenomenological considerations. In fact, a plot of the compliance versus $\log t/a_T$ (where a_T is the shift factor of the WLF equation) would have yielded a similar curve. Accordingly, if the time of duration of the excitation is kept constant, a plot of the compliance (isochronous) as a function of the temperature would also show the two transitional states of Fig. 7.1; but since there is no direct proportionality between a_T and the temperature (i.e.

$$\log(a_T) = \frac{C_1(T - T_g)}{C_2 + (T - T_g)}$$

where C_1 and C_2 are constants and T_g is the glass transition temperature), the curve loses the symmetrical sigmoidal shape of the exponential functions obtained from the spring and dash-pot models. Both the variations of a_T and of the isochronous compliance as a function of the temperature are shown in Fig. 7.2.

The dotted curve shows the typical secondary transitions exhibited by crystalline polymers, which arise from relaxations within the amorphous phase. Such transitions are, obviously, absent at temperatures above the

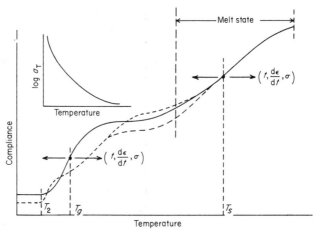

Fig. 7.2. Phenomenological description of the deformational behaviour of polymers.

rubbery state in view of the one-phase morphology acquired by the polymer after the crystals have been destroyed by the melting process.

In this diagram the glass transition temperature and the plateau corresponding to the rubbery state are clearly recognisable. The transition at higher temperatures corresponds to what is generally known as the 'melt' state, and its position in relation to the glass transition temperature may be established by identifying a midpoint in the curve corresponding to the maximum exhibited by the first derivative plotted as a function of the temperature. This may be conveniently termed the 'softening point' or 'flow temperature' (T_s).

The horizontal arrows on the transition portions of the curve indicate that their relative position in relation to the temperature is dependent on the conditions shown in brackets; i.e. longer periods of excitations and higher stress (directional types) levels would displace the curve towards lower temperatures, and vice versa for the case in which the compliance is measured at constant straining rate. Similarly, hydrostatic (compressive) stresses, which hinder the relaxation mechanisms, displace the transitions towards higher temperatures.

7.3 THE GLASSY/RUBBERY TRANSITIONAL STATE

The controversy over the kinetic character of the glass transition temperature has been resolved by Gibbs and DiMarzio in their pronunciation of the existence of a true equilibrium state, called the 'glassy state', in which all molecular chains are at their lowest conformational energy level. In this state only one possible conformation for the entire chain exists; i.e. the conformation entropy is zero (see Chapt. 2).

If this is to be interpreted in terms of response of a polymeric material in its glassy state to the action of stresses, it would simply mean that the resulting deformations would arise purely from dilatations.

This does not violate the concept of a constant unoccupied volume of polymers in the glassy state but simply makes it dependent on the state of external stresses acting on the material.

The associated internal energy change corresponds to the work of expansion to overcome solely the intermolecular forces, i.e. there is no segmental motion; hence $\Delta S = 0$, therefore

$$\Delta G = \Delta E - T\Delta S = \Delta E$$

$$\Delta E = \Delta H - \bar{\sigma}_m \bar{\varepsilon}_v$$

where σ_m = mean stress, i.e. $\frac{1}{3}(\sigma_1 + \sigma_2 + \sigma_3)$; ε_v = volumetric strain = $\varepsilon_1 + \varepsilon_2 + \varepsilon_3$; and $\bar{\sigma}_m \bar{\varepsilon}_v$ = dilatation work/unit volume.

The lack of chain rotations brings about the nucleation and subsequent growth of cracks leading to brittle fractures. Hence the glassy state *per se* is of no interest in shaping operations in the processing of thermoplastics.

At temperatures above T_2 (Fig. 7.2) the larger free volume and the higher internal energy acquired by the polymer allow the chains to undergo rotations (i.e. the conformational entropy increases), which are biased by the direction of the stresses. Because of the kinetic nature of the chain rotations, the deformation resulting from a given stress increases with time as the chains seek to find new equilibrium configurations.

By raising the level of the excitation stress, the length of segments involved in the molecular chain rotations becomes increasingly large, so that each increment in stress produces an increasingly large increment in the resulting deformation (i.e. the deformational behaviour becomes non-linear (see Chapt. 5).

Eventually a level of stress is reached at which catastrophic rotations occur, accompanied by large-scale molecular uncoiling (even if only in localised regions such as the amorphous domains of a semi-crystalline polymer) and resulting in very high rates of deformations. The new 'oriented' configurations of the polymer chains are stabilised by the overall increase in intermolecular forces resulting from the net reduction in average distance between dipoles.

The above phenomenon corresponds to the 'yield point' of the material.

The stress level at which yielding occurs decreases as the elapsed time increases. In other words, the longer the history of the stress, the lower the energy barrier which must be overcome to induce the catastrophic rotations for the yielding process.

Raising the temperature results in an increase in internal energy of the molecules and reduces the potential energy barrier for the onset of yielding, so that the yield stress eventually becomes zero. This also means that the deformations occur without going through a yield point and that there will be no stabilisation of the new configurations of the chains; so that the reverse process of deformation occurs readily when the stress is removed. When this occurs the material is said to be in its 'rubbery state'. The range of conditions under which the material is capable of yielding defines the 'glassy/rubbery' state. In terms of mechanics, this can also be called the 'viscoelastic–plastic' state. The term 'viscoelastic' indicates the time dependence of the phenomenon, whereas the term 'plastic' denotes the irreversibility of the deformations upon removal of the stress.

From a processing point of view the properties that are of interest for a material in its glassy/rubbery state are:

(1)   Yield criteria in terms of stresses and strains (i.e. effects of multiaxial stresses).
(2)   Temperature and strain rate dependence of the yield stress.
(3)   Extensibility limits of the material.
(4)   Recovery characteristics of post-yielding deformations.
(5)   Structure/properties relationships for polymers in their plastically deformed state.

7.3.1 Yield Phenomena in Relation to Processing

The modification of established yield criteria, e.g. Tresca and von Mises criteria, to include a compressibility coefficient has been discussed in Chapt. 6.

In view of the large yield strain values of polymers (relative to metals), criteria based on strain, however, would be more useful than those based on stress. In other words, the yield strain is the critical parameter to consider for polymers (rather than the stress), in shaping operations involving plastic deformations. For the case of metals, of course, the values of the strains at yield can be calculated directly from the corresponding stresses through their modulus and Poisson ratio. This is not so for polymers, in view of the non-linear relationship between stress and strain.

As already discussed in Chapt. 5, it is even more difficult to determine the level of strain at which the material may be considered to have reached the yielding conditions, since the transition from the viscoelastic to the plastic state is very diffuse: a phenomenon which is compounded also by localised temperature increments. Obviously, upon approaching the rubber state the yield strain becomes increasingly large and more diffuse.

A further difficulty arises from controversial reports about the volume changes occurring during yielding[1] in view of the various forms of microcavitations that may take place, e.g. crazing, microvoiding, etc.

7.3.2 Temperature and Strain Rate Dependence of the Yield Strength of Linear Polymers

A knowledge of the yield strength of polymers as a function of temperature and strain rate in relation to processing is useful insofar as it gives a good indication of the state of the material in relation to the rubbery state. A rapid change of yield stress with temperature is, obviously, advantageous in processing, as it would require only a slight amount of cooling from the rubbery state (say) to ensure that permanent deformations are set up so that the stresses can be removed (i.e. a component can be removed from the mould).

Similar deductions can be made with respect to the strain rate sensitivity of the material.

As already indicated in Chapt. 6, both effects can be quantified through Eyring's theory on the stress activation of flow processes. By considering, in fact, yielding as a rate process involving movements (or rotations) of segments of molecular chains, the effects of the applied stress can be deduced by the following reasoning.[2]

The stress, σ, decreases the energy barrier for the movements in the direction in which it acts, i.e. it creates a bias for the process (Fig. 7.3). Therefore

$$-\Delta E_0 = \sigma \times A \times l$$

where A = cross-sectional area of the entity considered, l = amount of displacement of the entity considered, and $A \times l = v$ = volume of material displaced by the action of the stress σ.

The new potential energy barriers are $E_0 - \sigma v$ in the direction of the stress and $E_0 + \sigma v$ in the opposite direction.

The probability of movements in each of these respective directions is:

$$P(\uparrow) = P_0 \exp\left[-(E_0 - \sigma v)/RT\right]$$

and

$$P(\downarrow) = P_0 \exp\left[-(E_0 + \sigma v)/RT\right]$$

where P_0 is the probability in the absence of the stress activation effect.

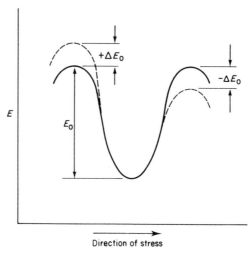

Fig. 7.3.   Change in energy barrier with the direction of the stress.

The net velocity of the movement of the structural entities involved in the deformation process is proportional to the difference of these two probabilities, i.e.

$$u = f \left\{ \exp \left(-\frac{E_0}{RT} \right) \left[\exp \frac{\sigma v}{RT} - \exp \left(-\frac{\sigma v}{RT} \right) \right] \right\}$$

Therefore

$$u = f \left\{ \exp \left(-\frac{E_0}{RT} \right) \sinh \frac{\sigma v}{RT} \right\}$$

If one considers that these entities move at a velocity corresponding to the rate of straining, then one obtains an expression which relates the stress at yield to both temperature and rate of strain, i.e.

$$\dot{\varepsilon} = A \exp \left(-\frac{E_0}{RT} \right) \sinh \frac{\hat{\sigma}_Y v}{RT} \qquad (7.1)$$

where v is known as the 'activation volume', and E_0 is the 'activation energy' (free energy if one neglects the entropy term).

Practical experience shows that the term $\hat{\sigma}_Y v$ is much larger than RT; so the sinh function becomes approximately equal to the exponential function. Hence the above equation can be written as

$$\dot{\varepsilon} \simeq A \exp \left[\frac{-E_0 + \hat{\sigma}_Y v}{RT} \right] \qquad (7.2)$$

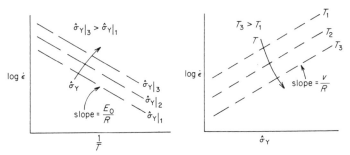

Fig. 7.4. Graphical estimation of activation energy and activation volume of yielding.

Therefore, the activation energy and activation volume can be obtained from the slope of the plots $\log \dot{\varepsilon}$ versus $1/T$ and $\log \dot{\varepsilon}$ versus $\hat{\sigma}_Y$, respectively (Fig. 7.4). Note that slight modifications to eqn. (7.1) have, occasionally, been suggested. These amount to no more than the introduction of constants in the second term and, therefore, do not alter the nature of the relationship between the various parameters.

Combining this information with the concepts expounded in Chapt. 6 it can be concluded that the parameters defining the yielding characteristics of polymers are yield strength ($\hat{\sigma}_Y$) and the associated dilatation or compressibility coefficient (μ), the activation energy (E_0), and the activation volume (v). The interaction of these parameters in Cartesian coordinates generates the paraboloids shown in Fig. 7.5. These have elliptical bases, the axes of which are functions of the dilatability or

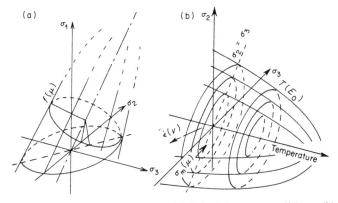

Fig. 7.5. Yield behaviour of polymers. (a) Triaxial stress conditions. (b) Plane stress conditions: effects of temperature, strain rate and hydrostatic stresses.

compressibility coefficient (μ) and also of the activation volume (v) when the third axis (the height) represents the temperature (Fig. 7.5(b)).

In discussing the processability of thermoplastics, one would like to have a knowledge of these parameters in order to predict the effects of processing conditions and to determine the adequacy of certain design features of the processing equipment. For the selection of a material or in the formulation of a plastics composition, it is also necessary to know the influence of the molecular and morphological structure of the base polymer and the effect of its interactions with additives.

At present these factors are poorly defined and their relationships are, at best, only speculative. For instance, it is known that for amorphous polymers the yield strength increases with molecular weight, chain stiffness and intermolecular forces (i.e. all the factors that increase T_g). However, while the yield strength can be reduced by addition of plasticisers and increased by incorporation of fillers, in the latter case T_g remains the same and the effect noted is due to the filler carrying higher loads than the matrix and, possibly, to the increased triaxiality of the stresses within the matrix. As to the activation energy (i.e. the temperature coefficient of the yield strength), it appears to be independent of the molecular weight; it is not affected by the filler (i.e. yielding is a matrix phenomenon with, possibly, little interfacial or interaction contributions). However, the activation energy is increased by plasticisers and, possibly, also by rubbery toughening agents. In other words, E_0 is affected primarily by the chain stiffness and intermolecular force balance within the polymeric composition. The activation volume and, obviously, the compressibility coefficient must be related to free volumes and chain stiffness.

7.3.3 Dimensional Stability and Recovery Characteristics of Post-Yielding Deformations

In assessing the processability of a material it is necessary to know its dimensional stability subsequent to the removal of the stresses, e.g. at the exit of a die, so that steps can be taken to remedy any undesirable effects that may ensue. For the case of metals, it is well known that post-yielding deformations are almost completely irrecoverable, i.e. the extent of recovery amounts to a very small fraction of the total deformation and corresponds to the elastic distortions of the crystals while they are forced to slide over one another under the influence of the applied stresses.

Recovery of plastic deformations of polymers, as already implied in previous discussions, can be substantial in view of the vibrational energy of the segments and end groups of the molecules occupying the amorphous regions. These will create free volumes and induce partial recoiling of molecular chains and some back tilting of crystals.

The extent of recovery, therefore, is dependent not only on the amount of deformation produced and type of stresses involved (i.e. dilatational hydrostatic stresses would increase the amount of recovery, whereas hydrostatic pressure would suppress recovery), but also on the time/temperature conditions of the deformations. The closer they are to the rubbery state conditions, the greater the extent of recovery.

The molecular and morphological structure of the polymer plays an important role in the deformational and recovery processes. The higher the molecular weight, the greater is the extensibility of the polymer system and, possibly, the higher is the post-yielding dimensional stability owing to the lower incidence of end groups. Similar conclusions can be drawn with respect to chain branches, although it must be borne in mind that straining of covalent bonds can also play an important role.

Crystalline polymers, blends, and composites (particulate and short-fibre types) exhibit a complex behaviour, which is governed by events involving both molecular and microscopic entities. In the case of crystalline polymers one has to consider the events that can take place inside the spherulites and in the interspherulitic regions. In other words, the molecular motions within the amorphous phase, which are the precursors to major large-scale deformations, affect the manner in which the spherulites deform, the tilting of crystals in the direction of the stresses, and any subsequent rearrangements of the crystals in lamellar (biaxial) deformations or fibrillar (monoaxial) deformations.

Normally, approximately 100% elongations for monoaxial deformations and, possibly, half this amount for plane stress deformations are required to break down the spherulitic structure. Much greater stability is to be expected after such an event because any retraction mechanism involves tilting of the crystals, which is much more difficult when these lie in the axial direction.

It is understood that recovery depends especially on the temperature of the polymer immediately after the deformation process is completed. Rapid application of the stresses, for instance, is conducive to producing adiabatic heating; consequently recovery is higher than when stresses are applied more slowly and isothermally. If the temperature is artificially raised, recovery will increase accordingly, to the point where, upon reaching the rubbery state conditions, total retraction takes place. In the case of crystalline polymers this occurs very rapidly once the thermodynamic melting point is reached.

Unpublished work by the author on the monoaxial drawing of polypropylene under conditions falling towards the upper regions of the glassy/rubbery state and followed by sharp cooling (i.e. as in the production of film yarns) shows that the recovery (i.e. post-shrinkage) at temperatures substantially below the melting point ($\simeq 100$–$120\,°C$) is a function of both draw ratio and temperature at which the deformation has been carried out, as shown in Fig. 7.6.

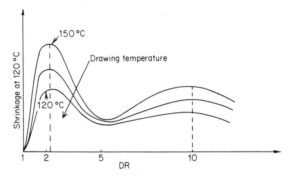

Fig. 7.6.   Overall recovery of uniaxially drawn polypropylene films. (Author's unpublished work.)

In these graphs the much higher shrinkage occurring for deformations below the level at which spherulitic breakdown takes place is clearly revealed. Presumably, at drawing ratios of the order of 5:1 the crystals are reasonably well aligned in the drawing direction; above this value it is possible that chain unfolding takes place and a chain-extended type of crystallisation ensues. At draw ratios greater than 10:1 shrinkage decreases because of the greater extent of molecular alignment. No similar data have been reported on the shrinkage of amorphous polymers, but it is to be expected that the 'knee' in the first portion of the curves would be absent. The microscopic events that may accompany the deformation process consist of microcavitations (i.e. crazes and voids) causing the so-called stress whitening phenomenon.

These microcavitations result from phase discontinuities, such as pigment particles, fillers, crystallites, and local structure inhomogeneities (i.e. 'nibs', gel spots, etc.), which cause the setting-up of dilatational stresses and nucleate the formation of cavities. It is interesting to note, for instance, that the type of stabilisers and lubricants used in rigid PVC formulations can determine whether the material will undergo stress whitening on yielding. The lower the compatibility of these additives, the greater their tendency to cause microcavitations.[3] Obviously, such events will reduce the extensibility of the polymer composition; crazing, in particular, will also affect the recovery characteristics, insofar as they constitute geometrical obstacles to the retraction of molecular chains in the surrounding regions.

7.4 THE RUBBERY STATE

According to the thermodynamic theory of rubber elasticity, when a level of internal energy is reached so that the internal movements within polymer molecules through the creation of large free volumes reduce the potential energy barriers to zero, the only forces necessary to extend the polymer chains are those required to reduce the configurational entropy. This assumes, of course, that there are no constraints to the lateral changes of dimensions under the influence of direct (or normal) stresses; i.e. there is no change in volume, so that the work of the expansion term ($\sigma_m \varepsilon_v$) can be neglected. Since the work necessary to extend the molecules in the direction of the stress depends only on the rotational energy of the backbone bonds, there can be no stress activation effect; i.e. the activation volume is zero, and consequently each increment in stress level produces equal increments in strain.

The linear relationship between stress and strain does not preclude, however, the time dependency of this behaviour. The rubbery state behaviour, in fact, is a special case of linear viscoelasticity, i.e. that state of the material for which the retardation and relaxation times of the deformation are of the same orders of magnitude as the time of duration or decay of the stress. However, since the contribution from intermolecular forces or from the hindered rotations is negligible, there is no elastic component in the deformational parameters. Consequently the rubbery state behaviour can be quite adequately represented by a series of Voigt or Maxwell elements. Normally a combination of four or five elements is sufficient to describe the time-dependent compliance and relaxation modulus.

I.e.

$$D(t) = \sum_{i=1}^{n=4} D_0|_i (1 - l^{-t/\lambda_c}) \quad \text{and} \quad E(t) = \sum_{i=1}^{n=4} E_0|_i l^{-t/\lambda_r} \quad (7.3)$$

In view of the fairly accurate representation of the behaviour by means of these simple models, one is in a reasonably easy position to find correlations between molecular features of a polymer composition and the parameters of the equation in (7.3).

For instance one can infer the following relationships:

(i) $D_0|_i$ and $E_0|_i$ are related to the rotational energy barriers of the molecular chains; i.e. the greater the chain stiffness, the higher the $E_0|_i$ values which will satisfy eqns. (7.3).

(ii) The retardation and relaxation times (λ_c and λ_r respectively), on the other hand, are related to the molecular weight of the polymer.

These relationships are illustrated, for the case of polystyrene, in Figs. 7.7 and 7.8.[4,5]

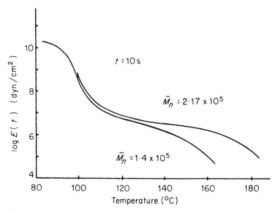

Fig. 7.7.   Effect of MW on the isochronous modulus of polystyrene in function of temperature. (After Aklonis et al.[4])

7.4.1 Dimensional Stability of Polymers Deformed in Their Rubbery State

From Fig. 7.8 it is quite clear that for a polymer of a given molecular weight the longer the time scale of the observation, the higher the λ value for which the rubbery state is observed; i.e. the polymer exhibits rubbery characteristics at lower temperatures. In Fig. 7.7, on the other hand, it is shown that with increasing molecular weight, the rubbery characteristics extend to higher temperatures, while the effect on the glass transition temperature is quite small.

Naturally, if the material is deformed in its rubbery state and the temperature is dropped immediately (ideally this process should be

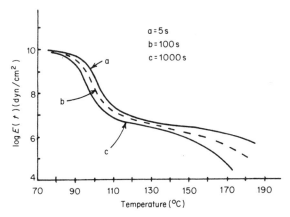

Fig. 7.8.   Effect of time of duration of load on the isochronous modulus of polystyrene as a function of temperature. (After Tobolski.[5])

instantaneous) so that it is brought into the glassy state, there will be no recovery of the deformation. Partial recovery will occur, on the other hand, if the temperature is reduced somewhat slowly, but it will cease immediately once the material has reached the glassy state conditions.

The cooling rate is also an important factor in the molecular packing of the polymer chains for a given level of deformation. A slower cooling rate enhances the degree of crystallinity in crystalline polymers and also produces a somewhat higher density in amorphous polymers. In either case the material would have a higher dimensional stability; i.e. it will exhibit lower shrinkage when subsequently heated to temperatures corresponding to the glassy/rubbery state of the material, and will have superior strength and stiffness. Consequently, any difference in properties of materials deformed in their glassy/rubbery state (plastic deformations) and their rubbery state is most probably due to the difference in thermal history to which the material has been subjected in the two deformational processes.

7.4.2 Relationship between Stress and Strain for Deformations Occurring in the Rubbery State

The important features of the rubbery state are the high undilatability, so that the Poisson ratio can be taken to be equal to 1/2, and the reasonable linearity in the relationships between stress and strain. However, the large deformations involved require a redefinition of strain in order that the appropriate constitutive equations can be cast into a formalism analogous to that of classical elasticity theory.

The concept of finite strain e_{ii} produces the following relationships[6]

$$\lambda_1^2 = 1 + 2e_{xx} \qquad \lambda_2^3 = 1 + 2e_{yy} \qquad \lambda_3^3 = 1 + 2e_{zz}$$

where λ_1, λ_2 and λ_3 are the extension ratios in the three principal directions (i.e. $\lambda = L/L_0$); while if the strain is infinitesimal

$$\lambda_1 = 1 + \varepsilon_{xx} \qquad \lambda_2 = 1 + \varepsilon_{yy} \qquad \lambda_3 = 1 + \varepsilon_{zz}$$

When shear strain components are zero and $v = \frac{1}{2}$ we obtain the following constitutive equations:

Classical (infinitesimal strain) elasticity	*Rubber (-like) elasticity*

$$\varepsilon_{xx} = \frac{3}{2E}(\sigma_{xx} - \sigma_m) \qquad\qquad \lambda_1^2 = \frac{3}{E}(\sigma_{xx} - \sigma_m)$$

$$\varepsilon_{yy} = \frac{3}{2E}(\sigma_{yy} - \sigma_m) \qquad\qquad \lambda_2^2 = \frac{3}{E}(\sigma_{yy} - \sigma_m)$$

$$\varepsilon_{zz} = \frac{3}{2E}(\sigma_{zz} - \sigma_m) \qquad\qquad \lambda_3^2 = \frac{3}{E}(\sigma_{zz} - \sigma_m)$$

where $\sigma_m = (\sigma_{xx} + \sigma_{yy} + \sigma_{zz})/3$.

It follows, therefore, that:

(1) For deformations under uniaxial stresses, i.e. $\sigma_1 = \sigma^*$, $\sigma_2 = \sigma_3 = 0$ and $\sigma_{12} = \sigma_{13} = \sigma_{23} = 0$, $\lambda_1^2 = (3/E)(\sigma^* - \sigma_m)$ and $\lambda_2^2 = \lambda_3^2 = -(3/E)\sigma_m$. Since for conditions of undilatability and incompressibility $\lambda_1 \lambda_2 \lambda_3 = 1$, then

$$\sqrt{\lambda_2^2}\sqrt{\lambda_3^2} = \frac{1}{\lambda_1} = -\frac{3}{E}\sigma_m \qquad \text{and} \qquad \sigma_m = -\frac{E}{3}\frac{1}{\lambda_1}$$

Therefore

$$\lambda_1^2 = \frac{3}{E}\left(\sigma^* + \frac{E}{3}\frac{1}{\lambda_1}\right)$$

and

$$\sigma^* = \frac{E}{3}\left(\lambda_1^2 - \frac{1}{\lambda_1}\right) \tag{7.4}$$

This equation becomes

$$\sigma = \frac{E}{3} \left(\lambda_1 - \frac{1}{\lambda_1^2} \right)$$

where σ is the nominal stress, i.e.

$$\sigma = \frac{F}{A_{(\text{original})}} = \frac{F}{A_{(\text{actual})}\lambda_1} = \frac{\sigma^*}{\lambda_1}$$

(2) For biaxial (plane) stresses, i.e. $\sigma_1 = \sigma_2 = \sigma^*$, $\sigma_3 = 0$ and $\sigma_{12} = \sigma_{13} = \sigma_{23} = 0$,

$$\lambda_1^2 = \lambda_2^2 = \frac{3}{E} (\sigma^* - \sigma_{\text{m}}) \qquad \lambda_3^2 = -\frac{3}{E} \sigma_{\text{m}}$$

Therefore

$$\sigma_{\text{m}} = -\frac{E}{3} \lambda_3^2$$

Hence

$$\lambda_1^2 = \lambda_2^2 = \frac{3}{E} \left(\sigma^* + \frac{E}{3} \lambda_3^2 \right) = \frac{3\sigma^*}{E} + \lambda_3^2$$

and

$$\sigma^* = \frac{E}{3} (\lambda_1^2 - \lambda_3^2) = \frac{E}{3} (\lambda_2^2 - \lambda_3^2)$$

Now if $\lambda_1 \lambda_2 \lambda_3 = 1$, then $\lambda_3 = 1/\lambda_1^2 = 1/\lambda_2^2$ and

$$\sigma^* = \frac{E}{3} \left(\lambda_1^2 - \frac{1}{\lambda_1^4} \right) = \frac{E}{3} \left(\lambda_2^2 - \frac{1}{\lambda_2^4} \right) \tag{7.5}$$

7.4.3 Deformability Limits of Polymers in Relation to Temperature and Strain Rate

Typical stress/strain curves for polymers tested at constant strain rate (or at constant stress, under isochronous conditions) over a range of temperatures between the glassy state and the melt state are shown in Fig. 7.9.

The extensibility of a polymer shows a maximum in the rubbery state and it rises very sharply within the melt state. In view of the time/temperature equivalence of the mechanical behaviour, an increase in strain rate would shift the curves so that they will overlap with those occupied at lower

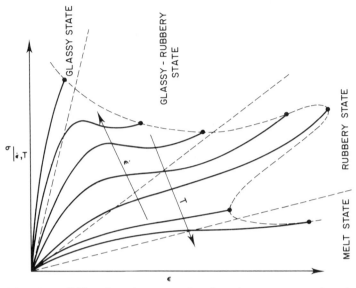

Fig. 7.9.   Extensibility of a polymer as a function of temperature and strain rate.

temperatures under low strain rate conditions. In other words a time/temperature superposition operates not only with respect to modulus and compliance but also for the strain at break.[7] Consequently one should be able to construct a master curve for the extensibility limits as a function of temperature and strain rate by a means of a shift factor a_T (Fig. 7.10).

Two phenomena may be largely responsible for the maximum extensibility within the rubbery state: (a) the tendency of the material to develop stronger intermolecular forces as a result of the stretching out of molecular chains (i.e. some form of crystallisation), and (b) the transfer of stresses to fewer and fewer molecular chains as the process of molecular recoiling and flow takes place. The first would provide a crack-stopping mechanism since the crack would have to grow in the directions normal to the molecular chains, while the second would probably favour the formation and propagation of cracks. With the approach of the melt state the number of chains which can build up stresses to the point of rupture of covalent bonds becomes negligible in view of the greater ability of the molecules to dissipate the strain energy into kinetic energy via translational motions and flow.

Other factors that affect extensibility include the molecular structure of the polymer system and the type of stress involved. Altering the molecular

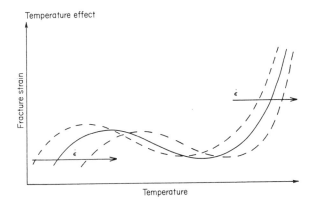

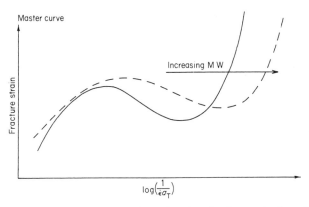

Fig. 7.10.   Deformability limits (strain at break) of polymers in the rubbery and rubbery/viscous states.

structure of the polymer system, i.e. increasing the molecular weight and widening the distribution by producing higher-molecular-weight tails, broadens the range of conditions (e.g. temperature) over which the rubbery state is exerted and displaces the melt state to higher temperatures (or longer relaxations or retardation times). This results in somewhat higher fracture strains at the peak region of the curve and a displacement of the curve to higher temperatures similar to the effect of the straining rate (Fig. 7.10). Consequently, by producing grafts and introducing some cross-links (as in the case of terpolymer blends) the master curve broadens towards higher temperatures.

7.5 ORIENTATION IN POLYMERS DEFORMED IN THE GLASSY/RUBBERY TRANSITIONAL STATE AND RUBBERY STATE

From the discussions in previous sections it can be easily inferred that deformations in the glassy/rubbery state and in the rubbery state produce morphological changes with a directional orientation along the axes of maximum stresses. This phenomenon is generally known as 'orientation'.

It is opportune at this stage to emphasise the difference between 'orientation' and 'internal stresses' within a polymer, since the two terms are sometimes erroneously used interchangeably.

Orientation is a permanent and stable structural state of the material at temperatures below the melting point or main transitions.

Internal stresses, on the other hand, imply that the structural entities of the material are subjected to forces that act in opposite directions to maintain equilibrium. When small sections of a component are removed this equilibrium of internal stresses is disrupted, causing distortions in the geometry of the component.

7.5.1 Definition of Orientation Functions

If one considers polymer molecules in their random coiled configuration, the vector $\bar{r}$, representing the mean length of a polymer molecule, forms equal angles with the reference cartesian coordinates. From statistical considerations of the configuration of polymer chains, it can be proved that

$$\frac{\bar{x}^2}{\bar{r}^2} = \frac{\bar{y}^2}{\bar{r}^2} = \frac{\bar{z}^2}{\bar{r}^2} = \frac{1}{3}$$

where $\bar{x}$, $\bar{y}$ and $\bar{z}$ are the projections of vector $\bar{r}$ on each of the three cartesian axes.[4] Hence the vector $\bar{r}$ can be taken to represent the 'unoriented or isotropic' state.

Obviously, this argument applies equally well to the case of crystalline polymers, since one is considering average configurations or molecular positions on a macroscale. When the molecules, on the other hand, are being stretched in one or two directions so that projections on each axis are no longer the same, the deviations of these projections from those for the isotropic state can be taken as a measure of the orientation of the molecules. (See Fig. 7.11.)

If the vector $\bar{r}$ (representing the axis of symmetry for the average position of the individual units of polymer molecules in space) lies along one of the

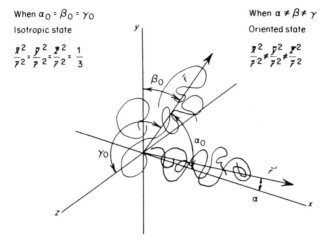

When $\alpha_0 = \beta_0 = \gamma_0$

Isotropic state

$$\frac{\bar{x}^2}{\bar{r}^2} = \frac{\bar{y}^2}{\bar{r}^2} = \frac{\bar{z}^2}{\bar{r}^2} = \frac{1}{3}$$

When $\alpha \neq \beta \neq \gamma$

Oriented state

$$\frac{\bar{x}^2}{\bar{r}^2} \neq \frac{\bar{y}^2}{\bar{r}^2} \neq \frac{\bar{z}^2}{\bar{r}^2}$$

Fig. 7.11. Change in orientation angle of polymer molecules subjected to a uniaxial stress in the x direction.

axes (say the x axis), the deviations are:

$$\Delta\left(\frac{\bar{x}^2}{\bar{r}^2}\right) = 1 - \tfrac{1}{3} = \tfrac{2}{3} \qquad \text{along the } x \text{ axis}$$

(i.e. $\cos^2 0 = 1$; $\cos^2 \alpha_0 = \tfrac{1}{3}$, or $\alpha_0 = \beta_0 = \gamma_0$)

$$\Delta\left(\frac{\bar{y}^2}{\bar{r}^2}\right) = 0 - \tfrac{1}{3} = -\tfrac{1}{3} \quad \text{along the } y \text{ axis} \qquad (7.6)$$

(i.e. $\cos^2 \pi/2 = 0$)

$$\Delta\left(\frac{\bar{z}^2}{\bar{r}^2}\right) = 0 - \tfrac{1}{3} = -\tfrac{1}{3} \quad \text{along the } z \text{ axis}$$

If the vector $\bar{r}$ lies in one of the planes (the xy plane, for instance) and forms equal angles with each of the two axes that define the plane, the deviations of the projections are:

$$\Delta\left|\frac{\bar{x}^2}{\bar{r}^2}\right| = \Delta\left|\frac{\bar{y}^2}{\bar{r}^2}\right| = \tfrac{1}{2} - \tfrac{1}{3} = \tfrac{1}{6} \qquad \text{along the } x \text{ and } y \text{ axes}$$

(i.e. $\cos^2 \pi/4 = \tfrac{1}{2}$; $\cos^2 \alpha_0 = \tfrac{1}{3}$) and

$$\Delta\left|\frac{\bar{z}^2}{\bar{r}^2}\right| = 0 - \tfrac{1}{3} = -\tfrac{1}{3} \qquad \text{along the } z \text{ axis} \qquad (7.7)$$

The deviations will be intermediate to those in (7.6) and (7.7) above in all other cases and will be related to the average angle that the mean vector $\bar{r}$ forms with each of the respective reference cartesian axes.

Hence, if

$$\frac{\bar{x}^2}{\bar{r}^2} = \overline{\cos}^2 \alpha \qquad \frac{\bar{y}^2}{\bar{r}^2} = \overline{\cos}^2 \beta \qquad \frac{\bar{z}^2}{\bar{r}^2} = \overline{\cos}^2 \gamma$$

where α, β, and γ are the angles formed by the chains with each of three cartesian axes; the deviations of the square cosine from $\frac{1}{3}$ (the value for the isotropic state), which are a measure of the orientation, are

$$\Delta\left(\frac{\bar{x}^2}{\bar{r}^2}\right) = \overline{\cos}^2 \alpha - \frac{1}{3} \qquad \Delta\left(\frac{\bar{y}^2}{\bar{r}^2}\right) = \overline{\cos}^2 \beta - \frac{1}{3} \qquad \Delta\left(\frac{\bar{z}^2}{\bar{r}^2}\right) = \overline{\cos}^2 \gamma - \frac{1}{3}$$

On the other hand, if one defines the degree of anisotropy in each plane as the difference of the deviations of the projection ratios, it follows that:

$$\left.\begin{array}{l} f(x/y) = \Delta\left(\frac{\bar{x}}{\bar{r}}\right)^2 - \Delta\left(\frac{\bar{y}}{\bar{r}}\right)^2 = \overline{\cos}^2 \alpha - \frac{1}{3} - \overline{\cos}^2 \beta + \frac{1}{3} \\[2mm] \qquad\qquad\qquad\qquad\qquad = \overline{\cos}^2 \alpha - \overline{\cos}^2 \beta \\[4mm] f(x/z) = \Delta\left(\frac{\bar{x}}{\bar{r}}\right)^2 - \Delta\left(\frac{\bar{z}}{\bar{r}}\right)^2 = \overline{\cos}^2 \alpha - \overline{\cos}^2 \gamma \\[4mm] f(y/z) = \Delta\left(\frac{\bar{y}}{\bar{r}}\right)^2 - \Delta\left(\frac{\bar{z}}{\bar{r}}\right)^2 = \overline{\cos}^2 \beta - \overline{\cos}^2 \gamma \end{array}\right\} \qquad (7.8)$$

where $f(x/y)$, $f(x/z)$ and $f(y/z)$ are called the 'orientation functions' of the vector $\bar{r}$.

(a) Monoaxial Orientation

When the random molecules are drawn from their isotropic state along one axis only (say the x axis), the vector $\bar{r}$ will form equal angles with the y and z axes; i.e. $\overline{\cos}^2 \beta = \overline{\cos}^2 \gamma$. Since $\overline{\cos}^2 \alpha + \overline{\cos}^2 \beta + \overline{\cos}^2 \gamma = 1$, substituting in eqns. (7.8) above we obtain:

$$f(x/y) = f(x/z) = \overline{\cos}^2 \alpha - \frac{1 - \overline{\cos}^2 \alpha}{2} = \frac{3\overline{\cos}^2 \alpha - 1}{2}$$

and

$$f(y/z) = 0$$

Hence one orientation function is sufficient to define the magnitude and direction of the molecules in their uniaxial oriented state; i.e.

$$f^* = \frac{3\overline{\cos^2}\alpha - 1}{2} \tag{7.9}$$

The function f^* varies from 0, when $\overline{\cos^2}\alpha = \frac{1}{3}$, i.e. the random or unoriented state, to 1 for $\alpha = 0$, i.e. when the vector lies on the x axis, denoting the maximum orientation.

When the vector $\bar{r}$ forms an angle of 90° with respect to the x axis, it lies in the y–z plane and the function f^* assumes the value of $-\frac{1}{2}$. The negative sign indicates that the change of angle is in the opposite direction, while the value $\frac{1}{2}$ corresponds to the conditions for which $\overline{\cos^2}\beta = \overline{\cos^2}\gamma$ and $\overline{\cos^2}\alpha = 0$; i.e. the conditions of balanced biaxial orientation in the y–z plane. (See later.)

(b) Biaxial Orientation

When the vector $\bar{r}$ is stretched simultaneously in two directions (say, in the x–y plane), the three angles α, β, and γ are different, except in the limiting conditions when the vector lies in one of the planes and forms equal angles with the axes that define the plane. From the definition of orientation functions in (7.8) above, it follows that

$$f(x/y) = \overline{\cos^2}\alpha - \overline{\cos^2}\beta$$

$$f(x/z) = 2\overline{\cos^2}\alpha + \overline{\cos^2}\beta - 1$$

$$f(y/z) = 2\overline{\cos^2}\beta + \overline{\cos^2}\alpha - 1$$

For reasons of consistency we would choose $f(x/z)$ and $f(y/z)$ as the reference orientation functions, bearing in mind that the third function, $f(x/y)$, can always be derived from the reference functions; i.e. $f(x/y) = f(x/z) - f(y/z)$. When the vector lies on the x–y plane, i.e. $\cos^2\gamma = 0$, the three functions become

$$\left.\begin{array}{l} f(x/y) = 2\cos^2\alpha - 1 \\[6pt] f(x/z) = \cos^2\alpha \\[6pt] f(y/z) = 1 - \cos^2\alpha \end{array}\right\} \tag{7.10}$$

When $\alpha = 0$ the two functions $f(x/y)$ and $f(x/z)$ are equal to 1 and $f(x/z)$ is zero, i.e. they become synonymous with the monoaxial orientation function.

7.5.2 Quantitative Relationships for the Molecular and Morphological Orientation of Polymers

The orientation or anisotropy functions of crystalline polymers can be expressed in exactly the same manner as for amorphous polymers, except that one takes the crystallographic axes as the reference vectors to define their orientation in relation to the Cartesian coordinates.

If the a, b, and c crystal axes are mutually perpendicular, i.e. the unit cell of the material is isometric, tetragonal or orthorhombic, the orientation functions have to be defined for each crystal axis. In the case of monoaxial orientation, one has to specify three orientation functions:

$$f(x/y)|_a = \tfrac{1}{2}(3\overline{\cos^2}\,\alpha_a - 1)$$

$$f(x/y)|_b = \tfrac{1}{2}(3\overline{\cos^2}\,\alpha_b - 1)$$

$$f(x/y)|_c = \tfrac{1}{2}(3\overline{\cos^2}\,\alpha_c - 1)$$

However, since $\overline{\cos^2}\,\alpha_a + \overline{\cos^2}\,\alpha_b + \overline{\cos^2}\,\alpha_c = 1$, the sum of the three orientation functions is zero.

Hence for uniaxially drawn materials only two functions are required to characterise the orientation or anisotropy state of the crystals, while for biaxial orientations *four* functions are necessary. It has been suggested[7] that the same treatment can also be used for crystals whose crystallographic axes are not perpendicular. However, since the orthogonality conditions no longer exist, the relationship between the various functions becomes more complex.

Crystalline polymers also contain an amorphous phase which must be characterised by its orientation functions in addition to those for its crystal orientation. Such a complete characterisation would be rather tedious, and probably inaccurate. It would, therefore, be much easier to measure overall orientation functions, i.e. average functions weighted by the respective volumetric fractions of the two phases.

For the crystal orientation function one would choose the orientation corresponding to the crystal axis which is parallel to the chain fold direction (i.e. along the major chain direction), since this is the most likely orientation to have an effect additional to that of the amorphous orientation on mechanical properties.

In the case of polypropylene, for instance, one would choose the c axis, since this is parallel to the helical axis of the isotactic molecules in the unit cell,[7] hence,

$$\bar{f} = \phi_{\mathrm{cry}} f(x/y)|_c + (1 - \phi_{\mathrm{cry}}) f(x/y)|_{\mathrm{am}}$$

where ϕ_{cry} is the volumetric fraction of the crystal phase which can be easily estimated from density measurements of the unoriented material.

Obviously, if fibrous fillers are also present there is an additional orientation function to take into account.

7.5.3 Relationships between Orientation and Mechanical Properties

Although theories have been advanced to relate molecular orientations directly to mechanical properties, the numerous uncertainties and computation difficulties involved make such an approach to prediction of properties rather unattractive in practice. Consequently, one normally relies on estimates of the molecular orientation through studies of the anisotropy of some other physical properties (e.g. optical birefringence, infrared dichroism, X-ray diffraction, thermal retraction, etc.) and relates these to some independently measured mechanical properties, either by means of empirical functions or graphically.

Within the scope of this text it is instructive to consider two simple methods used for the assessment of orientation, both of which provide estimates of the overall orientation in the case of crystalline polymers.

(a) Birefringence Methods

Birefringence is the term used to define the difference between the refractive index of the material measured in two perpendicular directions. In terms of polymer molecules stretched out in one direction it corresponds to the difference in refractive indices between the chain's axis direction and that transverse to the chain's axis, i.e. $\Delta_T = \eta_{\parallel} - \eta_{\perp}$.

Birefringence is, therefore, a measure of the anisotropy of the velocity of light arising from the difference in electron density along the chain axis of the molecules (intramolecular electron density) and between adjacent molecules (intermolecular electron density). This in turn produces a corresponding difference in induced dipoles and in polarisability in the two directions considered.

On the basis of the equation in Section 7.5.1 one can equate, therefore, the orientation functions to the ratio $\Delta_T / \Delta_T|_{max}$ since, in the case of monoaxial orientation, for instance, this varies from 1 (when all the molecules are fully oriented to *zero* (for the unoriented state).

It, obviously, becomes negative when the reference axis is rotated in the opposite direction, i.e. when there is a switch in direction of measurement relative to direction of molecular alignment; and it reaches the minimum value of $-\frac{1}{2}$ when the $\eta_{\parallel}$ is taken exactly transverse to the chain axis. In this case two equal directions would have to be considered for the measurement of the new $\eta_\parallel$ value. In other words, $\Delta_T/\Delta_T|_{max}$ becomes the same as the orientation functions defined earlier.

One can use exactly the same argument for the case of biaxial orientation, where there will be two birefringence values corresponding to the two orientation functions defined in eqn. (7.10). (See Fig. 7.12.) The $\eta_\parallel$ values correspond to those measured in the plane of the orientation, while $\eta_\perp$ corresponds to that in the direction transverse to the plane of orientation.

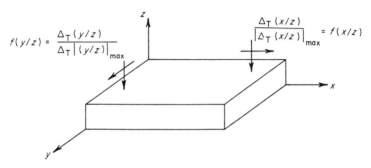

Fig. 7.12.   Birefringence of biaxial orientated materials.

In any case, $\Delta_T|_{max}$ cannot be measured directly because of the practical difficulties in obtaining fully extended chain configurations. Hence, it would have to be calculated or estimated by means of extrapolation procedures. Since neither of these two propositions is useful, birefringence measurements are not normally converted into the corresponding orientation functions, but simply used to provide an indication of the 'degree' of orientation.

This procedure can produce misleading results, however, in the case of crystalline polymers when an attempt is made to relate birefringence to mechanical properties.

Assuming that a simple additivity law exists for the birefringence arising from the orientation of the amorphous phase and that of the crystalline phase, one could write an equation for the overall birefringence as:

$$\Delta_T = \phi_{cry} \Delta_{cry}^0 f_{cry} + (1 - \phi)_{cry} \Delta_{am}^0 f_{am}$$

where ϕ_{cry} = volume function of crystals, Δ^0 = maximum birefringence, f = orientation function, cry = subscript referring to the crystalline phase, and am = subscript referring to the amorphous phase.

Samuels[8] has suggested an experimental procedure for the graphical determinations of Δ_{cry}^0 and Δ_{am}^0 and was able to differentiate the amorphous from the crystal contribution to the total measure birefringence (Fig. 7.13).

The negative amorphous contribution at low elongations indicates that the amorphous chains are oriented towards the perpendicular direction of the axial stresses used to obtain the desired elongation. Only at above 100 % elongations do the molecules in the amorphous region begin to align themselves in the draw direction. This is rather interesting insofar as a plot

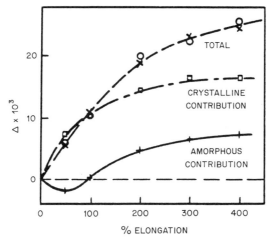

Fig. 7.13. Crystalline and amorphous contributions to the birefringence of deformed, isotactic polypropylene films: $\bigcirc$, $\Delta = \Delta_T$ (experimental); $\times$, $\Delta = \Delta_T$ (calculated); $\square$, $\Delta = \beta \Delta_c$ (calculated); $+$, $\Delta = (1 - \beta) \Delta_{am}$ (calculated). (After Samuels.[8])

of the modulus (short-time measurements) versus percentage elongation or draw ratio reveals that little change occurs up to about 100 % extension. Amorphous materials, on the other hand, show a monotonic increase right from the early stages of the drawing process.

It is instructive to compare the behaviour of polypropylene with that of natural rubber (dynamic modulus, at room temperature and above, measured while the sample is held in tension at various draw ratios)[9] as there are some similarities in the two curves at 22 °C and −20 °C, respectively. (See Figs. 7.14 and 7.15.) This seems to confirm the findings of Samuels. At even lower temperatures, natural rubber shows, in fact, a decrease in modulus, which tends to indicate that the transverse alignment

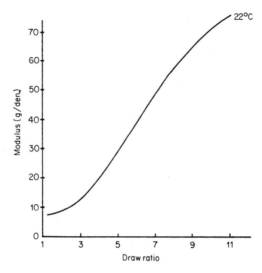

Fig. 7.14.   Modulus of polypropylene film yarns as a function of the draw ratio. (Author's unpublished work.)

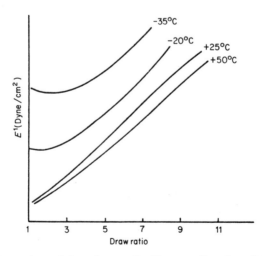

Fig. 7.15.   Dynamic modulus of natural rubber as a function of the draw ratio. (After Mason.[9])

of the molecules to the direction of drawing has a greater influence than the orientation of the crystal on the short-term modulus. At room temperature and above, on the other hand, rubber behaves like a normal amorphous material, and the modulus increases almost linearly with increasing draw ratio. One does not expect, however, to find similar anomalous behaviour in the case of biaxial orientation of crystalline polymers.

In the case of monoaxial orientation, special attention must be given to the variation of modulus with increasing draw ratio in relation to the angle formed by the applied stress with the direction of draw.

In the case of amorphous polymers it is relatively easy to perceive that the increase in modulus in the direction of the axis of alignment of the molecules (draw direction) will be appreciably greater than that measured in the perpendicular direction. Furthermore, the orthotropic plate analysis can be used to describe the variation of the modulus as a function of the angle formed by the direction of the stress to that of the molecular alignment.

Such behaviour is maintained at all levels of draw ratios; and, naturally, the degree of anisotropy (i.e. $E_{\parallel}/E_{\perp}$) increases with the extent of draw.

For the case of crystalline polymers, on the other hand, it is not a simple matter to predict the variation of modulus with draw ratio and the angular dependence of the modulus, in view of the complexity of the orientation state of the two phases and its relationship to the draw ratio. Raumann and Saunders,[10] for instance, have obtained very high values of the modulus in the direction transverse to the draw direction for low density polyethylene sheets stretched to relatively small draw ratios (4·6:1). At very low draw ratios (2·4:1) the modulus in the transverse direction is even higher than that in the direction of draw. The striking difference between the effects of orientation on crystalline and amorphous polymers is shown in Fig. 7.16.

The situation with respect to strength and toughness is different insofar as both crystalline and amorphous polymers show overall similarities in their behaviour; i.e. in the transverse direction, strength and toughness show definite decreases up to fairly low draw ratios (or levels of birefringence), while in the axial direction these properties show a substantial and monotonic increase up to fairly high orientation levels. This increase is accompanied by a gentle reduction in the steepness of the curve as the drawability limits of the material are approached.

Owing to the disparity in orientation direction between the amorphous and crystalline phases at low draw ratios, the crystalline polymers exhibit much milder changes at these low levels of orientation (Fig. 7.17).

It is instructive to note that often tensile strength measurements on both

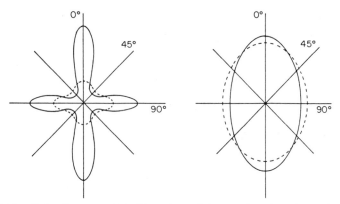

Fig. 7.16.   Polar diagrams of the Young's modulus as a function of draw ratios for crystalline and amorphous polymers respectively. Left: low density polyethylene. Right: polymethyl methacrylate. – – – Draw ratio = 2·4, —— draw ratio = 4·6. (After Raumann and Saunders.[10])

crystalline and amorphous polymers show a maximum at fairly high draw ratios. This is associated with the fibrillar mode of fracture which probably arises from the fact that the crystalline phase densifies to a greater extent than the amorphous phase. As a result, fracture of these fibrils occurs in a stepwise fashion, leaving each consecutive surviving structural element to carry the applied load. Hence the anomaly is not in the actual material property but in the manner in which the strength is calculated (i.e. merely using the initial cross-sectional area value in the calculation of the stress to fracture, $\hat{\sigma}$).

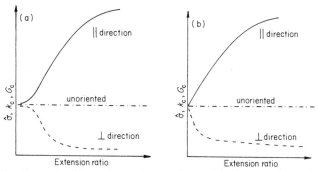

Fig. 7.17.   Strength ($\hat{\sigma}$) and toughness (k_c = critical stress intensity factor; G_c = critical strain energy release rate) of uniaxially drawn materials. (a) Typical crystalline polymer. (b) Typical amorphous polymer.

(b) Shrinkage Methods

When orientation is introduced by means of axial stresses (tensile stresses), the overall state of orientation can be deduced with good approximation from the draw ratio (since plastic deformations are totally recoverable). However, under complex stresses, particularly when these vary considerably over the entire component, as is frequently the case in processing, it is not always possible to map the state of draw or 'global orientations' of the component by the normal 'graphic' methods, e.g. by examining distortions and other geometrical changes of grids painted onto the component and/or traces incorporated in the bulk of the material. Furthermore, there could be numerous immediate, localised retractions, especially at low draw ratios which would create appreciable error in the estimates.

Thermal retraction, on the other hand, will provide accurate estimates of the orientation state of the material. Measurements are often made easier if the specimen is irradiated by γ-rays or with an electron beam prior to subjecting it to a source of heat.

7.6 THE MELT STATE

The melt is a state of the polymer in which the action of an external stress produces flow. This requires translational motions of complete molecules, which can only take place through recoiling of molecular chains from the oriented configurations acquired in the initial steps of the deformation.

On the basis of the discussions about the rubbery state it is expected that, if the intermolecular forces are negligible, the only competitive processes resulting from the action of the stresses will be the uncoiling of molecular chains and the translation of entire molecules.

In order to explain the occurrence of orientation, the theory has to allow for the existence of factors which hinder the molecular translation process. These restrictions are normally known as 'entanglements', owing to the early visualisations of polymer molecules as bundles.

The true physical explanation of this phenomenon is not known as yet but could be considered to consist of 'dispersive' interactions or momentary associations between portions of molecular chains, as in a liquid composed of small molecules, atoms or ions.

As the molecules become stretched out, the length of chain segments involved in such interactions grows increasingly large; so that the stress must increase in order to produce further molecular extensions. However, as this process continues, some local perturbations result from the vibrational energy of the molecules, especially at the chain ends, where less steric hindrance occurs. These perturbations initiate the recoiling process, which is favoured thermodynamically by the decrease in entropic energy (configurational) which the system seeks to achieve. Consequently if the stress increases at a much lower rate than that at which the recoiling process can take place, flow will result.

Under these conditions the material is in a 'viscous state' since at any two consecutive instants the molecules are always in their coiled configurations, but in a different position. Note, therefore, the contrast in deformation mechanism between the rubbery state and the viscous state. If, however, the stress increases at a much higher rate than that of the molecular recoiling process, at any given instant the molecular chains are stretched out by a certain amount; i.e. the extent of molecular orientation observed at any given time is a function of the ratio of the two concomitant rate processes.

Similarly under the influence of a step input of the excitation stress, i.e. when applied instantaneously and kept constant, the configuration and position of the molecular chains are dependent on the time at which the observation is made. In other words, each subsequent observation reveals an increased amount of recoiling until the random coiled configuration is reached.

Consequently this situation is characterised by features of both the rubbery state and the viscous state, hence the melt state represents the 'rubbery–viscous' state of the polymer.

Obviously, until the viscous state is reached, the polymer still retains some characteristics of solids; i.e. under the influence of directional stresses some energy remains stored in the material as potential energy, which corresponds to the loss of configuration entropy. This property of melts can be characterised in terms of a modulus or compliance, which will be discussed in a later section.

Since the influence of temperature is self-explanatory (see Figs. 7.1 and 7.2), it is instructive to enquire whether the nature of the stress may have an effect on the behaviour of the material in this state. One can easily foresee that isotropic pressure would not promote any flow and that the resulting volumetric strain would be a function of the bulk modulus of the material, which in turn would also be related to rotational energy of the backbone chains. It is to be expected, therefore, that the hydrostatic components of the stresses would hinder both the uncoiling of the molecular chains and their subsequent recoiling; i.e. there would be a decrease in the compliance and an increase in the viscosity of the polymer melt. This effect can be considered to be equivalent to a displacement of the curve in Fig. 7.2 to lower temperatures.

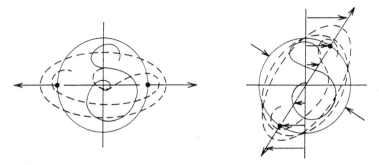

Fig. 7.18.   Movements of a polymer molecule in tensile (left) and simple shear
(right) fields.

An analysis of the behaviour in simple shear compared to that under the
influence of tensile stresses can also be quite instructive (Fig. 7.18).

When a polymer melt is subjected to a tensile field, the molecular chains
stretch out readily, while their recoiling is hindered by the neighbouring
molecules moving into closer proximity to one another. Compare, for
instance, the relative distances of planes parallel to the equatorial plane in
the deformation of a sphere to an ellipsoid (Fig. 7.19).

Under the influence of shear stresses (simple shear), the molecular chains
will undergo rotations and translations, so that spherical agglomerates will
become ellipsoids in the rotations from 0 to $\pi/4$ and from $-\pi/4$ to 0 but will
regain their spherical configuration in the rotation from $\pi/4$ to $\pi/2$ and from
$-\pi/2$ to $-\pi/4$ (Fig. 7.18). Although the molecular chains can still undergo
elongations, the cyclic nature of the process favours their recoiling since the
average time over which the molecular chains would be in close contact with
each other to stabilise the oriented configuration is only a fraction of that in
a tensile field.

The recoiling process can be considered to be equivalent to a stress

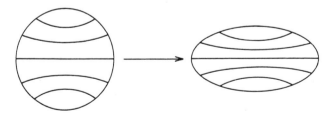

Fig. 7.19.   Deformation of a sphere in a tensile field.

relaxation process. Although the stress acting on the system may be constant during the course of a deformation, if the two simultaneous processes of molecular chain uncoiling and recoiling are isolated, one can imagine that the stress associated with the uncoiling process increases from zero to a finite value within the time interval considered, while that connected with the recoiling process decays from a finite value to zero. Thus the sum of the stresses connected with the two processes is equal to the external stress.

Having separated the two processes, one should be able to assign to them a characteristic retardation time (λ_U) and a relaxation time (λ_R). Hence it can be inferred that for a given time of observation t, the relative importance of the two processes is determined by the ratio λ_R/λ_U and that the melt state occurs for λ_R/λ_U values ranging from 1 (the viscous state) to ∞ (the rubbery state).

7.6.1 Modelling the Melt State

As pointed out earlier, in the melt state polymers acquire the characteristics of liquids; i.e. they acquire the capability of undergoing irreversible deformations or flow. However, on the basis of previous discussions the characteristic relaxation time λ_R for the recoiling of molecular chains is greater than the retardation time for their uncoiling (λ_U), so that during any time interval one observes both flow and orientation; i.e. if the stress is suddenly removed, one observes partial recovery of the imposed total deformation.

Consequently, the characteristics of a polymer in the melt state have to be described both in terms of viscosity (properties of liquids) and their 'recovery' compliance (properties of a solid); i.e.

$$J_{(rec)} = \frac{\varepsilon_{(recovery)}}{\sigma_{(input)}} \tag{7.11}$$

Note that the compliance in Fig. 7.1 refers to an arbitrarily defined total compliance; i.e. it includes both recoverable and irrecoverable strain, which obviously approaches infinity upon reaching the viscous state.

Obviously, the Voigt model used to represent the rubbery state behaviour does not satisfy the requirements of the melt state, since it does not have an element which would produce irreversible flow.

A Maxwell element (i.e. a spring connected in series to a dash-pot), on the other hand, would show both flow and recoverable strain, but the latter process would take place instantaneously through the timeless retraction of the elastic spring.

A more accurate description of the overall behaviour, i.e. one which allows a time-dependent recovery, can be obtained by means of Voigt models in series with a dash-pot. This would also account for a strain rate response which is a function of the time of observation, i.e. the history of the deformation.

For simplicity it is preferable to assume that a single Voigt element in series with a dash-pot (i.e. the Jeffrey model, Fig. 7.20) is adequate to represent the behaviour of the material and to derive the underlying constitutive equation between strain rate ($d\varepsilon/dt$) and the excitation stress (σ_0). Given that the general constitutive equation for linear viscoelasticity reduces to:

$$\frac{1}{E_M}\left(\frac{\eta_M+\eta_v}{\eta_M}\right)\frac{d\sigma}{dt}+\frac{\sigma}{\eta_M}=\frac{d\varepsilon}{dt}+\frac{\eta_v}{E_M}\frac{d^2\varepsilon}{dt^2}$$

When the excitation stress, σ_0, is constant, the solution of the differential equation provides the desired constitutive equation,

$$\frac{d\varepsilon}{dt}=\frac{\sigma_0}{\eta_M+\eta_v}\left\{1+\frac{\eta_M}{\eta_v}\exp\left[-t\left(\frac{E_M}{\eta_M\eta_v}\right)(\eta_M+\eta_v)\right]\right\} \qquad (7.12)$$

where the various coefficients correspond to the parameters defining the behaviour of the Jeffrey model shown in Fig. 7.20.

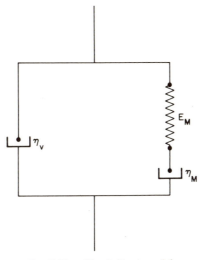

Fig. 7.20. The Jeffrey model.

According to the above equation, when t is very small

$$\frac{d\varepsilon}{dt} \simeq \frac{\sigma_0}{\eta_v}$$

and when t is very large

$$\frac{d\varepsilon}{dt} \simeq \frac{\sigma_0}{\eta_M + \eta_v}$$

Both of these are typical constitutive equations for time-independent Newtonian liquids.

The constitutive equation, (7.12), predicts, therefore, that a steady (constant) state strain rate is established only after a certain period of time, i.e. when $t \gg (\eta_M \eta_v)/[E_M(\eta_M + \eta_v)]$. Consequently it is expected that, under normal testing conditions, the viscosity (defined as the ratio of excitation stress to strain rate) will increase with time until the steady state is reached. This effect has been confirmd experimentally and is shown in Fig. 7.21 for two grades of polystyrene.[11]

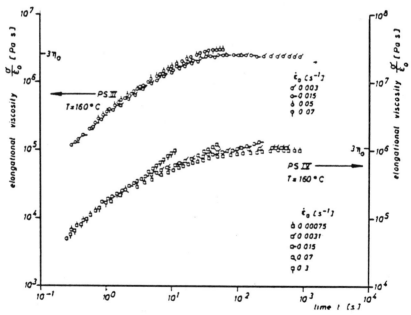

Fig. 7.21.   Time dependence of the elongational viscosities of two polystyrenes with a higher molecular weight. (After Munstedt.[11])

The Jeffrey model, however, predicts a linear behaviour; i.e. the viscosity (η_a) is independent of the excitation function σ_0. In practice one observes a linear region only at low stresses, above which there is a disproportionate increase in strain rate and, therefore, a reduction in viscosity. The stress level for the onset of the non-linearity is substantially higher for normal stresses than for shear stresses. However with tensile stresses, when the temperature is appreciably near the thermodynamic melting point of the polymer (if the polymer is crystalline), the viscosity can increase with increasing the stress (typically polypropylene). This is a phenomenon, often known as melt structure,[12] which is attributed to stress-induced crystallisation. However, the phenomenon of 'melt structure' can occur also with shear stresses under the influence of superimposed isotropic compressive stresses.

The difference between shear viscosity and tensile viscosity behaviour as a function of the stress input level is shown in Fig. 7.22 (bottom). Figure 7.22 (top), on the other hand, shows that after a certain time the steady state strain rate conditions are reached (note, in fact, the small deviations from the straight lines at short times).

The decrease in viscosity with stress level can, obviously, be explained in terms of Eyring theory of stress-activated flow, in a manner similar to the yield behaviour discussed in Section 6.1.3 which leads to the following types of constitutive equations:

$$\dot{\varepsilon} = C \sinh\left(\frac{\sigma}{A}\right) \qquad or \qquad \dot{\varepsilon} = C \sinh\left(\frac{\sigma}{A}\right) + \frac{\dot{\sigma}}{G}$$

where C, A and G are constants for a given material at a certain temperature, while the dot on the strain and stress terms denotes the usual first derivative with respect to time.

Sometimes the reduction in viscosity as a function of the excitation function, i.e. straining rate or stress rate, is viewed as a general phenomenon of linear viscoelasticity.

Using, in fact, a Maxwell model, the general linear viscoelasticity equation becomes

$$\frac{\sigma}{\eta_M} + \frac{1}{E_M}\frac{d\sigma}{dt} = \frac{d\varepsilon}{dt}$$

which for a sinusoidal stress input $\sigma = \sigma_0 \sin \omega t$ gives the solution

$$\eta_M^*(\omega) = \frac{E_M \lambda}{1 + \omega^2 \lambda^2} \qquad \text{where } \lambda = \frac{\eta_M}{E_M}$$

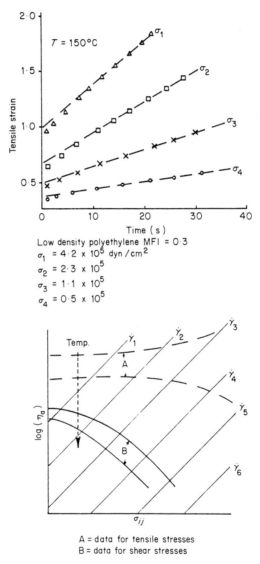

A = data for tensile stresses
B = data for shear stresses

Fig. 7.22.   Time dependence and non-linearity of the melt viscosity of polymers.
(After Cogswell.[13])

η_M^* is the 'complex' dynamic viscosity, i.e. $\eta_M^* = \eta_M' - i\eta_M''$. If it is expressed in terms of distribution of relaxation times, the equation of the viscosity as a function of the angular velocity becomes

$$\eta^*(\omega) = \int_{-\infty}^{+\infty} \frac{\lambda H(\lambda)\,d\lambda}{1 + \omega^2\lambda^2}$$

By assuming that the average strain rate is equivalent to the angular velocity of the strain output, so that $\omega = \dot{\varepsilon}/2$, then

$$\eta^*(\omega) = \int_{-\infty}^{+\infty} \frac{\lambda H(\lambda)\,d\lambda}{1 + (\dot{\varepsilon}/2\lambda)^2} \qquad (7.13)$$

and, therefore, a plot of $\log \eta^*$ versus $\log \omega$ is equivalent to that of $\log \eta$ versus $\log \dot{\varepsilon}$ (Fig. 7.23).

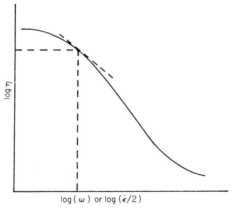

Fig. 7.23. Equivalence of angular velocity and strain rate in dynamic measurements of the viscosity.

By taking the slope of the curve at various intervals of the angular velocity or strain rate, one obtains the value of the distribution of relaxation time $H(\lambda)$ for conditions of $\lambda = 2/\dot{\varepsilon} = 1/\omega.$, i.e.

$$H(\lambda)|_{\lambda = 2/\dot{\varepsilon}} \simeq -\dot{\varepsilon}/2\,\frac{d(\log \eta)}{d(\log \dot{\varepsilon})} \qquad (7.14)$$

In dynamic tests one normally measures the amplitude ratio σ_0/ε_0 and the phase angle ϕ formed by the stress (σ) and strain (ε) vectors, as a function of the frequency (or angular velocity) of the excitation function.

From these one calculates the two dynamic moduli (see also Section 5.2.1(e)), i.e.

$$E' = \sigma_0/\varepsilon_0 \cos \phi \qquad \text{(storage modulus)}$$

$$E'' - \sigma_0/\varepsilon_0 \sin \phi \qquad \text{(loss modulus)}$$

(7.15)

By assuming that the 'losses' (i.e. the dissipated energy) occur entirely as a result of *flow*, one can define the 'real' component of the viscosity as

$$\eta' = \frac{E''}{\omega} \qquad (7.16)$$

This assumption creates, however, the dilemma of establishing whether the losses result entirely from flow, i.e. net displacement of molecules, or whether there are other contributory energy dissipation mechanisms.

Secondly, studying the viscosity as a function of the strain rate by conventional dynamic tests, where the stress and strain amplitudes are extremely small, does not provide the information required in processing, where the stress level can be very high. If, on the other hand, large values of stress and strain amplitudes are used, the non-linearity of the relationships between stress and strain rate (or vice versa) in relation to the stress, for example, creates serious problems in the estimation of the phase angle ϕ,[14] since this parameter is not constant throughout the entire period of the oscillations.

In view of the difficulty of using models to derive the constitutive equations for the flow behaviour of polymers, it is common practice to display the experimental data in graphical form, such as in Fig. 7.24.

7.6.2 Empirical Relationships for the Non-Newtonian Behaviour of Polymer Melts

When stress versus strain rate is plotted on a log–log scale, there is a region (about 1–2 decades) over which the data fit reasonably well onto a straight line, or sometimes the whole range of experimental results will fit two or three straight lines of different slopes. (See Fig. 7.24.)

This has led to the use of empirical power law relationships of the type

$$\tau = \eta_k \dot{\gamma}^n \qquad \text{and} \qquad \dot{\gamma} = m_i \tau^\beta \qquad (7.17)$$

where τ = the shear stress (input function), $\dot{\gamma}$ = the shear rate, η_k = viscosity index, m_i = fluidity index, and n and β = empirical constants, which define the deviation from Newtonian behaviour.

However, in the above plots the values of η_k and m_i represent the

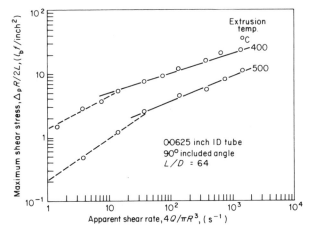

Fig. 7.24. Typical $\log \sigma$ versus $\log \dot{\varepsilon}$ of polymers for data obtained by capillary rheometer extrusion. Polymer:polystyrene melt index (ASTM D1238) = 0·53. (After Bernhardt.[15])

intercepts that cannot be obtained when using the logarithmic scale, but since at low stresses the behaviour is Newtonian, it is possible to make the following amendments.

In the low stress region there should be a portion of the curve with slope = 1, i.e. Newtonian behaviour (by definition) and, therefore, one can define a Newtonian viscosity for the material (η_0). If this region is not apparent from the experimental data available one can take η_0 as being a 'reference state' viscosity, i.e. the viscosity at $\dot{\gamma} = 1\,\text{s}^{-1}$, so that the intercept at $\log \dot{\gamma} = 0$ becomes η_0.

One can now define apparent viscosity η_a as the ratio of the shear stress to shear rate at a given shear stress or shear rate value. Substituting in the left-hand equation above (which seems to be the one preferred by rheologists) one obtains

$$\eta_a = \frac{\tau}{\dot{\gamma}} \qquad \eta_a = \eta_k \dot{\gamma}^{n-1} \qquad \text{i.e. } \eta_a = \eta_0 \left(\frac{\dot{\gamma}}{\dot{\gamma}_0}\right)^{n-1} \tag{7.18}$$

or simply

$$\eta_a = \eta_0 \dot{\gamma}^{(n-1)} \qquad \text{and} \qquad \eta_a = \eta_0 (\tau/\tau_0)^{n-1/n} \tag{7.19}$$

It is not certain, however, to what extent such power law curves are only a manifestation of the applicability of the more general sinh type of relationships within a certain range of values of the excitation function.

Furthermore, the majority of data published so far have been obtained by means of equipment in which the polymer is sheared between metallic surfaces, where one cannot guarantee that the magnitude of error is not greater than the actual effects that one is attempting to measure. The errors in this case arise from the assumption of a zero velocity at the interfaces and of a constant shear rate gradient across the thickness of the sample. When the distance between the shearing surfaces is reduced to eliminate the latter source of error, one cannot be too sure of the effect of size (i.e. the relative contribution of the bulk of the sample in relation to changes in mechanisms) on the phenomena at the interface.

7.6.3 Additional Considerations about the Viscosity of Polymer Melts

So far the problem of defining and measuring viscosity has been dealt with in general terms. For a more complete understanding of the behaviour of polymers, one must also have a knowledge of the relationship between viscosity measured in a shear field and in tension and of the effect of temperature and pressure on viscosity.

(a) Relationship between Shear Viscosity and Tensile Viscosity

For Newtonian fluids it can be demonstrated, by simple analogy to elasticity theory (where the strain tensor is replaced by the strain rate tensor) that for unidirectional laminar flow, i.e. when $\sigma_2 = \sigma_3 = 0$, $\tau_{13} = \tau_{23} = 0$,

$$\eta_{ii} = 3\eta_{ii} \tag{7.20}$$

That is, the normal stress viscosity (or Trouton viscosity) is three times the shear viscosity (or Newtonian viscosity) when incompressibility is assumed.

In other words, if $E = 2(1 + v)G$, then for conditions of incompressibility (i.e. $v = \frac{1}{2}$)

$$E = \frac{\sigma_1}{\varepsilon_1} = 3\frac{\tau_{12}}{\gamma_{12}}$$

and consequently

$$\eta_T = \frac{\sigma_1}{\dot{\varepsilon}_1} = 3\frac{\tau_{12}}{\dot{\gamma}_{12}} = 3\eta_N$$

Similarly for biaxial stresses, i.e. $\sigma_1 = \sigma_2 = \sigma^*$, $\sigma_3 = 0$, for elastic solids one has the relationship:

$$\varepsilon_1 = \frac{\sigma_1}{E} - \frac{v}{E}\sigma_2$$

which becomes

$$\varepsilon_1 = \frac{1}{2E}\,\sigma^* = \varepsilon_2 \qquad \left(\text{i.e. } \frac{\sigma^*}{\varepsilon^*} = 2E\right)$$

when $v = \frac{1}{2}$ and $\sigma_1 = \sigma_2 = \sigma^*$.
Therefore, for fluids

$$\frac{\sigma^*}{\dot{\varepsilon}^*} = \eta_b = 2\eta_T$$

where η_b is the biaxial stress viscosity. Hence,

$$\eta_b = 6\eta_N \qquad (7.21)$$

For linear viscoelastic solids the same relationship applies if it is assumed that the relaxation or retardation times under normal stresses are the same as for shear stress. Using, for instance, the equations of the viscosity for a Maxwell model, one obtains:

$$\frac{\eta_T(\dot{\varepsilon})}{\eta_N(\dot{\gamma})} = \frac{E_M\lambda_T}{1 + (\dot{\varepsilon}/2\lambda_T)^2}\,\frac{1 + (\dot{\gamma}/2\lambda_N)^2}{G_M\lambda_N} \qquad (7.22)$$

and if $\dot{\varepsilon}$ and $\dot{\gamma}$ (respectively tensile strain rate and shear strain rate) have also the same numeric value, i.e. under iso-straining rate conditions, then

$$\frac{\eta_T(\dot{\varepsilon})}{\eta_N(\dot{\gamma})} = \frac{E_M}{G_M} = 3$$

and for biaxial stresses

$$\frac{\eta_T(\dot{\varepsilon})}{\eta_N(\dot{\gamma})} = \frac{2E_M}{G_M} = 6$$

However, at a given temperature (since the rubbery state in tension is more prevalent than in shear) the value of λ_T can be much higher than λ_N. (See discussions in Section 7.4.) Consequently the ratio $\eta_T(\dot{\varepsilon})/\eta_N(\dot{\gamma})$ at equal strain rates can be substantially greater than 3 for monoaxial stresses and 6 for biaxial stresses. In practice, values of up to 30 000 have been found.[16] To what extent the non-linearity of the flow behaviour is responsible for these very high values of the viscosity ratio is difficult to say. If at any given temperature, however, the behaviour in tension is typical of the rubbery state, while in shear the polymer is already in the rubbery/viscous state, the viscosity ratio (η_T/η_N) is expected to be infinity. This statement has been formalised mathematically by Coleman and

Noll,[17] who derived an expression for the relationship between the tensile and shear viscosity by taking the so-called 'memory function' of a Maxwell model, i.e.

$$\eta_T = \frac{3\eta_N}{(1 + 2\dot{\varepsilon}\lambda)(1 - \dot{\varepsilon}\lambda)} \tag{7.23}$$

where the rubbery state for tensile deformation implies that $\dot{\varepsilon}\lambda = 1$.

Regarding the effect of the formulation constituents of a given polymer composition on the viscosity ratio, the lack of experimental data makes it impossible to provide any clear indications. One can, however, speculate that any factors that promote the so-called 'chain-entanglements' or hindrances to molecular recoiling, such as high-molecular-weight fractions, long branches and 'gel-fractions', would increase the η_T/η_N ratio. On the other hand, any formulation feature that promotes the recoiling process, e.g. addition of plasticisers or internal lubricants would have the opposite effect.

(b) Effects of Hydrostatic Stresses and Temperature on the Viscosity of Polymer Melts

The assumption that a melt is incompressible implies that the isotropic (or hydrostatic) stress component of the total state of stress has no effect on the shear and tensile viscosity coefficients of the material.

In other words, for the simple case of steady state unidirectional laminar flow, the total state of stresses acting on the melt comprises one isotropic (pressure) component and five non-zero deviatoric components, i.e.

$$\sigma_{ij} = - \begin{vmatrix} p & 0 & 0 \\ 0 & p & 0 \\ 0 & 0 & p \end{vmatrix} + \begin{vmatrix} \sigma_1' & \sigma_{12} & 0 \\ \sigma_{21} & \sigma_2' & 0 \\ 0 & 0 & \sigma_3' \end{vmatrix} \tag{7.24}$$

where the two shear components of deviatoric stresses are equal, i.e. $\sigma_{12} = \sigma_{21} = \tau_{12}$, and the sum of the three normal deviatoric stresses is zero, i.e. $\sigma_1' + \sigma_2' + \sigma_3' = 0$ (since $-p = \sigma_m = (\sigma_1 + \sigma_2 + \sigma_3)/3$). Hence the system is fully defined if the isotropic pressure, two normal stresses and the shear stress are known. The coefficients that relate each of these stresses to the corresponding deviatoric strain rate components define, respectively, the tensile viscosity and shear viscosity of the material.

The pressure does not enter into these relationships since it is assumed, from the definition of incompressibility, that it does not cause any change in volume. Incompressibility, therefore, also means that the volumetric viscosity is infinity and that flow results only from the action of the deviatoric stress components. Obviously, in practice all materials can be compressed, but the volumetric strain rates are usually sufficiently small to be neglected in the underlying constitutive equations.

In fact, measurements of the viscosity coefficients which involve pressure are normally based on a monodimensional analysis of flow through capillaries, where the excitation shear stress at the capillary wall is calculated from the pressure drop along the length of the capillary. The excess exit pressure (i.e. that in excess of the discharge pressure) is considered to be equivalent to the first normal stress difference (i.e. $-p_{ex} = \sigma_1 - \sigma_2 = \sigma_1' - \sigma_2'$) at that point.[18] Hence, this is responsible only for the recovery of the strain and cannot contribute to the flow process. On the basis of this assumption, several authors have reported values of the shear viscosity as a function of the input pressure[19] (Fig. 7.25).

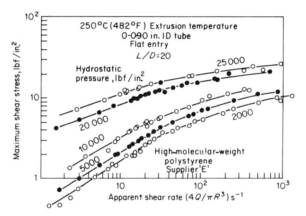

Fig. 7.25.   Effects of hydrostatic pressure on shear stress/shear rate relationship.
(After Westover.[19])

Obviously, a level of pressure is eventually reached where the shear viscosity becomes infinity (i.e. the 'melt' ceases to flow); therefore, any deformation induced will be fully recoverable. This would correspond to a shift from the rubbery/viscous state to the rubbery state.

In other words, the hydrostatic pressure has an effect equivalent to a decrease in temperature. This statement supports the observation[20] that if the volume is constant the temperature dependence of the viscosity is negligible. On the basis of this, Powell[21] has constructed log-linear plots for the viscosity ratio of several polymers as a function of the temperature difference and pressure difference, relative to some reference value. Both plots, in fact, are drawn as straight lines, which is exactly what is expected from considerations of rate processes. The 'zero' shear rate (or Newtonian) viscosity can be related to temperature and pressure through the use of the Arrhenius and Doolittle equations respectively:

Arrhenius rate equation   *Doolittle free-volume-fraction equation*

$$\eta = \eta|_{T=\infty}\, e^{\Delta H/RT} \qquad\qquad \eta = A e^{B/f}$$

where $f = (V - V_f)/V_f$, and V_f = free volume.

In other words, a decrease in temperature in the rate equation is equivalent to a reduction in the free-volume fraction, f, that is brought about by the increase in pressure in the Doolittle equation. An exact equivalence or superposition between temperature and pressure, however, would imply linear behaviour and direct proportionality between pressure (or hydrostatic stresses) and fractional free volume. In practice although linearity can be found in the viscosity/temperature relationship[22] (see Fig. 7.26) there is little indication of direct proportionality between pressure and changes in free volumes.

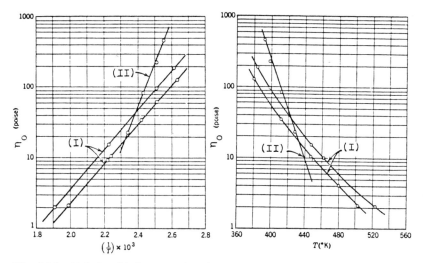

Fig. 7.26. Relationship between viscosity and temperature. Plots of log (viscosity at zero shear rate, η_0, versus (left) reciprocal absolute temperature and (right) direct temperature for (I) two grades of polyethylene and (II) polyvinyl butyral. More linear relationships are obtained using reciprocal temperature. (After McKelvey.[22])

7.6.4 Phenomena Related to Recoverable Deformations of Polymer Melts

Rubbery deformations can be analysed by means of a Voigt or Maxwell model (or a series of these). In view of the equivalence of angular velocity (in dynamic tests) to strain rate, one can describe the 'recovery' functions in terms of elastic components of the complex modulus. However, as already discussed with reference to the viscosity, the non-linearity of the relationships could render the above analyses unattractive for the solution of processability problems.

Therefore, for the case of simple unidirectional laminar flow, the current tendency is to express the recovery characteristics in terms of normal stress differences by assuming (a) that recovery of the volumetric strain resulting from the compressibility of the melt can be neglected and (b) that normal stress differences result entirely from the rubbery component.

In other words, the stress differences must be a manifestation of the potential energy stored by the molecular chains, having been forced to assume higher energy (oriented) configurations.

Consequently, since in many situations $\sigma_2 \simeq \sigma_3$, the recovery characteristics can be related to $(\sigma_1 - \sigma_2)$ and τ_{12}. The ratio $(\sigma_1 - \sigma_2)/\tau_{12} = S_R$ (i.e. the residual stresses per unit stress input) describes the recovery characteristics of the material, often referred to as the 'recoverable shear'. However, a knowledge of the normal stresses difference at a given shear stress (or shear rate) would automatically give an indication of the recovery compliance, since the two are inversely related. (This can be inferred directly from the relationship in eqn. (5.23).)

The assumption that the melt is incompressible implies also that $\sigma_1' - \sigma_2' = \sigma_1 - \sigma_2$. Therefore, $(\sigma_1 - \sigma_2)/2$ is the maximum shear stress associated with the recovery process. If the behaviour is linear, the 'recoverable shear' corresponds to the ratio of the deviatoric strain (maximum shear strain) to γ_{12} (the principal shear strain). This ratio, therefore, is a measure of the geometric distortion of an element of the melt during recovery. Since geometric distortions are associated with shear strain, the ratio $(\sigma_1 - \sigma_2)/\tau_{12}$ is also known as the 'recoverable shear strain'. (Note that a factor 2 is often included in the denominator.)

(a) Criteria for Fracture Phenomena and Flow Instabilities

For a long time it has been observed that increasing the flow rate of polymer melts through either capillaries of extrusion rheometers or the dies of processing extruders results in the formation of grossly distorted extrudates (bambooing effects), surface roughness (sharkskin and orange peel), and other low incidence surface defects (grain). The severity of these defects in terms of dimensions decreases in the order written above.[23] (See Fig. 7.27.)

Fig. 7.27.   Fracture   phenomena   in   extrusion.   Left: melt   fracture.   Right:
sharkskin.

The gross distortions are associated with die entry instabilities, which create pressure fluctuations for constant velocity extrusions or pulsating flow under constant pressure flow conditions. They are observed at some critical value of the pressure[24] input. The resulting fluctuations in either pressure or draw down force reach some characteristic amplitudes and frequencies depending on the nature of the material, on the geometric features of the die and on the temperature of the melt.

The waviness phenomenon resulting from an increase in the force at which the extrudate is pulled out of the capillary is also known as 'draw down resonance'. Sharkskin and orange peel are phenomena which originate, apparently, at the die exit at some critical velocity and increase in severity with increasing output rate. Grain, on the other hand, is associated with melt heterogeneities (microscopic entities) having relaxation characteristics that are different from those of the surrounding bulk. These appear below a certain critical 'pull down' force and cooling rate of the extrudate.

It is now well established that all these phenomena are associated with the recovery characteristics of the material as explained below.

(b) Gross Distortions and 'Melt Fracture'
Spencer and Dillon were the first to report this phenomenon.[25] It seems to result from some instability or 'turbulence' at die entry giving the impression that the melt stream has fractured and that the recovery of the outer skin strain has caused some periodic perturbation in the velocity profile. (See Fig. 7.28.)

The phenomenon of draw down resonance as a manifestation of melt fracture occurring at some critical 'pull-down force' or normal stress (Fig. 7.29) was reported by Bergonzoni and DiCresce.[26] However, the idea that viscoelastic fluids may fracture under the influence of large local concentrations of strain energy was first introduced by Reiner and Freudenthal in the 1930s.[27]

Although a variety of mechanistic interpretations have been suggested

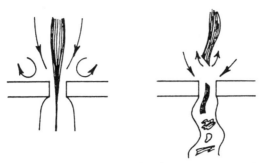

Fig. 7.28. Onset of melt fracture for cylindrical extrudates. Left: low pressure.
Right: high pressure.

since the early observations, the concept of a critical dimensionless number proposed by several other authors[28-30] seems to be the most attractive in terms of material properties. Basically this concept stipulates that for a simple laminar unidirectional shear flow, melt fracture occurs when the Weissenberg number reaches a critical value; i.e.

$$|N_{We}|_{crit} = |\bar{\lambda}(\bar{V}/R)|_{crit} \qquad (7.25)$$

where $\bar{\lambda}$ is a characteristic retardation time for the recovery component. Note that the phenomenon of unconstrained die swelling (i.e. free discharge to the atmosphere) is synonymous with the recovery of strains as a result of the sudden decay of the normal stresses; i.e.

$$\bar{\lambda} = \eta_N J_{(rec)}$$

where η_N = shear viscosity of the melt and $J_{(rec)}$ = compliance. $\bar{V}$ is the mean

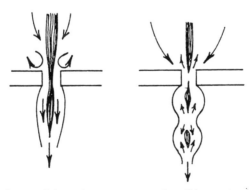

Fig. 7.29. Onset of draw down resonance for ribbon extrudate (profile).

velocity of the melt through a capillary of radius R. Substituting in (7.25) above each parameter of the respective constitutive equation, i.e. $J_{(rec)} = \gamma_{12}/\tau_{12}$, $\eta_N = \tau_{12}/\dot{\gamma}_{12}$ and $\bar{V}/R = \dot{\gamma}_{12}$, one obtains

$$|N_{We}|_{crit} = |\gamma_{12}|_{crit} \qquad (7.26)$$

since the terms τ_{12}, γ_{12} and η_N cancel out.

In other words, the critical Weissenberg number corresponds to a critical value of the shear strain, which can only correspond to the fracture shear strain if the phenomenon involves an actual fracture of the melt. This criterion is very attractive since it allows at least qualitative predictions for the occurrence of melt fracture from the behaviour of the polymer, as described previously and in Fig. 7.10. Obviously, the strain to fracture in Fig. 7.10 can have a general denomination and could easily represent the shear strain. This is reproduced in Fig. 7.30 for easy reference.

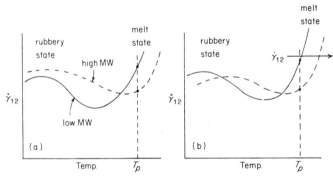

Fig. 7.30.   Factors affecting fracture strain at high temperatures. (a) Effect of structural parameters. (b) Effects of processing variables.

At a given processing temperature, T_p, within the melt state, higher-molecular-weight polymers can exhibit a lower fracture strain than lower-molecular-weight polymers, in view of the shift of the rubbery/viscous transition to higher temperatures (see Fig. 7.30(a)). This is in agreement with practical observations that higher-molecular-weight polymers are more prone to melt fracture.

Similarly, for a given polymer, increasing the shear rate at a certain processing temperature reduces the recovery strain at which fracture occurs; consequently the polymer becomes more prone to melt fracture (see Fig. 7.30(b) and later discussions about corresponding normal stresses). In any case, increasing the temperature raises the value of the strain to fracture; reducing, therefore, the susceptibility of the polymer to undergo melt fracture.

For a given length/diameter ratio (L/D) and geometry of the capillary entrance (flat entry, for example), the greater the ratio of the reservoir diameter to that of the capillary (D_R/D), the greater the acceleration of the outer particles to reach a certain velocity through the capillary. This increases the straining rate at the die entry, which automatically makes the melt more prone to exceed the critical shear strain than when the D_R/D ratio is lower. Han and Kim[31] have found that the swell ratio of the extrudate, which is related to recoverable strain components, increases in fact with increasing the D_R/D ratio, which means in effect, that the state of the polymer has been shifted more towards the rubbery state.

If the L/D ratio of the capillary is increased, on the other hand, the elapsed time for the recovery of the strain is also increased, so that, although at the die entry the value of the recoverable strain may reach the critical value (normally it cannot be expected that the entire flowing element would actually undergo fracture nor that this would occur instantaneously), the severity of the distortions observed at the exit will be reduced, owing to some loss of 'memory' by the time the melt emerges from the die. If the capillary is sufficiently long, in fact, the fracture can be expected to 'heal' completely before it reaches the exit and no distortions of the extrudate will be observed. Thus pressure fluctuations are a manifestation of the velocity discontinuities within the flowing mass only at the point at which the measurement is made, while the appearance of the fracture (or distortions) on the extrudate depends on its history from the inception of melt fracture to the time it reaches the die exit.

The arguments about the effects of the D_R/D ratio apply equally well to the observations of the die-entry angle. A conical entry would reduce the acceleration of the outer particles in their passage from the pressurised reservoir to the capillary, which will be accompanied by a concomitant reduction in shear rate on the outer surface. No observations have been reported on the effects of increasing the pressure of the melt prior to entering the capillary for a given pressure drop along the length of the capillary (i.e. by discharging the extrudate against a back pressure). However, the above arguments would intuitively predict that increasing the pressure at the die entry would raise the value of the 'wall shear stress' for the onset of the fracture. Conversely, if the extrudate were discharged against a negative pressure, i.e. by pulling the extrudate down the capillary, the phenomenon, which in this case would be associated with 'drawn down resonance', would occur at lower values of the die entry shear stress.

It is instructive at this stage to inquire whether any relationship exists between the maximum wall shear stress as a fracture criterion for melts (based on the assumption that the constitutive equation for stress as a function of strain is linear up to the point of fracture) and the Tresca criterion for yield phenomena of polymers in their glassy/rubbery state. Since σ_2 is taken to be equal to σ_3, the value of $(\sigma_1 - \sigma_2)/2$ (as already mentioned) coincides exactly with the maximum shear stress in the melt, it suggests that the two fracture criteria are equivalent. (Possibly the von Mises criterion may provide even more accurate predictions of melt fracture.) Hence it would be expected that the effects of pressure (mean stress) can be taken into account in the same manner.

Since in this case (i.e. for $\sigma_2 \simeq \sigma_3$), $\tau_{oct} = \sqrt{\frac{2}{3}}(\sigma_1 - \sigma_2)$ one can write

$$|\tau_{oct}|_{crit} = |\tau_{oct}|_0 + \mu\sigma_m$$

which predicts that the wall shear stress for the onset of melt fracture would increase with pressure.

This would, therefore, provide a rational explanation for the intuitive statement made above. Furthermore, this type of interpretation would also support the more general conception of melt fracture as a phenomenon of flow instability resulting from excessively high shear deformations in localised regions.

In comparing this criterion with that proposed earlier, one notices that the Weissenberg number concept does not take into account normal stress differences. If, however, the $(\sigma_1 - \sigma_2)/\tau_{12}$ ratio is constant up to the point of fracture, and the relationships between stresses and strains are linear up to the fracture limits, the two criteria are equivalent except for a factor of 2 which arises in converting strain terms into stresses.† This seems to be confirmed by the results of Han and Lamonte[32] for polystyrene (Fig. 7.31).

Although they reported that their results for low density polyethylene were too highly scattered to support this conclusion, Husbey and Matsuoka[33] calculated the value of the strain rate for which the recoverable strain becomes a maximum (critical value for the occurrence of melt fracture) using linear models (Fig. 7.32). Clearly, the experimental values of Han and Lamonte could fit the theoretical curve of Husbey and Matsuoka for linear PE.

† This has recently been explained by B. Hlavacek, P. Carreau and H. P. Schreiber in *Science and Technology of Polymer Processing*, Nam P. Suk and Nak-Ho Sung (eds.), MIT Press, Massachusetts, 1979, pp. 497–507.

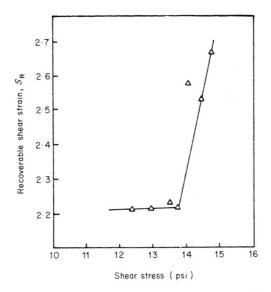

Fig. 7.31. Critical recoverable shear for polystyrene (Styron® 686 at 187°C). (Experimental values.) (After Han and Lamonte.[32])

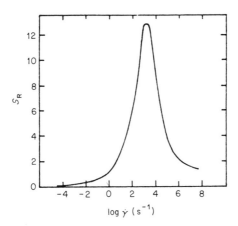

Fig. 7.32. Critical recoverable shear for a linear polyethylene. (Calculated values at 190°C.) (After Husbey and Matsuoka.[33])

(c) Sharkskin and Orange Peel Phenomena

These phenomena are believed to take place at the die exit because fracture is restricted to the surface. The surface roughness assumes regular patterns, hence the only differences between the two phenomena are the greater severity and lower frequency of the surface irregularities in sharkskin. These two phenomena have not been explained as yet, although Howells and Benbow[23] have observed that they originate at some critical flow velocity and increase in severity with an increase in output rate. However, other observations (such as the increase in critical velocity when the temperature of the die is raised, and the greater severity of these surface defects with higher-molecular-weight polymers), suggest that melt fracture, sharkskin and orange peel are part of a spectrum of the same phenomenon and that any differences arise purely from the prevailing conditions at the point at which the phenomenon occurs. In fact, the major problem mentioned in discussions of melt fracture and other surface fracture phenomena is the difficulty of separating interactions with the equipment from those resulting purely from material properties.

If sharkskin and orange peel are indeed die-exit phenomena, the only plausible explanation for their occurrence is that the material at the surface of the extrudate accelerates when emerging from the die as a result of the swelling of the extrudate, thereby rearranging the velocity profile from a parabolic type to a plug type. On the other hand, the material in the centre is expected to undergo a deceleration in the process of adjustment of velocity profile.

The acceleration of the material at the outer surface (Fig. 7.33(a)) increases the local strain rate. This in turn reduces the strain to fracture of the material at that point, while the deceleration of the material in the

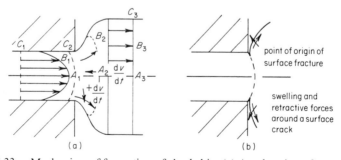

Fig. 7.33. Mechanism of formation of sharkskin. (a) Acceleration of material at the outer surface layers. (b) Surface tractions at the onset of the fracture at the die exit.

central regions has the opposite effect. A fracture might appear therefore, at the outer edge of the die, thereby producing a growing ridge (Fig. 7.33(b)) around the circumference of the extrudate as a result of the surface retractions associated with release of molecular orientation. Propagation of the fracture would be arrested until at some later instant a new fracture appears as the unrelaxed material in the capillary reaches the exit. Howells and Benbow[23] have observed that these circumferential ridges form thread-like ridges, and that their angle in relation to the extrusion direction decreases with an increase in draw down of the extrudate. The fact that these ridges can break up to form characteristic discontinuous protrusions or orange peel, indicates that some vibrations are set up at the exit, causing small undulations of the extrudate, as in the case of melt fracture phenomena.

The suggestion has often been made that surface fractures occur as a result of a slip–stick effect at the melt/wall interface, which involves a periodic breaking of the interfacial adhesion. Howells and Benbow have shown this by drawing a metal slider over a pool of silicone rubber.

This is primarily a fracture phenomenon that must be associated with a critical stress or strain; the periodicity of the event must result from the recovery of deformations in the time period between the formation of a fractured region or surface and the establishment of new critical conditions. However, the fact that increasing the temperature increases the level of the shear strain or extrusion velocity at which the phenomenon occurs, supports the critical strain criterion on the basis of the behaviour described in Fig. 7.30.

This, however, does not exclude the possibility that sharkskin, like 'melt fracture', originates from excessively high shear deformations on the surface of the extrudate at the die-exit when the $(\sigma_1 - \sigma_2)/\tau_{12}$ value reaches critical levels. Such localised deformations leave the characteristic ridge-like pattern since the strains in the regions beneath remain below the critical value. In other words, these ridges will be much smaller at all times, unlike the melt fracture and draw-down resonance phenomena which may involve large shear deformations across a major part or all of the extrudate.

7.6.5 Effects of Polymer Structure and Additives on the Rheological Characteristics of Thermoplastics in the Melt State

(a) *Effects on Melt Viscosity*
Practical experience shows that the melt viscosity of polymers correlates better with weight-average than number-average molecular weight, and it is strongly influenced by molecular-weight distribution, chain branches and

chain stiffness; the same factors have a pronounced effect also on the recovery characteristics. Much less conclusive evidence exists with respect to melt fracture characteristics. Although no systematic studies which clarify the influence of all the above parameters have been reported, some fairly conclusive guidelines have emerged from consideration of various disconnected reports.

Substantial agreement exists between molecular theories and experimental results for the relationship between melt viscosity and weight-average molecular weight, i.e.

$$\eta_N = k\bar{M}w^x$$

where k is a factor which depends on temperature, and is strongly influenced by chain stiffness of the polymer; and x is a parameter which depends on shear rate and varies from 3·4 at zero shear rate to about 1·5–2 at very high 'apparent' rates of shear. When considered as functions of temperature and shear rate, k and x seem to change in the opposite manner in relation to molecular weight distribution and chain stiffness.

In the case of flexible chain polymers (such as polyolefins) and broad molecular-weight distributions, when the temperature is increased, k decreases much less rapidly than in the case of stiff chain polymers and those with narrow molecular-weight distributions. The factor x, on the other hand, shows the opposite trend when considered as a function of shear rate. This effect is illustrated by the idealised diagrams in Fig. 7.34 and substantiated by experimental evidence given in Table 7.1.

Furthermore, Mendelson[36] finds that for polyethylene (flexible chain polymer), at a given molecular weight and molecular-weight distribution, the melt viscosity decreases with the number of long chain branches. This

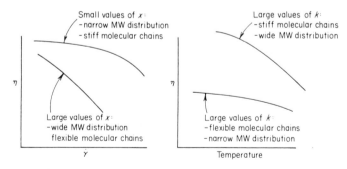

Fig. 7.34. Influence of molecular structure on viscosity characteristics of polymer melts.

Table 7.1: Typical Viscosity Characteristics of Polymers in Relation to Chain Flexibility

Material		Viscosity (poise)		$\left|\dfrac{\eta_{T_1}}{\eta_{T_2}}\right|_{\dot{\gamma}} = 10^2 s^{-1}$	$\left|\dfrac{\eta_{\dot{\gamma}_1}}{\eta_{\dot{\gamma}_2}}\right|_T$
		$\dot{\gamma}_1 = 10 s^{-1}$	$\dot{\gamma}_2 = 10^3 s^{-1}$		
Low density polyethylene MFI $\simeq 1 \cdot 07$[34]	150 °C	$4 \cdot 1 \times 10^4$	$2 \cdot 0 \times 10^3$	2·5	20
	200 °C	$1 \cdot 25 \times 10^4$	$0 \cdot 95 \times 10^3$		13
Acetyl copolymer[34]	180 °C	$7 \cdot 1 \times 10^4$	$4 \cdot 5 \times 10^3$	1·5	18
	200 °C	$3 \cdot 0 \times 10^4$	$3 \cdot 5 \times 10^3$		10
Polycarbonate[35]	250 °C	$6 \cdot 0 \times 10^3$	$3 \cdot 0 \times 10^3$	60	2
	315 °C				
Polysulphone[35]	300 °C	$1 \cdot 6 \times 10^4$	$5 \cdot 0 \times 10^3$	150	1·1
	360 °C				

suggests that the aspect ratio of a polymer molecule (assuming a cylindrical configuration for the fully stretched state) is a major contributory factor (in addition to the chain stiffness), rather than the actual molecular weight. Obviously, the more compact the molecular chains, the less the steric hindrance in the recoiling process and the lower the degree of orientation at any given strain rate.

The stress history of the polymer also seems to play a substantial role. Howells and Benbow[37] have shown that the melt viscosity of low density polyethylene is reduced considerably by repeatedly reprocessing the polymer at 190 °C. The original behaviour, however, is regained after prolonged annealing treatments at much higher temperatures or simply by 'recovering' the polymer through a dissolution process. Hanson and Cogswell[38] on the other hand, have reported a 100-fold increase in the melt viscosity of polypropylene after extrusion at 180 °C; the original behaviour was regained by means of thermal treatments at much higher temperatures (e.g. 260 °C). No such effects have been reported to date in respect of amorphous polymers.

It seems, therefore, that melt viscosity is also influenced by the 'morphological' structure of the polymer. It is possible, for instance, that in the case of polypropylene the crystal nuclei are not destroyed until very high temperatures are reached and that they grow to sub-spherulitic sizes under the influence of the stresses encountered during processing. This phenomenon would subsequently create hindrances or otherwise raise the energy barriers for the molecular recoiling process. The results for polyethylene, on the other hand, were obtained at temperatures considerably higher than the melting point of the polymer; consequently the effect of repeatedly reprocessing the polymer was the gradual destruction of these sub-spherulitic nuclei.

To what extent the difference in the behaviour discussed above hinges on the difference in pressures and environments used in the polymerisation processes (i.e. high pressure bulk polymerisation for low density polyethylene, and very low pressure in presence of diluents for polypropylene) has not yet been resolved.

With regard to the influence of the additives of a given thermoplastics material, it is quite clear (not taking into account colateral effects of morphology) that the additives likely to have a major effect are plasticisers, processing aids (high temperature plasticisers), fillers and rubbery 'gel' particles as in an ABS composition. Fillers seem to enhance substantially the viscosity at low shear rates, depending on particle size and aspect ratio; i.e. the larger the particles and the lower their aspect ratio, the smaller the enhancement of viscosity. At high shear rates the viscosity approaches that of the unreinforced material. This phenomenon is related to the orbits and 'flow' path described by the filler particles within the polymer matrix within a given velocity field. Possibly, the filler particles (especially those that are fibrous or cylindrical) travel in a streamline fashion, i.e. via axial displacements, above certain shear rates; while, simultaneously, the layers at the metal interface become increasingly depleted of filler particles until 'plug flow' conditions are established. At this point the behaviour is determined entirely by the constitution of the polymer matrix in the interfacial regions. The effect of plasticisers is to depress the viscosity at all shear rates, but the nature of the plasticiser affects the thermal activation energy.[39] Grafted rubbery inclusions, on the other hand, seem to exhibit a behaviour which is completely the reverse of that of plasticisers.[40]

(b) *Effects on Recovery Behaviour and Fracture Phenomena*

To a large extent the recovery component of the 'flow' of a polymer melt can be related to degree of non-Newtonian behaviour of the polymer.

In other words the smaller the power law index n, the larger the recovery compliance. Consequently the qualitative effects of molecular structure could be inferred directly from earlier discussions regarding the factor x in the viscosity/molecular weight relationship. For instance, for a given molecular weight, the wider the molecular-weight distribution and the more flexible the chains, the greater the value of the recovery compliance at any given temperature. This is in accordance with the values obtained on swell ratio measurements on polypropylene by Coen and Petraglia[41] and by normal stress measurements by Mendelson,[36] who also confirms the expected reduction in recoverable shear strain expected from an increase in the number of chain branches. The molecular weight itself, however, is also expected to increase the compliance at any given temperature by virtue of the displacement of the 'softening' transition to high temperatures, a phenomenon inferred from the discussions on recoiling relaxation time.

A similar conclusion may be drawn for the effects of plasticisers, fillers and rubbery gel inclusions.

7.7 INTERFACE PHENOMENA IN POLYMER RHEOLOGY

7.7.1 The Coefficient of Friction of Polymers
Some aspects of friction between polymers and metal surfaces have already been discussed in Chapt. 6 and its importance in processing has already been highlighted in relation to melt fracture phenomena. In this section attention is focused on the manner in which interface phenomena affect the analysis of the deformational behaviour of polymers which are contained between rigid surfaces. To do this it is convenient to recall the concept of coefficient of friction, μ, defined as

$$\mu = \frac{\tau_{ij}}{\sigma_{ij}} = \frac{\tau_{ij}}{-p + \sigma'_{ii}}$$

i.e. μ is the coefficient of proportionality between the shear stresses (the excitation function for the sliding of the two surfaces in contact) and the normal stresses that oppose the sliding process. The normal stresses are compressive in nature since the interface is a discontinuity and, therefore, cannot support tensile forces.

As for the case of bulk mechanical properties, the coefficient of friction of polymers is expected to be time-dependent and, possibly, a non-linear function of the excitation stress. Also a limiting value of the stress (i.e. the interfacial shear strength) is expected to exist. This is, in fact, what has led to the distinction between 'static' and 'dynamic' coefficient of friction.

Our knowledge of the friction behaviour of polymers has not reached the same degree of sophistication as that of bulk properties and it is, therefore, difficult to make accurate judgements.

It can be stated, however, that while the static coefficient of friction is expected to be a measure of 'an average stiffness or compliance' of the two sliding members as well as of the degree of interfacial contact, the dynamic coefficient involves more complex phenomena, such as the transfer of material from one surface to another, as well as the formation of debris. Normally, the discussions about the coefficient (dynamic) of friction have to exclude the effects of interfacial debris in order to avoid excessive computation difficulties.

Practical experience seems to indicate that the curve for the dynamic coefficient of friction of thermoplastics as a function of temperature follows a relationship similar to that of strain to fracture; moreover, the effect of sliding velocity is similar to that of strain rate. Although it could be argued that the dissipation factor of the bulk of the material (a non-fracture property) exhibits a similar type of behaviour, this only points to a qualitative relationship between bulk and interfacial dissipation mechanisms and indicates that the separation of static and dynamic coefficients of friction may be more difficult to conceive in practice than is foreseeable from theoretical considerations.

7.7.2 Effects of Additives on the Frictional Behaviour of Polymers

The main factors that affect the coefficient of friction (dynamic) of a polymer composition at low temperatures, are the external lubricants and fillers, but the order of magnitude of the effects considered may well be dependent also on the interactions with other additives. If it is visualised that the sliding of metal surfaces occurs when the interfacial layers reach their shear strength value, a lubricant such as stearic acid at room temperature would fail at about $2 \cdot 6 \text{ MN/m}^2$. This is less than $\frac{1}{10}$ of the shear strength of polyamides or PVC, for instance. Admittedly, non-polar polymers such as polyolefins may show only a limited adhesion towards metals; therefore, shear failure could occur at even lower shear stress values. Fillers, on the other hand, have a much higher shear strength than the polymer (possibly 100 times greater) and are prone to form debris; hence the coefficient of friction would be increased over that of the virgin polymer.

Arguments similar to the above apply also for the rubbery state, except that if a lubricant is present, this may melt and reduce the coefficient of friction to such an extent that it may even assume values lower than in the glassy/rubbery state. In other words, the behaviour may change from a 'pseudo boundary' lubrication to a 'hydrodynamic' lubrication.

At even higher temperatures, i.e. in the melt state of the polymer, with increasing the shear stress at the interface there can be a sudden jump in flow rate. This phenomenon is reversible, but it would seem that the value of the critical shear stress is reduced (Fig. 7.35).[42]

At these temperatures the external lubricant and other low-molecular-weight additives (e.g. plasticisers) can diffuse very rapidly at the interface

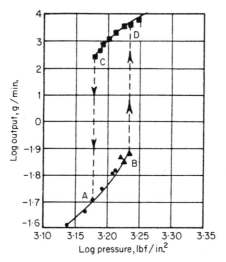

Fig. 7.35. Onset of slip at a critical pressure in the extrusion of high density polyethylene. (After Bagley *et al.*[42])

and produce a thin layer of low viscosity fluid. This will reduce the value of the critical shear stress at which the slip phenomenon takes place and even eliminate it, if hydrodynamic lubrication takes place.

To account for the effect of the lubricant it is possible to regard the flow rate as the sum of two components: a plug or slip (constant velocity) component and a shear (velocity gradient) component.

For the case of unidirectional laminar shear flow of a polymer melt between two metal plates, the true shear rate is equivalent to the velocity gradient between the velocity at the interface and the maximum velocity of the melt. If this is simulated by means of a plate shearing the polymer over a stationary surface (Fig. 7.36), we have

$$\int_{V_s}^{V_{max}} dV = \int_0^h \dot{\gamma}\,dx$$

or

$$V_{max} - V_s = \int_0^h \dot{\gamma}\,dx$$

where $\dot{\gamma} = dV/dx$ (shear rate). Hence the assumption of $V_s = 0$ can produce considerable errors in estimating the shear rate in rheometric measurements.

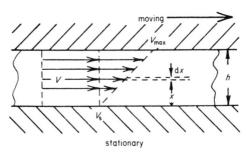

Fig. 7.36.   Slip velocity component in laminar shear flow between parallel plates.

Corrections can be made, however, through appropriate manipulation of the data. Assuming, for instance, that for a given recorded shear stress the velocity gradient is constant, i.e. independent of the gap distance h, the apparent shear rate can be plotted as a function of h (Fig. 7.37). The intercept becomes the 'true shear rate', which can be used for the computation of the viscosity, while the slope of the curves is the slip velocity. In many cases the slope is actually constant.

Note that when this procedure is used for high density polyethylene,[43] the slip velocity increases very rapidly at the critical shear stress (Fig. 7.38),

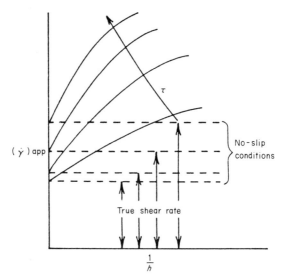

Fig. 7.37.   Principle of evaluation of the slip velocity in laminar shear flow in a rheometer.

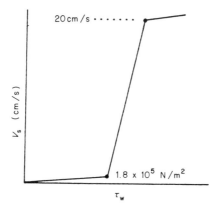

Fig. 7.38. Critical shear stress for the onset of slip flow for high density polyethylene as measured in a capillary rheometer. (Compare with data in Fig. 7.37.) (After Worth and Helmy.[43])

while for low density polyethylene the slip velocity begins to increase at much lower shear stress values but much less rapidly.

Although the procedure described above is not widely used in practice (possibly because the majority of commercially available rheometers do not permit varying of the relevant geometrical parameters, such as the gap between coaxial cylinders or the radius of the capillary), it is clear that it offers tremendous scope for the study of interfacial effects, particularly with respect to determining the efficiency of lubricants. An external lubricant will produce an increase in slip velocity, while an internal lubricant will show an increase in the apparent shear rate. It is also expected that an external lubricant will reduce the amount of die swelling at a given flow rate.

7.8   METHODS FOR THE CHARACTERISATION OF THE DEFORMATION BEHAVIOUR OF POLYMERS IN RELATION TO PROCESSING

7.8.1 Evaluation of the Yielding Characteristics by the Plane-Strain Compression Test

Although tensile tests are normally used to measure the yield strength of materials, including the effects of strain rate and temperature, these have severe limitations when data that can be related to processing are required.

First, most processes involving yielding operate under compression stress conditions (or a combination of compression and shear) and, therefore, the compressibility/dilatability of polymers does not allow the data to be used indiscriminately in all situations.

Second, on approaching the yield point, the instability caused by the formation of a 'neck region' makes it difficult to calculate the true value of the stress, since the extension ratio is not uniform and the undilatability conditions cannot be applied.

On the other hand, ordinary compression tests on cylindrical specimens are rather difficult to carry out with any degree of accuracy, in view of the problems associated with buckling, barrelling and misalignment. An effective way to overcome these difficulties is to use the 'plane-strain compression test'.[44] In this test a narrow band across the width of a rectangular specimen is compressed by narrow platens that are considerably wider than the strip (i.e. $w/b > 4$ and $h/b = \frac{1}{4}$ to $\frac{1}{2}$, Fig. 7.39).

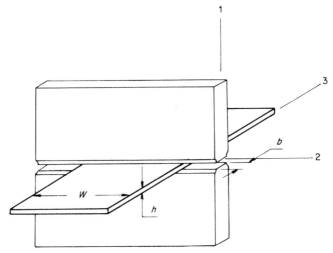

Fig. 7.39.   Loading arrangement for the plane strain compression test.

The dies and specimen surfaces are lubricated to give negligible friction, so that the specimen is not constrained to extend in the normal direction to the applied load. On the other hand, the deformed material under the die is restrained from moving parallel to the tool by the constraints imposed by the undeformed material on either side of the contact area ($\varepsilon_2 = 0$, i.e. plain strain conditions).

The stress/strain relationships for this type of loading arrangement can be obtained by assuming that the stress along the length of the sheet is very small due to the absence of any constraints resulting from the well-lubricated surface, i.e. $\sigma_3 = 0$.

Since around the yield point the material can be assumed to be undilatable, i.e. $v = 0.5$ the relationship between stress and strain can be written as[45]

$$\varepsilon_1 = \frac{\bar{\varepsilon}}{\bar{\sigma}} \left[\sigma_1 - \tfrac{1}{2}(\sigma_2 + \sigma_3) \right] \qquad (7.27)$$

$$\varepsilon_2 = \frac{\bar{\varepsilon}}{\bar{\sigma}} \left[\sigma_2 - \tfrac{1}{2}(\sigma_1 + \sigma_3) \right] \qquad (7.28)$$

$$\varepsilon_3 = \frac{\bar{\varepsilon}}{\bar{\sigma}} \left[\sigma_3 - \tfrac{1}{2}(\sigma_1 + \sigma_2) \right] \qquad (7.29)$$

where $\bar{\sigma}$ and $\bar{\varepsilon}$ are the equivalent stress and equivalent strain, as defined in eqns. (6.16) and (6.17).

From eqn. (7.28) $\sigma_2 = \tfrac{1}{2}\sigma_1$ and from the definition of $\bar{\varepsilon}$ we obtain $\varepsilon_3 = -\varepsilon_1$. Substituting in the equations defining $\bar{\sigma}$ and $\bar{\varepsilon}$ gives

$$\bar{\sigma} = \frac{\sqrt{3}}{2} \sigma_1 \qquad \text{and} \qquad \bar{\varepsilon} = \frac{2}{\sqrt{3}} \varepsilon_1 \qquad (7.30)$$

where $\varepsilon_1 = \ln h/h_0$, which can be measured directly from the displacement of the dies. Dividing the force exerted on the die by the area of contact gives the value of σ_1.

The data can be presented, therefore, as plots of $\bar{\sigma}$ versus $\bar{\varepsilon}$, from which the yielding conditions can be easily identified since the value of $\bar{\varepsilon}$ increases very rapidly for very small increments in $\bar{\sigma}$.

7.8.2 Evaluation of the Extensional Properties of Polymers in the Rubbery State and in the Melt State

(a) Uniaxial Tensile Tests

Although several authors have devised test methods for the measurement of the rheological properties of polymers under uniaxial tension,[46–48] the simplest apparatus has been developed by Cogswell.[49]

The specimen is submerged in an oil bath, fixed at one end and pulled at the other extremity under the action of a dead weight linked to the specimen clamp by means of a chain. The chain runs over a spiral-shaped cam, profiled in such a way so that the stress remains constant while the specimens are being stretched. In other words the cam acts as a variable-arm lever whose mechanical advantage (the force F acting on the specimen) changes inversely proportional to the length of the specimen, i.e.

$$F = F_0 \frac{L_0}{L} \tag{7.30}$$

Since the material deforms at approximately constant volume it follows that $L/L_0 = A_0/A$, which can be substituted in (7.30) to give $F/A = F_0/A_0$.

In this way the elongation of the specimen can be measured as a function of time for a series of stress inputs, from which the compliance and viscosity can be calculated. Retraction experiments can also be performed, from which the recovery compliance can be calculated using eqn. (5.9) in Chapt. 5. The rubbery state will be identified by conditions under which the strain recovers completely when the stress is removed.

(b) Biaxial Extensional Tests

Symmetric biaxial tensile tests are carried out on sheets clamped between two plates, both of which contain circular concentric holes. An inflation medium, gas or liquid, is introduced under pressure to a chamber mounted on one side of the plate.[50] The pressure difference, Δp, between the two sides of the sheet and the deformation of the sheet are monitored as functions of time, from which viscosity and compliance can be calculated.

The biaxial extension stress σ_b is obtained from classical shell analysis, i.e.

$$\sigma_b = R \frac{\Delta p}{2B} \tag{7.31}$$

where R is the radius of curvature of the shell and B is the thickness of the sheet.

The biaxial stretch ratio, λ_b, is related to the change in arc length from the pole to the edge of the sheet at the clamp, i.e.

$$\lambda_b = 2\alpha R/D \tag{7.32}$$

where D is the diameter of the hole in the clamping plate and is defined in Fig. 7.40.

According to the analysis by Rhi-Sausi and Dealy[50] the definition of the sine function can be used to show that

$$\lambda_b = \frac{\alpha}{\sin \alpha} \tag{7.33}$$

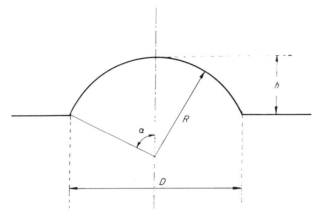

Fig. 7.40.   Parameters used in the analysis of the deformations of a sheet by inflation.

which can be calculated from measurements of the height of the bubble through the following trigonometric relationships.

$$h = R(1 - \cos \alpha)$$

and

$$D/2R = \sin \alpha \qquad (7.34)$$

Assuming that the deformations occur at constant volume the change in thickness is obtained from the following:[50]

$$B = B_0 \left(\frac{D^2}{8Rh} \right)$$

which can be used to calculate the stress from eqn. (7.31). Since the clamping plates prevent the sheet from being stretched uniformly over the entire surface only the regions near the pole may be considered to be subjected to symmetric biaxial extension (Fig. 7.41).

The stretch ratio in the longitudinal direction is given by

$$\lambda_{\parallel} = \frac{2\beta r}{d_0} = \frac{d\beta}{d_0 \sin \beta}$$

while the stretch ratio in the lateral direction ($\lambda_{\perp}$) is simply d/d_0.

If β is small the two extension ratios are equal since $\sin \beta \simeq \beta$ and, therefore, the extension can be considered to be biaxially symmetrical.

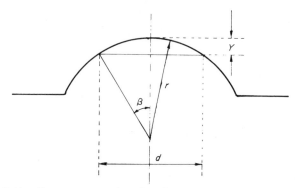

Fig. 7.41. Parameters used to describe the deformations at the pole.

Under these conditions it follows also that the change in thickness is related directly to the change in the diameter d, i.e.

$$B = B_0 \left(\frac{d_0}{d} \right)^2$$

By inflating the sheet with a liquid the hydraulic pressure can be used to calculate the stress, while the displacement of the piston that transfers the fluid to the chamber can be used to calculate the strain. The relationship between the volume displaced by the piston and the biaxial extension ratio at the pole of the bubble is obtained as follows.

The volume of the bubble is

$$V = \frac{\pi h}{6} \left(\frac{3D^2}{4} + h^2 \right)$$

which can be related to α and, therefore, to λ_D through the equations in (7.34), i.e.

$$V = XA_p = \frac{\pi D^3}{48} \left[3 \left(\frac{1 - \cos \alpha}{\sin \alpha} \right) + \left(\frac{1 - \cos \alpha}{\sin \alpha} \right)^3 \right] \qquad (7.35)$$

where X and A_p are the displacement and area of the piston respectively.

Equation (7.35) can also be written as a function of λ_b, through eqn. (7.33); i.e.

$$V_{(\lambda)} = \frac{\pi D^3}{48} f(\lambda_b)$$

and

$$X_{(\lambda)} = \frac{\pi D^3}{48 A_p} f(\lambda_p) \qquad (7.36)$$

where $f(\lambda_b)$ can be solved for a wide range of λ_b values.[50]

Obviously the apparatus can be used to measure the extensional viscosity of polymer melts by programming the rate of movement of the piston to produce a constant strain rate. In this case the strain ε_b will be taken as the natural logarithm of λ; therefore, a constant strain rate can be arranged in such a way that any given time t, the ratio ε_b/t remains constant, where ε_b is related to the movement of the piston by (7.36).

7.8.3 Evaluation of the Properties of Polymer Melts in Laminar Shear Flow

Although several techniques have been used to characterise the deformational behaviour of polymer melts in a shear stress field, the most commonly used methods are the 'cone-and-plate (Weissenberg) viscometer' and the 'capillary rheometer'.

(a) Cone-and-Plate Viscometer

Mooney and Ewart[51] showed that when a polymer melt is continuously sheared in the gap between a cone rotating over a stationary plate, the shear stress, τ_{12}, that induces laminar flow of the melt in the rotation direction is approximately constant when the gap angle is less than $10°$.

The shear rate of the melt, $\dot{\gamma}_{12}$, is given by

$$\dot{\gamma}_{12} = \Omega/\Delta\theta \tag{7.37}$$

where Ω is the angular velocity of the cone and $\Delta\theta$ is the gap angle.

The shear stress is calculated directly from the torque, T, developed by the rotating cone to maintain a constant angular velocity, i.e.

$$T = \int_0^R 2\pi r \tau_{12} r \, \mathrm{d}r \tag{7.38}$$

where $\tau_{12} r \, \mathrm{d}r$ is the shear force acting across the gap over the distance $2\pi r$.

Integration of (7.38) produces the relationship between torque and shear stress, i.e.

$$\tau_{12} = \frac{3T}{2\pi R^3} \tag{7.39}$$

The viscosity is obtained from the ratio of the shear stress from (7.39) and the shear rate in (7.37).

The Weissenberg rheogoniometer is also equipped with a load cell to measure the thrust, F, which pushes the cone and plate apart. From the measured value of F the normal stress difference can be calculated as

$$\sigma_1 - \sigma_2 = \frac{2F}{\pi R^2} \tag{7.40}$$

and the 'recoverable shear', SR, as

$$SR = \frac{\sigma_1 - \sigma_2}{2\tau_{12}} = \frac{2FR}{3T} \tag{7.41}$$

The cone-and-plate viscometer, however, operates at low shear rates and can only provide data relevant to processes such as compression moulding and calendering.

(b) Capillary Rheometer

The capillary rheometer operates at higher shear rates than the cone-and-plate viscometer, hence the two techniques are often used to obtain data over the whole range of shear rates encountered in polymer processes involving laminar shear flow. The capillary rheometer simply consists of a heated cylinder terminated by a fine capillary die (normally cylindrical). The polymer melt is discharged through the die by the action of a spherical ball under constant gas pressure or by means of a plunger driven at constant speed by an electrical motor.

The analysis proposed by Mooney,[52] which includes the effect of slippage at the wall, is as follows.

For a fully developed laminar shear flow the pressure drop, Δ_p, over a length, L, of the capillary may be related to the wall shear stress, τ_{12}, by a force balance, i.e.

$$\Delta p \, \frac{\pi D^2}{4} = \pi D L \tau_{12}$$

(Driving force) = (Resistive force)

and, therefore,

$$\tau_{12} = \frac{\Delta p D}{4L} \tag{7.42}$$

Since the pressure in the reservoir is measured at the point of contact with the piston, the total pressure drop recorded experimentally is the sum of various terms, i.e.

$$\Delta p_{(total)} = \Delta p_{(cylinder)} + \Delta p_{(die\ entry)} + \Delta p_{(capillary)} + \Delta p_{(exit)}$$

Note that $\Delta p_{(exit)}$ is the residual pressure that is responsible for the swelling of the extrudate at the die exit (see Section 7.6.3).

Of these terms $\Delta p_{(cylinder)}$ and $\Delta p_{(exit)}$ are very small and can be neglected. The value of $\Delta p_{(die\ entry)}$, on the other hand, can be determined empirically by measuring the pressure as a function L/D, i.e. using dies of different length. The intercept of this plot corresponds to $\Delta p_{(entrance)}$ and can be used to calculate τ_{12} using eqn. (7.42). Alternatively the value of τ_{12} can be obtained directly from the slope of the curve.

The shear rate at the wall is calculated from the flow rate through the capillary, on account of the equivalence of the shear rate and velocity gradient across the capillary, i.e.

$$\dot{\gamma}_{12} = -\left(\frac{du}{dr}\right)_w$$

Since the flow rate through a cylindrical die is

$$Q = \int_0^{D/2} 2\pi r u_{(r)} \, dr \tag{7.43}$$

integration by parts gives

$$Q = |\pi r^2 u|_0^R - \pi \int_0^R r^2 \left(\frac{du}{dr}\right) dr \tag{7.44}$$

For the conditions of non-slip at the wall we have $u = 0$ at $r = R$ and, therefore, the first term of eqn. (7.44) becomes zero.

Since the shear stress is directly proportional to the radius (i.e. it increases linearly from zero at the centre of the capillary to its maximum value, τ_w, at the wall surface), transformation of the variables in eqn. (7.44) gives

$$\frac{4Q}{\pi R^3} = \frac{4}{\tau_w^3} \int_0^{\tau_w} \left(\frac{du}{dr}\right) \tau^2 \, d\tau \tag{7.45}$$

Differentiation with respect to τ_w leads to

$$-\left(\frac{du}{dr}\right)_w = \dot{\gamma}_w = \frac{3}{4}\left(\frac{4Q}{\pi R^2}\right) + \frac{1}{4}\tau_w\left[\frac{d(4Q/\pi R^3)}{d\tau_w}\right] \tag{7.46}$$

which can be transformed to give

$$-\left(\frac{du}{dr}\right) = \frac{3}{4}\left(\frac{4Q}{\pi R^3}\right) + \frac{1}{4}\left(\frac{4Q}{\pi R^3}\right)\left\{\frac{d[\log(4Q/\pi R^3)]}{d[\log(\Delta p R/2L)]}\right\} \tag{7.47}$$

Since a log–log plot of $\Delta p R/2L$ versus $4Q/\pi R^3$ yields the value of n in the power law relationship between shear stress and shear rate, the above equation can be also written as

$$-\left(\frac{du}{dr}\right)_w = \frac{3n+1}{4n} \frac{4Q}{\pi R^3} \tag{7.48}$$

where the term $(3n+1)/4n$ is the Rabinowitsch correction factor.[53]

If, on the other hand, the slip velocity at the wall, u_s, is taken into account, eqn. (7.45) becomes

$$\frac{4Q}{\pi R^3} = \frac{4}{\tau_w^3} \int_0^{\tau_w} \dot{\gamma} \tau^2 \, d\tau + \frac{4u_s}{R} \qquad (7.49)$$

In accordance with the analysis in Section 7.7.2, we can see that the slip velocity may be evaluated from the slope of the plot $4Q/\pi R^3$ versus $1/R$, at a given pressure, i.e.

$$u_s = \left[\delta \left(\frac{4Q}{\pi R^3} \right) \bigg/ \delta \left(\frac{1}{R} \right) \right]_{\tau_w} \qquad (7.50)$$

Hence the correction for the shear rate in eqn. (7.48) becomes

$$-\left(\frac{du}{dr} \right)_w = \frac{3n+1}{4n} \left(\frac{4Q}{\pi R^3} \right) - \left[4 + \frac{d \log (u_s/R)}{4d \log \left(\dfrac{\Delta p R}{2L} \right)} \right] \frac{u_s}{R} \qquad (7.51)$$

A capillary rheometer can also be fitted with a series of needle load cells along the length of the capillary so that the wall pressure (stress in the radial direction) can be recorded as a function of the distance from the die entry. The plot of the wall pressure versus distance, L, is a straight line which can be extrapolated to obtain the residual pressure at the die exit.

According to the analysis of Han et al.,[54] this corresponds to the first normal stress difference (see Section 7.6.3), i.e.

$$\sigma_1 - \sigma_2 = \tau_w \left(\frac{\delta P_{R,L}}{\delta \tau_w} \right) = -P_{(\text{exit})} \qquad (7.52)$$

Since the first normal stress difference is responsible for the swelling of the extrudate, it is to be expected that an estimation of the value of $(\sigma_1 - \sigma_2)$ can be obtained from measurements of the swell ratio ($B = D_{(\text{extrudate})}/D_{(\text{die})}$).

Both the exit pressure and die swell have been found to decay with increasing the L/D ratio of capillary and to reach a constant value (i.e. fully developed flow) for an L/D ratio in the region of 20.

The above authors were able to obtain a straight line for a semi-log plot of the die swell versus exit pressure suggesting that the following relationship should apply

$$\sigma_1 - \sigma_2 = -P_{\text{exit}} = k \exp B$$

where k is a constant that depends on the nature of the material.

In the absence of suitable instrumentation to measure the normal stress by the above method an approximation estimate of $(\sigma_1 - \sigma_2)$ can be obtained from the analysis of Metzner et al.[55]

I.e.

$$\sigma_1 - \sigma_2 = \frac{\rho D^2}{64n} \left(\frac{8u}{D}\right)^2 \left\{ (n+1)\frac{3n+1}{2n+1} - B^2 \left[n + 1 + \frac{d \log B}{d \log \left(\frac{8u}{D}\right)} \right] \right\}$$

where ρ = density, D = capillary diameter, n = power law index, u = velocity at the die exit, and B = swell ratio.

REFERENCES

1. R. N. HAWARD, *The Physics of Glassy Polymers*, Applied Science, London, 1973, Chapts. 5 and 6.
2. H. EYRING, *J. Can. Phys.*, **4** (1936) 283.
3. P. I. VINCENT, F. M. WILLMOUTH and A. J. COBBOLD, *Second International Conference, Yield, Deformation and Fracture of Polymers*, Cambridge, England, March 26–29 1975.
4. J. J. AKLONIS, W. J. McKNIGHT and M. SHEN, *Introduction to Polymer Viscoelasticity*, Wiley Interscience, New York, 1972, p. 42.
5. A. V. TOBOLSKI, *Properties and Structure of Polymers*, Wiley, New York, 1960, p. 74.
6. A. E. GREEN and J. E. ADKINS, *Large Elastic Deformations of Non-Linear Continuum Mechanics*, Clarendon Press, Oxford, 1960.
7. T. L. SMITH, *J. Polym. Sci.*, **32** (1955) 99.
8. R. G. SAMUELS, in *The Science and Technology of Polymer Films*, O. J. Sweeting (ed.), Wiley Interscience, New York, 1968, pp. 272–87.
9. P. MASON, *J. Appl. Polym. Sci.*, **5** (1961) 428.
10. G. RAUMANN and D. W. SAUNDERS, *Proc. Phys. Soc.*, **77** (1961) 1028.
11. H. MUNSTEDT, *J. Rheol.*, **24** (1980) 854.
12. P. LAMB and F. N. COGSWELL, *International Plastics Congress*, Amsterdam, 1966.
13. F. N. COGSWELL, *Trans. Plast. Inst.*, **36** (1968) 110.
14. J. M. DEALY, *Polymer Rheology and Plastics Processing Conference*, Loughborough, Reprints, 17–19 September 1975, pp. 35–48.
15. E. C. BERNHARDT, *Processing of Thermoplastic Materials*, Krieger, New York, 1959.
16. A. B. METZNER and A. P. METZNER, *Rheol. Acta.*, **9** (1970) 174.
17. B. D. COLEMAN and W. NOLL, *Arch. Rat. Mech. Anal.*, **6** (1960) 355.
18. C. D. HAN and W. PHILIPPOFF, *Trans. Soc. Rheol.*, **14** (1970) 393; **13** (1969) 455.
19. R. F. WESTOVER, *Polym. Eng. Sci.*, **6** (1966) 83.
20. J. A. BRYDSON, *Flow Properties of Polymer Melts*, Iliffe, London, 1970, p. 51.

346   THERMOPLASTICS: MATERIALS ENGINEERING

21. R. M. ORGORKIEWICZ (ed.), *Thermoplastics: Properties and Design*, Wiley, London, 1974, pp. 185 and 187.
22. J. MCKELVEY, *Polymer Processing*, Wiley, New York, 1957, p. 42.
23. E. R. HOWELLS and J. J. BENBOW, *Trans. Plast. Inst.*, **30** (1962) 247-9.
24. C. D. HAN and R. R. LAMONTE, *Polym. Eng. Sci.*, **11** (1971) 381.
25. R. S. SPENCER and R. E. DILLON, *J. Coll. Sci.*, **3** (1948) 163.
26. A. BERGONZONI and A. J. DiCRESCE, *Polym. Eng. Sci.*, **6** (1966) 45-59.
27. M. REINER and A. M. FREUDENTHAL, *Proc. Fifth Intern. Cong. Appl. Mech.*, 1938, p. 228.
28. D. C. BOGUE and J. O. DOUGHTY, *Ind. Eng. Chem. Fundamentals*, **V** (1966) 243.
29. N. TOKITA and J. L. WHITE, *J. Appl. Polym. Sci.*, **11** (1967) 321.
30. A. B. METZNER, J. L. WHITE and M. M. DENN, *AIChE J.*, **XII** (1968) 863.
31. C. D. HAN and K. J. KIM, *Polym. Eng. Sci.*, **11** (1971) 396.
32. C. D. HAN and R. R. LAMONTE, *Polym. Eng. Sci.*, **10** (1970) 385-94.
33. T. W. HUSBEY and S. MATSUOKA, *Mat. Sci. Eng.*, **1** (1967) 335.
34. Results from ICI Ltd, Personal communication.
35. H. D. BASSETT, A. M. FAZZARI and R. B. STAUB, *Plast. Tech.*, **11** (1965) 50.
36. R. A. MENDELSON, *Polym. Eng. Sci.*, **9** (1969) 350-5.
37. E. R. HOWELLS and J. J. BENBOW, *Trans. Plast. Inst.*, **30** (1962) 253.
38. D. E. HANSON and F. N. COGSWELL, *Polymer Rheology and Plastics Processing Conference, Loughborough, Reprints*, 17-19 September 1975, pp. 244-50.
39. H. P. SCHREIBER, *Polym. Eng. Sci.*, **9** (1969) 311-18.
40. A. CASALE, A. MORONI and C. SPREAFICO, *Advances in Chemistry*, Series 142, Am. Chem. Soc., 1975, pp. 172-84.
41. A. COEN and G. PETRAGLIA, *Materie Plastiche ed Elastomeri*, **31** (1965) 1270.
42. E. B. BAGLEY, I. M. CABBOTT and D. C. WEST, *J. Appl. Phys.*, **29** (1958) 109.
43. R. A. WORTH and H. A. A. HELMY, *Polymer Rheology and Plastics Processing Conference, Loughborough*, 1975, p. 157.
44. A. B. WATTS and H. FORD, *Proc. Inst. Mech. Eng.*, **69** (1955) 1141.
45. J. G. WILLIAMS and H. FORD, *J. Mech. Eng. Sci.*, **6** (1964) 412.
46. H. MUNSTEDT, *J. Rheol.*, **23** (1979) 421.
47. R. L. BALLMAN, *Rheol. Acta*, **4** (1965) 137.
48. J. MEISSNER, *Rheol. Acta*, **8** (1969) 78.
49. F. N. COGSWELL, *Plastics and Polymers*, **36** (1968) 109.
50. J. RHI-SAUSI and J. M. DEALY, *Polym. Eng. Sci.*, **21** (1981) 228.
51. M. MOONEY and R. H. EWART, *Physics*, **5** (1934) 350.
52. M. MOONEY, *J. Rheol.*, **2** (1931) 210.
53. B. RABINOWITSCH, *Z. Physik. chem. (Leipzig)*, **145A** (1929) 1.
54. C. D. HAN, M. CHARLES and W. PHILIPPOFF, *Trans. Soc. Rheol.*, **13** (1969) 455.
55. A. B. METZNER, W. T. HOUGHTON, R. A. SAILOR and J. L. WHITE, *Trans. Soc. Rheol.*, **5** (1961) 133.

8

Processing Performance of Thermoplastics in the Temperature Range between the Glassy State and Melt State

The classification of plastics processes given in page 273 can now be refined by grouping together the processes falling within the four states in which thermoplastics can undergo large deformations. In those cases where the process considered takes place in stages, each involving different deformational states, the process will be broken down accordingly.

For the analysis of the processing performance, observations of typical aspects of the process are considered in terms of the deformational characteristics described in Chapt. 7.

In some cases the transient conditions prevailing in a particular process may involve a change of state of the polymer. In these cases a particular process will be classified on the basis of its 'most favourable or prevalent' state.

Finally some examples will be given of how to quantify the processability of thermoplastics by defining a processability index which groups together the major relevant parameters.

The processability index will be a function of the type:

$$\alpha_p = f\left(\frac{\text{output, quality}}{\text{power requirement}}\right)$$

8.1 PROCESSES IN THE GLASSY/RUBBERY STATE

Within this group are included those processes which are akin to metal processing. Although many of these have not yet achieved wide commercial acceptance, the purpose of the analysis is to consider the implications of these processes for thermoplastics in terms of economic feasibility and performance potential. The analysis will include deviations from metallurgical practice to take into account the difference in behaviour between metals and thermoplastics. Processes in this group include the following:

(1)   Machining, drilling, blanking and grinding (buffing), etc.
(2)   Forging, rolling, stamping, cold (impact) extrusion, deep drawing and cold forming.
(3)   Pelleting, preforming.

8.1.1 Analysis of Machining Operations

A successful machining operation relies on the production of localised plastic deformations by point loading, followed by plane strain fracture. A simple model of two-dimensional (orthogonal) cutting is shown in Fig. 8.1. For the case of cellulose nitrate this process is used for the production of foils.

The tool is characterised by the rake angle, α, and the clearance (or relief) angle, δ. Cutting takes place after the material to be removed yields along the xy plane, which is called the 'shear plane'. The removal of material occurs by the formation of a continuous shaving, called the 'chip', the outer surface of which and of the machined component contain layers of

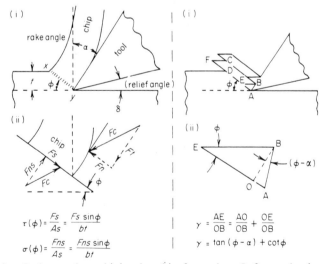

$$\tau(\phi) = \frac{Fs}{As} = \frac{Fs \sin\phi}{bt}$$

$$\sigma(\phi) = \frac{Fns}{As} = \frac{Fns \sin\phi}{bt}$$

$$\gamma = \frac{AE}{OB} = \frac{AO}{OB} + \frac{OE}{OB}$$

$$\gamma = \tan(\phi - \alpha) + \cot\phi$$

Fig. 8.1.   Orthogonal machining by chip formation. Left: mechanism of chip formation (i) and stresses on shear plane (ii). Right: shear planes (i) and calculation of shear strain (ii).

plastically deformed material. The amount of plastic deformation in these layers decreases in moving away from the cut surfaces; consequently, internal stresses are set up at some distance from the surface (see Fig. 8.2), since the strained material below these layers is prevented from recovering by the more stable layers deformed plastically. These could affect, therefore, certain properties, such as crazing resistance, in materials like polystyrene, polymethyl methacrylate, polycarbonates, etc.

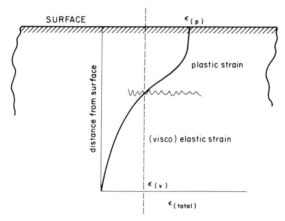

Fig. 8.2.   Strain at the surface of a machined thermoplastic surface.

In selecting machine tools for thermoplastics, it is normally recommended that the rake angle, α, be very small (sometimes even negative and always much smaller than for metals) and that the relief angle, δ, be large (typically 3–5° for metals and 15–20° for thermoplastics). Cutting speed should normally be lower than for metals.

These facts can be explained in terms of the difference in deformational behaviour between thermoplastics and metals in the following manner. Owing to the much higher yield strain of thermoplastics (at actual cutting temperatures it can reach values of 20 % or more) in comparison to metals (typically less than 0·2 %), it is essential that the shear strain in the yield plane xy be as large as possible. From geometrical considerations in Fig. 8.1, right (i), it can be deduced that the shear strain can be calculated by noting the distortion of an element ABCD to EBCF resulting from the forces exerted by the cutting tool. The shear strain γ (engineering) is the shear offset divided by the perpendicular distance between planes;[1] i.e.

$$\gamma = \tan(\phi - \alpha) + \cot\phi \qquad (8.1)$$

Although due to compressibility effects the shear plane angle ϕ for thermoplastics is expected to be somewhat lower than for metals, producing a somewhat larger value of γ, a much greater increase in shear strain is obtained by reducing α or making it negative.

The relief angle, on the other hand, has an enormous influence on tool wear and, possibly, also on the quality of the surface finish of the machined component.

From an analysis of the forces exerted by the cutting tool (see Fig. 8.1, right), it can be deduced that the normal stress (compressive) acting on the shear plane has a large component in the direction perpendicular to the surface being cut. Since plastics can undergo large viscoelastic deformations prior to reaching the yield strain, considerable recovery of the material will occur after the cut has been made. This creates compressive stresses along a substantial portion of the tool surface from the tip and causes considerable wear. Increasing the relief angle δ increases the length of the wear 'land' and decreases the amount of wear accordingly. This is particularly important when the thermoplastics material contains pigments or fillers, since the cutting tool produces debris which is trapped between the lower face of the tool and the surface of the machined part, thus increasing the amount of wear by abrasion.

The quality of the surface finish is also determined by the recovery of plastic deformations resulting largely from the temperature rise in the cutting area. Furthermore, if the material reaches the rubbery state, an increase in friction on the wear land of the tool surface and a much larger value of the fracture strain will result. Both of these factors contribute to 'slip–stick–tear' effects on the surface of the plastics material with the subsequent formation of localised asperities and roughness.

(a) *Formalisation of a 'Machinability Index'*
For metal cutting the machinability index is given by the expression[2]

$$\alpha_m = \frac{V_m}{P} \left(\frac{1}{V_w \times S_f} \right)$$

where V_m = volume of material cut per unit time, V_w = volume of tool wear per unit time, P = power consumption, and S_f = surface roughness factor.

The surface roughness factor is equivalent to the reciprocal of the 'quality' parameter given in the general formula of the processability index at the beginning of this chapter. In the case of plastics one must add another term to express the quality of the machined component, i.e. one that represents the level of residual stresses R_s. (These stresses exist also for metals but are not likely to affect the quality of the component.) Hence an expression for the machinability of thermoplastics could be cast in the following terms:

$$\alpha_m = \frac{V_m}{P} \left(\frac{1}{V_w \times S_f \times R_s} \right) \tag{8.2}$$

We can assume that V_m and P are parameters that can be evaluated only by trials carried out on the particular machining process considered, while the other terms are generalisable and can therefore be related to the deformational characteristics of the material.

From the previous discussions it can be inferred that surface roughness is mostly associated with localised deformations (tear and recovery) occurring when the material reaches the rubbery state through heating effects. Since the difference in thermal diffusivity between the various thermoplastics materials is not sufficient to produce large differences in heating rate, it can be concluded that surface roughness is inversely related to the glass transition temperature. Hence the classical engineering thermoplastics, e.g. nylons, acetals, polycarbonates, polysulphones, etc., are expected to exhibit a surface finish on machined components superior to materials such as low density polyethylene, high impact polystyrene, etc.

At the same time it should be remembered that the more brittle thermoplastics, e.g. unmodified polystyrene, polymethyl methacrylate, etc., can produce rough surfaces through brittle fractures, i.e. these materials remain in their glassy state if the rate of heat generation ahead of the cutting tool is insufficient to bring the material into the glass/rubber transitional state. Furthermore, the argument does not apply to thermosets since these are not capable of undergoing large-scale yielding; i.e. even if yielding can occur, as in the case where the cross-linking density is low, the deformations up to the point of fracture are very small.

Residual stresses and tool wear, on the other hand, are related to the recovery characteristics of the material upon removal of the load exerted by the cutting tool, i.e. after the cut has been produced.

It can be easily inferred, therefore, that a high degree of tool wear occurs if the modulus of the material is low, since this would produce a large amount of recovered strain and create friction in rubbing against the surface of the tool after the cut has been produced.

As already explained earlier the level of residual stresses on the surface of the machined component is related to the difference in strain recovered in the surface layers that have been plastically deformed, and in those beneath that have been deformed to strain levels below the yield strain and are, therefore, susceptible to full recovery. Unlike the case of metals (which exhibit an elastic/plastic behaviour and of which, therefore, the recovery of the deformation above the yield point is zero, while that below the yield point can be readily calculated from the modulus), the difference in recovered strain cannot be easily related to fundamental characteristics of the material with the present state of knowledge. We could introduce, however, an empirical factor $(\overline{\Delta\varepsilon})_{Y/E}$ in the expression for the machinability, which will give an indication of the level of residual stresses likely to be found on the surface of the machined component. Hence the machinability index of thermoplastics can be expressed as follows:

$$\alpha_m = f\left\{\frac{V_m}{P}, \left[\frac{T_g}{(\Delta\varepsilon)_{Y/E}}\right]\right\} \tag{8.3}$$

where the factor $(\Delta\varepsilon)_{Y/E}$ could be measured experimentally over a range of temperatures up to the rubbery state of the material from simple uniaxial tensile tests. The average value $(\overline{\Delta\varepsilon})_{Y/E}$ would then be used to calculate the machinability index.

8.1.2 Analysis of a Forging Process

The easiest analysis of a forging process is that for the compression of a flat circular disc (Fig. 8.3). Under these conditions $\sigma_\theta = \sigma_r$ and for an incompressible solid the conditions of yielding reduce to:

$$\hat{\sigma}_Y = \sigma_r - \sigma_z \tag{8.4}$$

By taking into account interfacial frictions and from the definition of the coefficient of friction $\mu = \tau_{zr}/\sigma_z$, an equation for the compressive stress σ_z required to produce yielding can be obtained,[3]

$$\sigma_z = \hat{\sigma}_Y \exp\left[\frac{2\mu}{h}(a-r)\right] \tag{8.5}$$

First of all one can see that since a combination of tensile and compressive stresses are acting on the material, eqn. (8.4) would have to be modified to account for the compressibility of thermoplastics, and becomes (see Chapt. 6):

$$\frac{\sqrt{2}\hat{\sigma}_Y|_c \cdot \hat{\sigma}_Y|_t}{\hat{\sigma}_Y|_c + \hat{\sigma}_Y|_t} = \frac{\sqrt{2}(\hat{\sigma}_Y|_c - \hat{\sigma}_Y|_t)3\sigma_m}{\hat{\sigma}_Y|_c + \hat{\sigma}_Y|_t} + (\sigma_r - \sigma_z) \tag{8.6}$$

This in turn produces further modifications in eqn. (8.5).

Secondly, setting $\sigma_r = \sigma_\theta$ means that the disc in its final dimension will be in a state of balanced biaxial stresses, hence the orientation function of the forged article will depend simply on the ratio a/r. Since the mechanical properties will be substantially improved over those of the undeformed disc, the forging process seems a very attractive proposition.

However, if considerable 'sticking' occurs at the tool/disc interface, so that 'fresh' material from the bulk has to feed in the circumferential areas

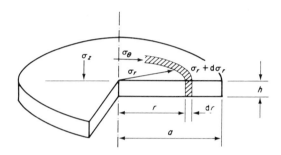

Fig. 8.3.   Stresses acting in compression on a flat circular disc.

outside the radius r, the final disc will be in a state of differential orientation along the diameters on the surface and across the thickness. This will create internal stresses, which would affect accordingly the dimensional stability of the component.

Even worse is the case where not a disc but a component of more complex geometry is deformed to acquire a geometry completely different from that of the feed stock. This produces situations of much greater differential orientation across the three 'major' axes and accentuates even more the problem of internal stress, differential recovery on removal of the stress and resultant distortions of geometry.

On the economics side, considerable doubts exist about the viability of such a process, since there has to be a preliminary process feeding into the forging operation. However, if the component has to have a considerable thickness across a major section, there is no alternative moulding process; hence forging may be preferable to machining. Bearing in mind that the feed stock may possibly be obtained by melt extrusion of solid-section rods or profiles.

8.2   PROCESSES IN THE RUBBERY STATE

Processes falling in this group can be conveniently subdivided into:

(a) *Monoaxial orientation processes*
These include the following:

(i)   monofilament and multifilament drawing processes
(ii)   drawing of nets for use in packaging
(iii)   drawing of tapes for weaving applications, strapping tapes, adhesive tapes, decorative tapes, etc.
(iv)   production of film yarns for bailer twines and ropes
(v)   production of fibres from films

(b) *Balanced biaxial orientation processes*
The major processes which introduce a balanced or pure biaxial orientation to the product are:

(i)   drawing of foils and sheets
(ii)   drawing of pipes under pressure

(c) *Unbalanced biaxial orientation processes*
In these processes are included not only those in which the orientation in the two major (plane) directions is unbalanced (i.e. $\lambda_1 \neq \lambda_2$), but also those in which the orientation produced may vary over the entire component. These include, therefore:

(i)   tubular film processes
(ii)   parison inflation for blow moulded containers (both injection and extrusion produced parisons)
(iii)   post-forming processes, i.e.

    (1)   embossing, corrugation, etc.
    (2)   shrink wrapping

(iv)   sheet forming processes, i.e.

    (1)   vacuum forming
    (2)   vacuum assisted drape forming
    (3)   plug-assisted vacuum forming
    (4)   bubble forming and air slip forming
    (5)   vacuum back-snap forming
    (6)   manual bending and other manual shaping operations
    (7)   deep drawing forging of containers

The reason why these processes are referred to as 'orientation processes', rather than as processes operating in the glassy/rubbery state, is that they represent the earliest and best-established processes involving molecular orientation.

8.2.1 Analysis of Monoaxial Orientation Processes†
The essence of this process is to draw widths of films to achieve substantial orientation in the machine direction and to develop fine splits (fiberise) by twisting the film once it has been cooled to room temperature.

A typical production line is shown in Fig. 8.4. Note, however, that the film can be extruded by either chill roll casting (as shown in Fig. 8.4) or by tubular film extrusion and that the heat setting oven is often omitted.

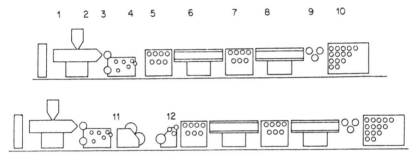

Fig. 8.4.   Production lines for stretched-film tape. Top: continuous production line for film tape. Bottom: discontinuous production lines for film and film tape. 1, Control cabinet; 2, extruder; 3, slit die; 4, chill roll and tape cutting tool; 5, septette; 6, drawing oven; 7, septette; 8, heat setting oven; 9, trio; 10, bobbin winder; 11, film winder; 12, film unrolling stand. (After Krassig.[4])

The purpose of this analysis is to consider the factors that affect the performance of this process with respect to specific requirements in order to derive an expression for the 'drawability index'.

The drawing of the film to the desired draw ratio is effected by arranging the second septet of rolls to rotate at a higher speed than the first set of rolls; the ratio of the peripheral speeds of the two sets of rolls (v_2/v_1) gives the draw (extension) ratio.

Heat setting is carried out in the second oven by allowing a certain amount of recovery (10–20%) between the drawing septet of rolls and the triplet of rolls at the exit of the second oven. The draw ratio is normally

† With special reference to the production of film yarns for bailer twines and ropes.

9–12:1, and the film yarns are cooled through contact with the metal rolls at the exit of each oven.

(a) Factors Affecting Production Rate
The output of a monoaxial orientation process can be expressed either in terms of velocity of the tapes leaving the triplet at the end of the heat setting stage (v_3), or in terms of the velocity of the second septet of rolls (v_2) when the drawing oven is omitted.

Alternatively, it can be expressed in terms of volumetric or mass rate, i.e.

$$V_m = v_3 A \qquad \text{or} \qquad Q_m = v_3 A \rho$$

where A is the total cross-sectional area of the film yarn and ρ is the density.

The maximum output (say v_3) is determined by the winding capacity of the machine on the one hand, and by maximum permissible drawing speed (i.e. without breakages) on the other. Considerations on materials processability will take into account only the latter, which is determined by extensibility limits (strain at break) of the material as a function of temperature and strain rate. At a given draw ratio the average strain rate along the length of the film is determined by the distance between the drawing rolls, which is a fixed machine parameter; hence the variation of strain at break as a function of temperature is the sole relevant parameter to consider. From the discussions in Chapt. 7, it follows that the range of temperatures affecting the extensibility (i.e. maximum draw ratio $\hat{\lambda}$) varies according to the average molecular weight of the polymer, as shown in Fig. 8.5.

Increasing the velocity v_2 at a given draw ratio reduces the residence time

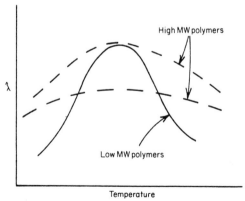

Fig. 8.5.   Effect of temperature on the extensibility of polymers.

in the oven and, therefore, the average (as well as the maximum) temperature reached by the film at a set oven temperature. As soon as the material enters the glassy/rubbery state as a result of the drop in temperature, drawing begins to take place by necking; while the shoulder of the drawn material gradually moves towards the inlet of the oven where the film temperature is lowest. This reduces the extensibility of the material and causes breakage of the tapes; moreover, if the oven temperature is increased to counteract this effect, at the exit regions of the oven the tapes will reach temperatures above those of the peak of the curve and breakages may also result (at the other end of the oven).

From such considerations it can easily be deduced that the highest production rates are obtained when the material exhibits a broad peak in the extensibility curve as a function of temperature, as in high molecular weight polymers. With crystalline polymers this is also achievable by producing extruded films with the lowest possible degree of crystallinity and, therefore, the film leaving the extruder die has to be quenched at the fastest possible cooling rate (as indicated in Fig. 8.6, for the case of polypropylene). Hence, in this respect the chill roll casting process is preferable to tubular film extrusion.

(b) Factors Affecting Quality
For the purposes of this analysis quality implies:

(1) Highest possible tenacity (tensile strength) at a given draw ratio.
(2) Highest dimensional stability of the tapes prior to their reaching the winding stage, in order to minimise the level of shrinkage stresses exerted by the tapes on the bobbins. These could in fact produce a permanent curl in the final tapes, since their inner surfaces will shrink to a lesser extent than their outer surfaces.
(3) Highest level of fiberisation in the twisting stage of the production of bailer twines. (Obviously, the opposite would be true if weaving tapes or strapping tapes were to be produced.)

All three factors seem to be largely affected by the thermal treatment to which the film is subjected in the overall drawing stage of the process, as well as, of course, by the draw ratio. It is for this reason that the majority of monoaxial orientation processes nowadays contain a heat setting stage.

The reasons why superior tenacity and dimensional stability can be achieved by suitable thermal treatments can be inferred from discussions in Chapt. 7, while the effect on fiberisation characteristics of the tapes can be interpreted as follows.

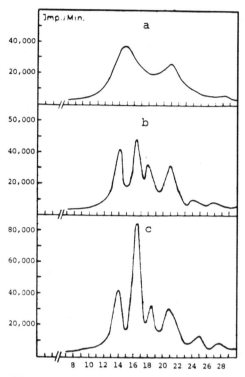

Fig. 8.6.   X-ray diffractograms of differently quenched PP film. (a) Chill roll cast (11 °C). (b) Chill roll cast (55 °C). (c) Blown. (After Krassig.[4])

Fiberisation results from the stresses acting over a wide range of angles from the drawing direction in the twisting process; these stresses produce a combination of crack-opening and shear deformations along the axis of the molecular orientation. Consequently, fiberisation must be related to the angular strength, $\hat{\sigma}_\theta$ (and fracture strain, ε_θ) of the tapes, which in turn means that crystalline polymers would exhibit fiberisation characteristics superior to those of amorphous polymers. In other words, it is both the fibrillar nature of the crystals and the inferior strength in the range of angles between 30° and 70° that create favourable conditions for the fiberisation process. However, the rigidity of the tapes also plays an important role insofar as this controls the level of stresses developed by the twisting operation; while the angular breaking strain, $\hat{\varepsilon}_\theta$, determines whether the fracture will take place and the number of splits formed for a given angle of twist. Consequently, it is to be expected that intrinsically tougher polymers, such as low density polyethylene, possibly nylons and others, will resist fiberisation to a much greater extent than materials like polypropylene (homopolymers).

(c) *Formalisation of a Drawability Index*

On the basis of the above discussions, therefore, an expression for the 'drawability index' can be cast in the following terms:

$$\alpha_D = f\left\{\frac{v_2}{P}, \frac{\hat{\sigma}_\parallel}{\sigma_\theta}\left[\frac{\lambda_e}{T_0}, \frac{\Delta T}{\lambda_0}\right]\right\} \tag{8.7}$$

where v_2 = the output velocity of the tapes; P = power consumption for the drawing operation; $\hat{\sigma}_\parallel$ = tenacity of the yarns at the operating draw ratio λ_0; $\hat{\sigma}_\theta \simeq E_\theta \varepsilon_\theta$ = the angular strength of the yarns (say, at 45°; hence E_θ and $\hat{\varepsilon}_\theta$ are respectively the modulus and strain at break of the yarns measured at 45° from the drawing direction); λ_e = extensibility limits, i.e. maximum draw ratio at the operating temperature, T_0; λ_0 = draw ratio of the process; and ΔT = temperature interval, over which $\lambda_e = \lambda_0$ at the process strain rate.

Obviously, the inclusion of a weighting factor in each term, either as powers or proportionality constants would add precision to the definition of α_D.

The power consumption would have to be measured experimentally, although an indication can be obtained by estimating the heat and mechanical energy requirements for the process, i.e.

$$P = v_1 A_1 \rho \bar{C}_p (T_0 - T_R) + v_2 A_2 \sigma_0 (\lambda_0 - 1) \tag{8.8}$$

where v_1 = velocity of the first drawing rolls septet; ρ = density of the material; A_1 = cross-sectional area of the film to be drawn; T_0 = processing temperature; T_R = room temperature; $\bar{C}_p$ = mean specific heat of the material over the temperature range $T_R \to T_0$; λ_0 = draw ratio of the process; A_2 = cross-sectional area of the drawn films (which can be calculated from A_1 and λ_0); v_2 = velocity of the second drawing rolls septet; and σ_0 = stress at λ_0.

Obviously, an expression for the drawability of tapes for weaving and strapping applications would contain the term $\hat{\sigma}_\theta$ in the numerator since fiberisation is an undesirable characteristic.

8.2.2 Analysis of the Tubular Film Process

Although many variants exist for tubular film extrusions, this analysis is concerned with 'normal' air-blown/air-cooled tubular films, and refers specifically to the blowing operation.

(a) Factors Affecting Output

At any given extrusion speed, v_1, the linear velocity of the film reaching the winding stage is v_2; hence the output that one is concerned with in this analysis is v_2/v_1. The film is blown and stretched biaxially in order to achieve a balanced orientation between the machine and transverse directions.

It can be easily shown that the velocity ratio (v_2/v_1) is equal to the blow ratio λ_B (i.e. D_B/D_D, where D_B is the diameter of the bubble and D_D is the diameter of the extrudate at the die). In fact, if it is assumed that blowing occurs entirely in the rubbery state, one can calculate the extension ratios in the two directions considered by means of eqns. (7.5); i.e., since blowing occurs under plane stress conditions, $\sigma_3 = 0$, then

$$\lambda_1^2 = \frac{3}{\bar{E}}(\sigma_T - \sigma_m) \qquad \lambda_2^2 = \frac{3}{\bar{E}}(\sigma_M - \sigma_m)$$

where σ_T = stress in the transverse direction, σ_M = stress in the machine (longitudinal) direction, σ_m = mean stress, $\sigma_m = (\sigma_T + \sigma_M)/3$, and $\bar{E}$ = average modulus of the material in the rubbery state.

The air pressure inside the bubble sets up stresses in the two directions of interest, i.e.

Axial direction (machine direction)

$$\sigma_M = \frac{p\bar{r}}{2\bar{t}} + \frac{T}{2\pi\bar{r}\bar{t}}$$

Circumferential direction (transverse direction)

$$\sigma_T = \frac{p\bar{r}}{\bar{t}}$$

where p = internal pressure, $\bar{r}$ = mean radius of the bubble, $\bar{t}$ = mean thickness of the film, and T = tension force in the axial direction.

Hence when no external forces are applied in the longitudinal direction, T can be ignored because it is very small (i.e. $T = mg$, where m = mass of the bubble and g = acceleration of gravity); therefore, the circumferential (hoop) stress is twice the axial stress. It follows that in order to achieve a balanced orientation, one must be able to exert a pull, T, in the axial direction such that $T/2\pi\bar{r}\bar{t}$ has the same magnitude of σ_M under free blowing conditions.

By setting $\lambda_1 = \lambda_2 = \lambda_B$, it follows that, since σ_M must be equal to $\sigma_T = \sigma^*$, the mean stress becomes $\sigma_m = \frac{2}{3}\sigma^*$. Therefore,

$$\lambda_B = \frac{3}{E}\,(\sigma - \tfrac{2}{3}\sigma) = \frac{\sigma}{E} = \frac{p\bar{r}}{tE}$$

i.e. the blow ratio can be calculated directly from the pressure, and vice versa. This gives, in turn, an estimate of v_2/v_1.

Consequently, the maximum output at a given extrusion velocity is determined by the extensibility limit of the materials at the die-exit temperature. This is the most likely region for the fracture since it coincides with the conditions where the fracture strain is lowest; i.e. it falls within the trough portion of the curve in Figs. 7.8 and 7.32. This also means that the higher the molecular weight of the polymer, the less the likelihood of die fracture. This is therefore, one of the reasons for preferring high-molecular-weight polymers for tubular film extrusion.

There are, however, other reasons why fracture is likely to take place at the die exit. One is associated with the fact that in this region surface fractures (sharkskin) occur and can easily propagate in the circumferential direction, if during processing the longitudinal stresses created by the pull action of the rolls exceed the hoop or transverse stresses generated by the pressure inside the bubble. This may also be one of the reasons why films produced with blow ratios greater than 3:1 tend to have a higher level of orientation in the transverse direction. In other words, the die face fracture phenomena fix an upper limit to the amount of pull in the machine direction.

Accordingly, the maximum extrusion velocity, v_1, is determined by the cooling rate achievable between the die face and the take-off nip-rolls. Above a certain extrusion rate the temperature of the film at the nip-rolls becomes sufficiently high to cause 'blocking' problems as a result of surface creep imposed by the compressive stresses at the nip. Admittedly, this problem can be alleviated by suitable additives and by reinflating the bubble after it has passed through the nip-rolls.

(b) Factors Affecting Quality
Excluding those properties that cannot be controlled by the blowing process, e.g. antistatic behaviour, etc., and accepting that others are also influenced by the degree of unavoidable unbalance of orientation, one defines quality of films in terms of optical properties (haze, gloss and see-through clarity) and mechanical properties (strength, toughness and rigidity).

Optical Properties. 'Haze' results from scatter of light at the surface and within the bulk for the case of crystalline polymers, assuming that all additives are present at levels below their compatibility limits (see Chapt. 3). The drawing (or blowing) process can reduce the amount of surface scatter by flattening the asperities at the surface. However, much of the success of this operation depends on whether the asperities have reached a 'relaxed' state prior to being drawn out and on their initial magnitude at the die exit. In other words, in view of the viscoelastic nature of the recovery of the deformations after the fracture has appeared on the surface, the surface retractions which are associated with the formation of the asperities will oppose locally the drawing forces and reduce the degree of flattening. Therefore, it is expected that a high recovery compliance (i.e. low value of $(\sigma_1 - \sigma_2)$) will produce a film with a lower degree of external haze. (Compare these predictions with the findings of Schroff *et al.*[5])

Internal haze, on the other hand, arises from the difference in refractive index between the amorphous phase and the crystalline domains (when these exceed the size of the wavelength of light). Consequently, internal haze does not arise with amorphous polymers, while with crystalline polymers it increases with the degree of crystallinity and size of spherulites or other forms of crystal aggregates, as well as with the ratio of the density between the crystalline phase and the amorphous phase.

'Gloss' is normally defined as the ratio of the intensity of light reflected at a certain angle (normally 45°) to the intensity of the incident light. It is, therefore, inversely related to surface haze.

'See-through clarity' refers to the ability of the film to resolve fine details of fairly distant images viewed through the film. This is due to forward scattering occurring at very small angles (i.e. between $\simeq 1$ and $1\cdot5°$) from the surface of the film. This arises from large surface irregularities, such as those produced from the flattening of severe sharkskin. It is also related, therefore, to the recovery compliance.

Mechanical Properties. Tensile modulus and tensile strength are mainly related to the blow ratio and, to a lesser extent, to the degree of crystallinity.

Tear strength and impact strength are influenced by both blow ratio and degree of anisotropy but decrease with the degree of crystallinity.

In general, however, the lower the degree of crystallinity, the higher the quality of the film. Consequently, those crystalline polymers that display the highest glass transition temperature and the lowest T_m/T_g ratio (e.g. polycarbonates, polychlorotrifluoroethylene and polyethylene tereph-thalate) can undergo the highest extent of supercooling and will develop a very small degree of crystallinity and spherulitic growth.

Formalisation of a 'Processability Index by Inflation'. If all the factors affecting output and quality are lumped together and the quite small power consumption required to supply air for blowing and cooling the film is ignored, an expression for 'the processability by inflation' can be written as

$$\alpha_B = f\left[\left(\frac{v_2}{v_1}\right), \left(\frac{\lambda_e}{T_R}\right), \left(\frac{T_g}{T_m}\right), \left(\frac{\rho_a}{\rho_x}\right), \frac{1}{J_{(rec)}}\right]$$

where $T_R \simeq$ temperature for the onset of the rubbery state, which corresponds to the temperature at which no further blowing occurs at the operating air pressure, which results from the rapid increase in modulus; λ_e = extensibility limit at the extrusion temperature under biaxial stresses (note that the curve for the extension ratio at break as a function of temperature is qualitatively similar to that in Fig. 8.5); ρ_x and ρ_a = density of crystalline and amorphous phase respectively; and $J_{(rec)}$ = recovery compliance at the extrusion temperature.

REFERENCES

1. G. E. Dieter, *Mechanical Metallurgy* (2nd edn.), McGraw Hill, New York, p. 705.
2. A. Kobayashi, *Encycl. Poly. Sci. and Tech.*, Vol. 8, Wiley Interscience, New York, 1968, p. 347.
3. G. E. Dieter, *Mechanical Metallurgy* (2nd edn.), McGraw Hill, New York, 1976, p. 593.
4. H. Krassig, *Plastics and Rubber Processing*, 1 (1976) 143.
5. R. N. Schroff, L. V. Cancio and M. Shida, *Modern Plastics*, 52 (1975) 62.

9

Processing Performance of Thermoplastics in the Melt State

An important characteristic of melt processes is the prevailing shear mode of deformation over tensile deformations (in contrast with rubbery state processes). Grouped in ascending order of maximum shear rate involved, these processes can be divided as below.

Compression moulding: At the rather low shear rates involved (typically in the region of 1 s^{-1}) in a compression moulding process, the rubbery to melt state transition occurs at relatively low temperatures. This means that at the upper processing temperature limit of the material (i.e. that temperature at which degradation reactions occur at an uncontrollable or unacceptable rate) the rubbery component is minimum. Consequently, there will be substantial relaxation of normal stresses or recovery of associated viscoelastic (rubbery) strains within the practical range of processing conditions (i.e. the characteristic time of the process).

In other words, if a polymer melt is compressed in a 'cold mould' (i.e. at a temperature above the main glass transition temperature of the material), by the time the 'ejection temperature' is reached the normal stresses will have relaxed and/or associated orientation strain will have recovered almost completely. Hence a component produced by compression moulding has a high dimensional stability and no post-moulding distortions occur. (See later discussions on injection moulding.) A typical application of compression moulding of thermoplastics is the production of PVC gramophone records, where it is essential to control the dimensions of the grooves to very fine tolerances. Any residual orientation within the grooves would cause changes in dimensions and a loss (or distortion) of the quality of the sound produced from the acoustic system.

Calendering processes: These operate at somewhat higher shear rates (nominally $10 \, s^{-1}$)† and are probably the only example of processes involving simple unidirectional laminar shear flow. A tension is developed, however, between the take-off cooling rolls and the last of the calendering rolls. Consequently, it is likely that the normal stresses within the mass of polymer melt at the point of leaving the last calendering roll are completely relaxed; while the axial stresses developed by the sheet while cooling in its travel towards the take-off rolls may introduce considerable monoaxial orientation, as a result of the very long relaxation time for the molecular chains, caused by the reduction in temperature and the prevailing tensile mode of deformation. (In other words, under these conditions the material may be within its rubbery state.)

Consequently, calendering is an example of processes which fall neatly within two stages: the first involves net flow, while the second assumes all the characteristics of a monoaxial orientation process, although the extension ratio can be extremely small.

Extrusion processes (*screw extrusion*): Extrusion processes are concerned with the continuous delivery of a melt by means of an Archimedean screw through a die, where it acquires the required geometry. The latter has the additional task of feeding granular or powder polymer into the heated barrel and bringing it into the melt state as a uniform mass. Up to the die assembly the extruder is a universal machine, i.e. suitable for a multitude of purposes; hence extrusion processes will differ from one another only with respect to the type of die, take-off and cooling assembly, which determine the ultimate geometry and dimensions of the cross-section of the products required. (Note that the types of screw or screw arrangements used are not determined by the type of product but by the material which is being extruded.)

Hence one classifies extrusion processes as follows:

(1) Extrusion of solid profiles, i.e. rods, strips, monofilaments, etc.
(2) Extrusion of hollow conduits, i.e. tubings, pipes, etc.
(3) Wire and cable covering.
(4) Tubular film extrusion.
(5) Slit-die film extrusion, i.e. water quenched, chill-roll cast, etc.
(6) Foil and sheet extrusion, i.e. calendering-roll/cooling take-off, stender take-off, etc.

Note that film and sheet lamination in extrusion processes must be

† Note, however, that the overall straining rate (i.e. including the elongational component) may achieve values many orders of magnitude higher.

considered separately, since these are calendering operations for sheet take-off systems. The flow of the melt through extrusion dies occurs at shear rates in the region of $10^3 \, \text{s}^{-1}$.

Injection moulding (reciprocating screw process): Injection moulding consists of delivering a melt at high velocity (shear rates in the region of $10^4 \, \text{s}^{-1}$) into a 'cold' cavity, where the material is allowed to cool to a temperature below that corresponding to the rubbery state. Normally feeding, melting and homogenisation of the melt are carried out by an Archimedian screw in the first stage of the process, while the 'injection' pressure is generated by the same in a second stage and is immediately reduced while the melt is allowed to cool in the mould cavity. There are, of course, variants to the 'plasticisation' units of injection moulding machines, but these are determined more by the nature of the material than by the type of component to be produced. The type of component to be produced does, however, determine to a great extent the injection programme, i.e. pressure and injection velocity profile during the transfer of the melt into the mould cavity.

Another aspect of injection moulding that the type of component may influence is the mechanism for the continuation of flow within the mould when large-area mouldings are produced. In the case of 'solid skin' cellular mouldings, for instance, the melt containing dissolved blowing agent is injected at high speed into the cavity, where the stresses required for filling the cavity are generated by the expanding action of the blowing agent.

In some special cases the depth of the mould cavity is increased by retracting (or withdrawing) by the required amount the insert which forms one face of the cavity, thereby allowing the expansion of the blowing agent to form a cellular structure. In others, when the melt viscosity is too high, cavity filling may be aided by arranging the cavity insert to move forward and to 'compress' the melt. The latter can be considered, therefore, to be an injection–compression moulding process.

9.1   ANALYSIS OF THE CALENDERING PROCESS

Calendering is a complementary process to film and sheet extrusion. It is used mainly for the following purposes:

(1) To produce films that are free of 'sharkskin' and 'orange peel', even at high output rates. This can be achieved by virtue of the very low shear stresses acting on the surface of the melt, which minimise the

normal stresses developed at the exit from the nip, i.e. to values well below the critical conditions for the occurrence of surface fractures.

(2)   To produce sheets from heavily filled compositions that cannot be successfully extruded in view of filler segregation problems in extrusion dies.

(3)   To produce decorative striations, e.g. marble patterns in floor covering sheets.

(4)   To achieve economic production of sheets when the melt viscosity of the polymer is too high, i.e. when extrusion processes are uneconomical.

9.1.1 Factors Affecting the Output of a Calendering Process

The volumetric rate of the sheet leaving the last nip of the calendering rolls is simply:

$$Q/2 = L \int_0^h V_x \, \mathrm{d}y$$

where L = length of contact with the rolls, h = distance from the mid-axis of the sheet to the roll surface at the nip, $\mathrm{d}y$ = infinitesimal layer-thickness of the sheet at a height y from the mid-axis, and V_x = linear velocity of the sheet reaching the cooling rolls. By ignoring the negative velocity component resulting from the recovery of the strains at the exit roll nips and from the increase in the density of the sheet while it cools on its way to the take-off rolls, one can equate the velocity V_x to the peripheral velocity of the exit roll(s). This assumes that the velocity of the take-off/cooling rolls is the same as the velocity of the calendering exit roll(s). However, to achieve this output it is necessary that the same feed rate be achieved at the first nip of the calendering rolls. This is determined by the shear rate exerted by the first set of rolls (the feed rolls) onto the material. The maximum achievable shear rate at the feed rolls corresponds to that at which slip occurs at the roll/melt interface and is a function of the static coefficient of friction of the material.

One has to bear in mind, however, that the feed velocity is the result of two components, one along the length of the rolls (i.e. the spreading component) and the other down the nip. It is important that the coefficient of friction increases when the melt comes into contact with the middle roll, in order to avoid formation of a continuous band around one or both the feed rolls. This may be difficult to overcome with some heavily filled compositions, insofar as it may not be possible to prevent cohesive failure within the melt and to allow the transfer of material around the feed rolls. Note, however, that the level of deformation just beneath the surface can easily exceed the rather low fracture strain of these materials.

The fulfilment of the above requirement is achieved in practice by ensuring that the coefficient of friction is very sensitive to small variations in shear rate and temperature, so that the roll on which the sheet must adhere can be run at somewhat higher speed and temperature than the opposite roll. Normally the melt in the feed bank is subjected to fairly wide variations of temperature because it is exposed to the atmosphere and subjected to fluctuations of residence time prior to passing through the nip, whereas the shear rate undergoes rapid changes in passing through the nip. (See Fig. 9.1.)

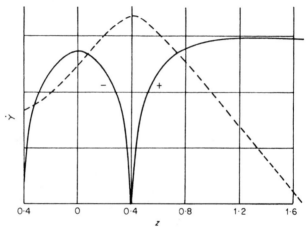

Fig. 9.1. Shear rate at roll surface versus distance along the z-axis. Note that negative and positive shear rates have been plotted on the same side of the z-axis. (After Glyde.[1])

This means that it is desirable for the viscosity (bulk property) to be insensitive to temperature and shear rate variations (low shear rate regions). The two opposing requirements of viscosity and coefficient of friction (surface property) pose considerable formulation difficulties with respect to polymer and additives selection, to the extent that few plastics materials can be successfully calendered at high speeds.

9.1.2 Factors Affecting Quality
The quality of a calendered sheet is normally judged in terms of its dimensional stability and anisotropy, thickness variations across the width

of the sheet, and freedom from surface blemishes, such as 'crow feet', 'bank marks' and variable gloss.

(a) Dimensional Stability and Anisotropy

These result from monoaxial orientation and are therefore related to the recoverable compliance function. In other words the 'frozen' longitudinal strain (in the machine direction) is related to the effective extension ratio between the take-off rolls and the exit calender rolls and to the combination of elongational and shear strains imposed on the sheet in passing through the calendering rolls. One has to bear in mind, in fact, that a tension must be maintained on the sheet passing through the rolls to guarantee adherence on each subsequent roll. This means that one is dealing with a stress relaxation situation and that the frozen orientation is a function of the level of normal stress components imposed on the sheet and the nominal average time between stressing and cooling the sheet to below the glass transition temperature. Therefore, the higher the calendering speed, the shorter the relaxation time and the greater the roll differential speed required. Obviously, the lower the temperature, the higher the level of normal stresses.

It can easily be appreciated that the magnitude of these effects is directly related to the recovery compliance (elongational) and the associated distribution of retardation times at an average temperature between the processing temperature and the glass transition temperature of the material. In other words, assuming linear behaviour, the elongational strain in the machine direction can be expressed in terms of the recoverable extension ratio (λ_1) using the rubber-like elasticity approach discussed in Chapt. 7, i.e. for $v = \frac{1}{2}$, assuming also that $\sigma_2 = \sigma_3$, one obtains:

$$\lambda_1^2 = \frac{2}{E(t)} (\sigma_1 - \sigma_2) = 2(\sigma_1 - \sigma_2)f(D(t))$$

Hence a linear transformation of the relaxation modulus into a creep compliance and a knowledge of the normal stress difference developed by the calendering process would permit an estimate of the frozen elongational strain. The complexity involved in extracting the relevant information from rheological data and the uncertainties arising from both the transient temperature conditions (i.e. while the sheet is cooling between the take-off and the calendering rolls) and the estimation of an appropriate elapsed time (between stressing and cooling the sheet) would, however, render such an approach untenable in practice.

It would be more appropriate, therefore, to make measurements directly on sheets produced under different conditions for the estimation of the effective (or average) recovery compliance for the material under calendering conditions. This would involve measuring σ_1 and σ_2 by quickly heating a constrained sheet to its rubbery state and observing the forces developed along the two principal axes in the plane of the sheet. The λ_1 value, on the other hand, is obtainable directly from shrinkage measurements.

(b) Thickness Variations
These result from deflection of the rolls and from differential recovery of the sheet in leaving the calendering rolls. In practice the variations in thickness profile across the width of the sheet can be reduced, but not eliminated, by a combination of roll-crowning, roll-bending and cross-axis alignment, which results in matching the gap distance and the recovery of the sheet to the true opening along the nip of the calendering rolls. This involves considerable tailoring of roll design to the actual material to be processed and, consequently, it lacks operational flexibility. Reducing the recovery component is, therefore, an effective way of minimising the problem of thickness variation. Hence, considerable improvements are to be expected, when processing heavily filled polymers and those containing 'rubbery-gel' fractions.

(c) Surface Defects
Crow-feet and bank marks are manifestations of differential rates of deformation at the roll/melt interface arising from variations in melt viscosity, which result from partial cooling of the melt or improperly homogenised melt. Crow-feet marks, which are retained on the final sheet, originate at the feed because of the low shear rates and of the lack of further restriction to flow in the subsequent passage through the middle and exit rolls. Bank marks, on the other hand, originate between the nips of the middle rolls. A small bank is always necessary to ensure adequate contact of the sheet with the rolls and to prevent air entrapment (see Fig. 9.2). However, if the size of the bank is too large and the roll speed is sufficiently low, when the exposed surface of the bank rotates and reaches the nip, it will be at a lower temperature than the material arriving directly at the nip from the preceding roll. As a result, periodic small depressions will be formed along the width of the sheet. Depressions may also result from slip–stick effects created by the periodic change in the coefficient of friction.

Dull patches and surface matness may result from 'plate-out' (jargon for deposits) of additives on the surface of the rolls thereby creating localised

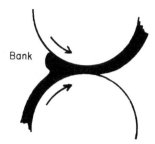

Fig. 9.2.   Formation of a 'bank' of polymer between the rolls of a calender.

friction on the surface of the sheet and a concomitant marring of the surface. Furthermore, if the temperature of the rolls is too low, sharkskin may result when the sheet leaves the nip, creating surface irregularities which decrease the gloss.

9.1.3 Formalisation of a Calenderability Index
In conclusion an equation for the 'calenderability index' of thermoplastics can be expressed in the following terms:

$$\alpha_c = f \left\{ \left(\frac{Q}{P}\right), \left(\frac{\Delta H_{(\mu)}}{\Delta H_{(\eta)}}\right), \left(\frac{1}{J_{(rec)}}\right) \right\} \qquad (9.1)$$

where Q = calendering output rate, P = power consumption, $\Delta H_{(\mu)}$ = activation energy for the static coefficient of friction, $\Delta H_{(\eta)}$ = activation energy for the melt viscosity, and $J_{(rec)}$ = recovery compliance of the material. P and Q are obviously related to the viscosity and coefficient of friction of the material at the processing temperature. In the calculation of P one would have to take into consideration also the rate of thermal energy input.

9.2   ANALYSIS OF EXTRUSION PROCESSES

As already mentioned an extruder basically consists of two parts: the plasticisation/pressure generating unit (screw section) and the shaping unit (die-assembly). Since the screw section is common to all extrusion processes, it can be treated independently of the type of extrudate produced. The die, in this case, will be considered simply as the flow restrictor unit (or flow valve) of a general extrusion operation, while the type of extrudate produced will be the parameter that determines the extrusion conditions.

As in previous cases the analysis will be concerned with factors affecting output, quality and power requirements. Screw design will be considered only inasmuch as it affects the conditions to which the material is subjected in an extrusion operation. For the purpose of this analysis only single-screw extruders will be considered.

9.2.1 Factors Affecting Output

The standard single-screw extruder receives cold powder (or granules) through a throat at the rear end of the barrel and delivers it to the compression zone of the screw. In this section the material is brought into its melt state and converted into a homogeneous mass.

From the compression (or transition) zone the melt passes through the metering zone, which has the function of stabilising the flow before it is pumped through the die. In practice the transition to the melt state may extend over the compression zone into either the feed zone or the metering zone. The continuous melt 'bed' in the metering section forms a barrier for the air occupying the interstitial regions of the particles of material arriving from the feed zone. The pressure developed at the die, on the other hand, ensures that the melt is fully homogenised (i.e. the interparticle boundaries are destroyed and the additives thoroughly mixed). This is achieved by virtue of the increased shear stress that acts on the material and is enhanced by a higher melt viscosity. Homogenisation is also aided by other flow restriction features in the die assembly (e.g. breaker plates), and at the tip of the screw (e.g. mixing torpedoes, rings, etc.).

In considering these events, it becomes quite obvious that it is necessary that the screw be so designed as to ensure that the conveying rate of the powder through the feed zone, and the rate of temperature rise and plasticisation in the transition zone, fully balance the flow rate through the metering zone. This is achieved by keeping the pitch of the screw constant and adjusting the depth of the screw channels, the relative length of each of the three zones, and the helix angle of the flights.

Other variants such as number of starts or leads and, occasionally, even a changeable helix angle are design features to be found in modern screw extruders. In order to analyse the factors that affect extrusion output, it is necessary to consider the flow rates in the feed and metering zone respectively and the plasticisation rate in the transition zone.

(a) Flow Rate in the Feed Zone

The velocity of powders or granules in the channel direction of the feed zone (V_f) can be considered to result from a rotational component (V_c) in

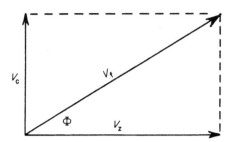

Fig. 9.3. Velocity components of powder particles in the feed zone of an extruder.

the circumferential direction (which produces no flow) and a translational component along the axis of the screw (V_z). Figure 9.3 indicates that the flow rate in the feed zone is, therefore, a function of the angle Φ.

By assuming that the flow of the granules can be simulated by a solid plug along the channels, equations for the flow rate in terms of geometry of the screw and of coefficient of friction against the barrel and screw surface have been derived.[2] These are:

$$Q_F = \pi^2 N H D_b (D_b - H) \frac{\tan \Phi \tan \theta_b}{\tan \Phi \tan \theta_b} \left(\frac{\bar{W}}{\bar{W} + e} \right) \qquad (9.2)$$

where $\bar{W}$ = average channel width, given by

$$\bar{W} = \frac{\pi}{\eta} (D_b - H) \sin \bar{\theta} - e$$

N = screw speed, H = height of channels, θ_b = helix angle of the screw flights at the root, D_b = diameter of barrel and outer diameter of screw, e = flight width, η = number of starts of flights, $\bar{\theta}$ = average helix angle from the root of the screw to the top of the flights, Φ = angle between the two velocity components, and

$$\cos \Phi = K \sin \Phi + 2 \frac{H}{W_b} \frac{f_s'}{f_b} \sin \theta_b \left(K + \frac{\bar{D}}{D_b} \cotan \bar{\theta} \right) + \frac{W_s}{W_b} \frac{f_s}{f_b} \sin \theta_b$$

$$\times \left(K + \frac{D_s}{D_b} \cotan \bar{\theta}_s \right) + \frac{\bar{W}}{W_b} \frac{H}{Z_b} \frac{1}{f_b} \sin \bar{\theta} \left(K + \frac{\bar{D}}{D_b} \cotan \bar{\theta} \right) \ln \frac{P_2}{P_1}$$

where

$$K \equiv \frac{\bar{D}}{D_b} \frac{\sin \bar{\theta} - f_s \cos \bar{\theta}}{\cos \bar{\theta} - f_s \sin \bar{\theta}}$$

Subscripts s and b are for screw and barrel respectively, f = coefficient of friction, Z_b = distance along the channel direction from the entry, and P_1 and P_2 = pressure at entry and end of feed zone.

These equations provide quantitative relationships for the qualitative predictions that can be made intuitively. It can be expected, for instance, that to maximise output the coefficient of friction on the screw surface should be as low as possible in order to favour translational movements of particles (or the sliding of the plug) along the axis direction of the screw, while that against the screw barrel should be as high as possible to minimise rotation in the circumferential direction. Since the coefficient of friction of the material up to the rubbery state increases with temperature, it can be anticipated that output increases with screw cooling and by heating of the barrel to the temperature corresponding to the rubbery state of the material. Higher temperatures would reduce once more the coefficient of friction against the barrel and would therefore tend to reduce output.

On the basis of these arguments, one can easily explain both the fairly high scatter in the results of Griffiths (see ref. 3) and the downward trend in output for polymethyl methacrylate granules for barrel temperatures above 200 °C.

Furthermore, it is to be expected that highly polished screw surfaces and rough inner barrel surfaces should also produce higher outputs, which is in perfect agreement with the results of Bernhardt.[4] The quantitative treatment of the above relationships, however, enables one to make accurate estimates of the optimum helix angle for various coefficients of frictions, so that, for the simplified case of $\mu_b = \mu_s$ and $P_2 = P_1$, one obtains the following relationship:

μ:	0	0·2	0·4	0·6	0·8	1·0
θ_b:	45°	26°	21°	17°	14°	12°

It is instructive to note also that the theory predicts the beneficial effect of increasing the entry pressure P_1, which would not be explained in terms of a pressure-dependent coefficient of friction since this would decrease the output by virtue of the increased friction against the screw. However, the theory does not allow for the following:

(i)   Tumbling effect of the granules (or powder), which produces a 30–50 % overestimate of the flow rate depending on particle size, geometry and polymer/polymer coefficient of friction. The larger the particles and the lower their aspect ratio and the smoother their surface, the greater the output.

(ii)   The temperature rise of the powder as it travels towards the compression zone, which not only affects the coefficient of friction but also the pressure as a result of the substantial thermal expansion. However, it can be anticipated that lubricants with a fairly high melting point, such as metal stearate, might help to keep the coefficient of friction constant over the desired temperature range and to minimise the effect of increased pressure by expansion.

(iii) The difficulty in defining the coefficient of friction of the mix in the presence of a substantial concentration of fillers.

However, by predicting that the flow rate through the feed zone will be approximately half that through the metering zone (see later) and recognising that there is a 30–50 % error in the calculation, it is quite remarkable that one should arrive at a compression ratio of about 4:1, which is common to most screws and was established long before any theory of extrusion had been formulated.

It is, on the other hand, more difficult to apply the same argument to materials like rigid PVC, for which a compression ratio of about 2:1 is normally used. This indicates that further errors may arise in theoretical predictions from other often-neglected factors, such as compressibility of the polymer melt and slip effects in the metering zone.

(b) Plasticisation or 'Melting' Rate in the Transition Zone

The melting mechanism of polymers in the transition zone of the extruder is reasonably well established. First, a thin layer of melt is formed by the increased friction of the granules against the hot barrel as the pressure increases. This may start, of course, well before the granules reach the compression zone. While they are still in the feed zone, the granules reach only their rubbery state and, consequently, are still free to rotate because the lack of flow at the interface prevents their sticking to one another. A bank of melt is then formed behind the consecutive flights and undergoes a circulatory flow (Figs. 9.4 and 9.5).

The ploughing action of the screw flights scrapes the molten layer and forces it downwards, mixing it with unmelted granules (zone C) and bringing other granules into the melt state through heat transfer. As the material moves along the compression zone, the melt region becomes increasingly large until the solid bed of granules breaks up and floats in a pool of melt. Some back flow takes place over the flights in the clearance region between screw flights and barrel (leakage flow). This back flow assists the circulation of the melt behind the flights of the screw.

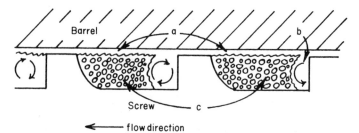

Fig. 9.4.   Melting mechanism in the transition zone of an extruder. a = Molten layer, b = circulating melt, c = melt mixed with granules.

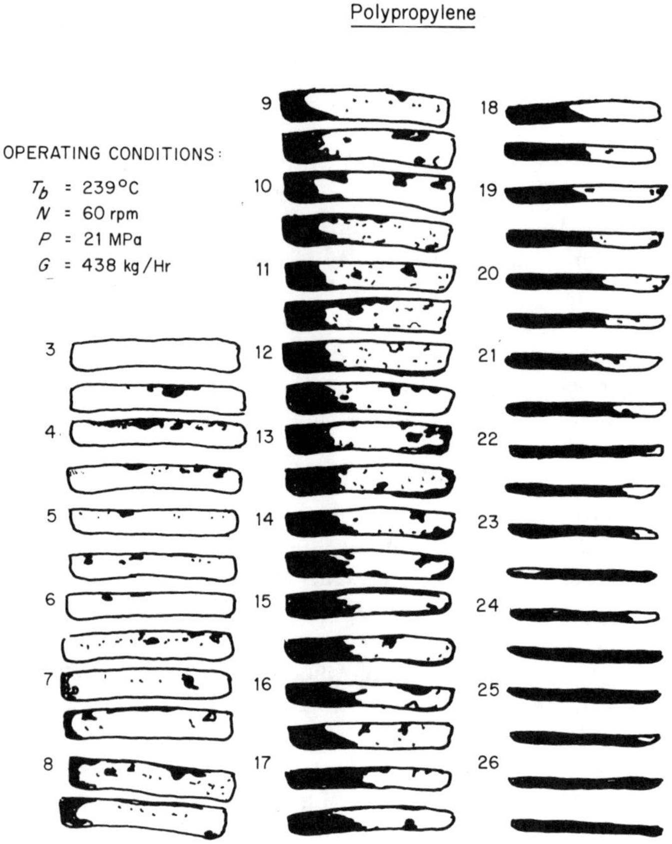

Polypropylene

OPERATING CONDITIONS:
T_b = 239°C
N = 60 rpm
P = 21 MPa
G = 438 kg/Hr

Fig. 9.5.   Typical melting mechanism along the flights of a screw. (Retraced from original photographs published by Tadmor et al.[5])

The actual melting should be completed before the material reaches the first channel of the metering zone, or at least the solid bed should have disintegrated so that individual granules float in the pool of melt. Otherwise, once an agglomerate of granules (or powder) reaches the metering zone, it is difficult to expel the entrapped air that appears in the extrudate as bubbles or porosity. Increasing the screw speed brings forward (i.e. towards the metering zone) the mixed solid bed/melt unless the rate of melting is increased at the same time by raising the barrel temperature and/or by increasing the compression ratio and shortening the length of the compression zone.

Prediction of the rate of melting of polymers in the transition zone is an important consideration for the prediction of maximum flow rate achievable. The theory in this area has now reached a very high degree of sophistication, thanks mainly to the work of Tadmor and Klein.[6]

The rate of melting per unit polymer/barrel interface area is obtained from a heat balance at the interface and a mass balance down the channel. These yield the following equations for the rate of melting:

$$\bar{\omega} = \frac{\Phi W^{1/2}}{2 - A/\Phi} \tag{9.3}$$

where $\bar{\omega}$ = mass rate (average) of melting; W = width of the channels; A = degree of tapering of the channels, i.e. the slope of the inner channel surface of the screw relative to the screw axis; Φ = a group of variables which gives a direct measure of the rate of melting (i.e. the higher Φ, the higher the rate of melting); this is given by the following expression

$$\Phi \equiv \left\{ \frac{V_{bx}\rho_m \left[K_m(T_b - T_m) + \dfrac{\eta}{2} - V_j^2 \right]}{2[C_p(T_m - T_s) + \Delta H_m]} \right\}^{1/2} \tag{9.4}$$

and $\Phi = (\Phi W^{1/2})/(G/H_0)$, where G = mass flow rate, and H_0 = depth of channel at the beginning of the compression zone.

It can be seen that all the material characteristics are included in the expression of Φ, which can be called, therefore, the 'melting rate index' of extrusion; where

ρ_m = density of the polymer melt;

T_m = temperature reached by the melt (or desired melt temperature);

T_s = temperature of the polymer granules at the entry of the compression zone, which can be taken to be approximately equal to the main glass transition temperature, since it is expected that the granules will remain in their rubbery state for some time to avoid premature particle fusion;

T_b = barrel temperature;

η = viscosity of the melt (assumed in this analysis to be independent of temperature and shear rate), these effects can be taken into account in the expression for Φ;

C_p = average specific heat of the polymer over the temperature range $(T_m - T_s)$;

ΔH_m = latent heat of fusion of the polymer;

V_{bx} = velocity of the material in the direction perpendicular to the flights, which can be calculated directly from the helix angle and screw speed, i.e. at 100 rpm and a helix angle of 18° and a screw diameter (flights diameter) of 62·5 mm

$$V_{bx} = \pi \times 6\cdot25 \times (100/60) \, \text{cm/s}$$

V_j = a velocity component of the material at the barrel interface, it is the vectorial difference between the peripheral screw velocity ($\pi N D_b$) and the velocity of forward movement of the solid bed (V_s).

Tadmore and Klein[7] give an example for the calculation of V_j for an extruder operating under the following conditions:

$$\text{Output} = 54\cdot5 \, \text{kg/h}$$
$$\text{Diameter of screw} = 63\cdot5 \, \text{mm}$$
$$\text{Helix angle} = 18°$$
$$\text{Flight width} = 6\cdot35 \, \text{mm}$$
$$\text{Channel depth in the feed zone} = 12\cdot7 \, \text{mm}$$
$$\text{Channel depth in the metering zone} = 3\cdot175 \, \text{mm}$$
$$\text{Density of the polymer (l.d. polyethylene)} = 0\cdot92$$
$$\text{Screw speed} = 82 \, \text{rpm}$$

$$V_j = V_b - V_{sz}$$

where $V_b = \pi \times 6\cdot35 \times (82/60) = 27\cdot25 \, \text{cm/s}$.

$$V_{sz} = \frac{G}{W H_0 \rho_s} = \frac{54\,500/3600}{[(6\cdot35 - 0\cdot635)\cos 18³] \times 1\cdot27 \times 0\cdot92} = 2\cdot4 \, \text{cm/s}$$

Therefore

$$V_j = 25 \, \text{cm/s}$$

Alternatively, one could isolate the effects of the machine from those of the material by eliminating the terms V_{bx} and V_j and redefining the extrusion melting rate index of the material as

$$\alpha_{er} \equiv f \left\{ \frac{\rho_m \left[K_m (T_b - T_m) + \dfrac{\eta}{2} \right]}{2[C_p (T_m - T_s) + \Delta H_m]} \right\}^{1/2} \tag{9.5}$$

One could also divide the above expression by G (the flow rate) and obtain a melting rate index for the material corresponding to the parameter ϕ for the complete machine/material interactions. In these expressions, however, one would have to take into account the temperature sensitivity of the material parameters, and also its shear stress and hydrostatic pressure sensitivity for the viscosity. On the basis of data presented by Powell,[8] this means that a decrease in temperature and/or an increase in pressure would produce a much larger increase in viscosity and, therefore, in melt rate index for acrylic polymers than for acetals, nylons or polypropylene.

An inspection of the melting rate index equation allows one to make several predictions about operating conditions and screw design requirements for different types of thermoplastics, as follows.

Those crystalline materials which display a narrow rubber state at quite high temperatures and have a low melt viscosity, e.g. Nylon 6, Nylon 6,6, acetals, poly 4-methylpentene 1, etc., are expected to require screws with a long feed zone, since it is necessary to have a long residence time in this section in order to bring the material to the rubbery state. They also require a rather short compression zone as well as a high compression ratio. The latter two features are necessary in order to maximise the mechanical energy input by the screw to generate the necessary latent heat of fusion. A short compression zone and high compression ratio means that there will be a rapid increase in shear rate developed by the screw to maintain a constant flow rate. Screws with these features are generally known as 'nylon screws' and have a compression zone which usually extends only over a $\frac{1}{2}$ turn of the flight around the screw.

Amorphous polymers that exhibit a broad rubbery state and have a high melt viscosity, e.g. polymethyl methacrylate, rigid and plasticised PVC, plasticised cellulose acetate, etc., are expected to perform better with screws having a short feed zone and long compression zone, owing to the absence of the latent heat of fusion. This prediction is made on the basis that it is not necessary to generate the same amount of heat by mechanical work as for crystalline polymers. Because of the high viscosity the rate of heat generation at a given shear rate is rather high, and, unless a small taper in the screw root diameter is used, an excessive temperature rise of the melt may occur, which could lead to decomposition of the polymer, e.g. PVC. For this reason, screws having a very short feed zone and a long compression zone (i.e. $\simeq \frac{3}{4}$ of the total length) are normally known as the 'PVC screws'.

Those screws in which the feed zone and the compression zone are approximately equal are known as 'general purpose' screws.

Despite the fairly high accuracy of predictions from the theory for the melting zone of extruders (see Fig. 9.6, for instance), it still fails to take into account important factors, such as the coefficient of friction of the polymer against the walls of the screw and barrel, as well as the effects of particle size and, possibly, of the aspect ratio of granules. When the solid bed in the compression zone breaks up, the circulatory flow of the melt takes place around the whole section of channel, thereby leaving a solid plug floating in the middle. This has the effect of creating a momentary reduction in the melt flow rate along the channel in view of the longer flow path that the melt must follow in the circulatory motion around the screw channels. In other words, a pulsating flow is expected to be set up over a certain section of the transition zone.

It is to be expected that the friction between the solid bed and the channel walls of the screw will affect the conditions in which the solid bed breaks up and, possibly, also the critical size of the solid bed for the onset of the phenomenon. Tadmor and Klein, in fact, have recorded the occurrence of the phenomenon (see Fig. 9.6) but have not stipulated a criterion for the conditions in which it takes place.

The breaking up of the solid bed affects in turn both the completion of the melting process (since when the solid bed is encapsulated in a pool of melt, the only heat source for the unmelted granules is by conduction and convection from the surrounding melt) and the 'degree' of instability of the flow which ensues, e.g. the amplitude and the frequency of the pulsations. Gale[10] has noted, in fact, that the level and type of lubricant in PVC not only affects the length of the melting zone but also can influence the whole fusion mechanism.

Regarding the effect of particle size, if one ignores the complications that may arise from a large gap between the screw flight and barrel, one can intuitively anticipate that powders will melt more slowly than granules in view of their lower packing density, which produces a lower rate of increase in pressure along the channels. This would have to be counteracted by a larger taper and a higher compression ratio to increase the levels of shear at the barrel surface as the powder advances along the screw channels.

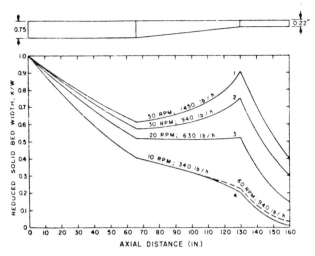

Fig. 9.6. Effect of frequency of screw rotation on solid bed profile without maintaining a constant throughput at 400 °F barrel temperature for low density polyethylene extruded in an 8 in diameter single screw plasticating extruder. Dotted line shows the effect of doubled screw speed alone on curve 2 maintaining throughput constant at 940 lb/h. The extrudate quality is markedly improved by the high screw speed and is equivalent to a 340 lb/h throughput at 10 rpm shown in curve 4. (After Tadmor and Klein.[9])

Consequently, the 'melting rate index' equation (eqn. (9.5)) could be suitably modified by inclusion of a dimensionless function $g(\bar{\mu}\bar{\phi})$, where $\bar{\mu}$ = average coefficient of friction in the temperature range corresponding to the rubbery state of the material, and $\bar{\phi}$ = a dimensionless shape factor which is dependent on particle size, surface roughness and aspect ratio, i.e.

$$\alpha_{er} \equiv f\left\{ g(\bar{\mu}\bar{\phi}) \left(\frac{\rho_m \left[K_m(T_b - T_m) + \frac{\eta}{2} \right]}{2[C_p(T_m - T_s) + \Delta H_m]} \right)^{1/2} \right\} \tag{9.6}$$

It is instructive to note at this point some of the opposing requirements in respect to the coefficient of friction against the screw and barrel surface in the feed zone and the compression zone respectively. From such considerations the advantages of cooling the screw up to the feed zone only and of external lubricant, which melts within the rubbery state temperature range and becomes compatible with the polymer at higher temperatures, can be easily appreciated (see Chapt. 4).

However, for considerations concerning choice of screw type, barrel temperature profile and also, possibly, type and level of lubricants, one has to take into account the nature of the extrudate to be produced. This can be best illustrated by analysing the recommendations of Basset *et al.*[11,12] (Table 9.1) in relation to polysulphones. Particularly relevant to these discussions is the difference in screw characteristics and temperature profile along the barrel between sheet and film extrusion. On the basis of the earlier discussion, it is clear why a long transition zone, a reasonably small compression ratio and a flat temperature profile have been recommended for sheet extrusion. In other words, one is dealing with a typical amorphous polymer of high melt viscosity; and, therefore, only a gentle increase in shear rate within the transition zone is required to supplement the heat input from the barrel.

The change to a nylon type of screw for film extrusion, therefore, would contradict the above arguments were it not for the fact that a much higher output is expected ($\simeq 475\,\mathrm{cm}^3/\mathrm{min}$ against $94\,\mathrm{cm}^3/\mathrm{min}$) with a smaller diameter screw and a smaller die gap. This can be achieved only by developing an appreciably higher temperature of the melt in order to reduce its viscosity. (Note the high temperature sensitivity of the viscosity of these materials from the discussions in Chapt. 7, Table 7.1.) Consequently the considerably higher rate of temperature rise required in the transition zone can be obtained only with a large taper and a high compression ratio. The actual melting process is likely to start in the feed section in view of the much higher barrel temperatures used.

A similar argument applies also to the case of nylon, especially for the production of solid profiles or thick section extrudates from glass filled grades. A longer transition zone and, possibly, a lower compression ratio may be more suitable than a conventional nylon screw in order to avoid excessive breaking of the fibres. Moreover, the generally lower extrusion rates required may ensure more efficient heat transfer by conduction from the barrel.

(c) Flow Stabilisation and Flow Rate in the Metering Zone
The flow instability created by the breaking up of the solid bed in the transition zone causes a fluctuation in pressure and temperature at the point and time of its occurrence. While the fluctuation in temperature disappears through equilibration of heat transfer between the vanishing solid bed and the surrounding melt, the fluctuation in pressure dampens out when the melt acquires uniform rheological properties across the channels of the screw. In other words, the damping of these fluctuations is the only evidence that melting is occurring. These fluctuations are

Table 9.1: Typical Conditions for Extruding Polysulphone Sheet, Film, Shapes[12]

	Sheet[a]	Film[b]	Shapes[c]
Screw diameter (mm)	115	63·5	63·5
Screw L/D ratio	20:1	28:1	16:1
Screw compression ratio	3:1	4:1	3·6:1
Screw turns			
feed	3	22·5	4
transition	11	0·5	6
metering	6	5·0	6
Channel depth (mm)			
feed zone	19·5	9·6	12·7
transition	—	—	—
metering	6·38	2·4	3·6
Screens	20–60–100	20–60–20	20–60–100
Die (mm) (not deckled)	660	1 120	—
Die setting (mm)	3·2	0·5	19·0; 12·7
Cylinder temp. (°C)			
feed	300	325	—
rear transition	310	340	300
front transition	330	360	300
metering	330	360	300
Adaptor temp. (°C)	330	310	300
Die temp. (°C)	330	340	300
Melt temp. (°C)			
barrel	—	—	—
adaptor	340	—	310
Melt pressure, (MN/m^2)			
barrel	—	32	11·5
adaptor	10·5	—	—
Screw speed (rpm)	25	27	15
Line speed (m/min)	0·61	3·70	2·5
Line power (kW)	74·6	30	20
Sheet width (mm)	610	—	—
Film width (mm)	—	1 000	—
Finishing roll temp. (°C)			
centre and top rolls	180–195	165	—
bottom roll	180–195	165	—

[a] For sheet 0·254 mm thick and 610 mm wide, die = 660 mm, screw = 115 mm diameter.
[b] For film 0·127 mm thick, die = 1120 mm, screw = 63·5 mm diameter.
[c] For tubing with 4·75 mm ID and 12·7 mm ID, screw = 63·5 mm diameter.

completely interdependent (see Fig. 9.7) insofar as the higher viscosity of the melting 'solid', which is naturally at a lower temperature, generates more viscous heating than the 'older' (and hotter) melt.

Consequently, in its capacity as a flow stabilising unit, the metering zone is only an extension of the compression zone; and in view of the melting process actually beginning in the feed zone, it is important not to use interchangeably the three zones of the screw with the three consecutive processes of 'solid conveying', 'thermal plasticisation' and 'flow stabilisation' taking place in the extruder barrel.

From the above discussions it can be deduced that the minimum length of the metering zone is determined by the molecular diffusion rate under the 'mixing' conditions imposed by the geometry of the screw. This intuitive

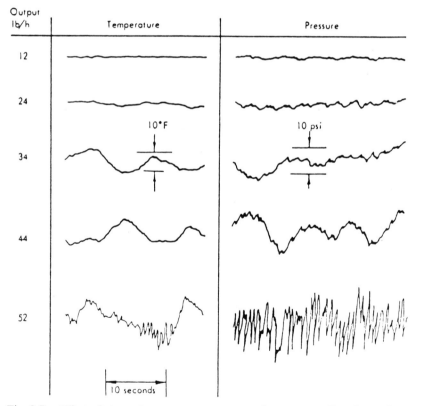

Fig. 9.7. Effect of throughput on temperature and pressure uniformity. (After Kessler *et al.*[13])

statement suggests a fundamental solution to the design of the metering zone, which has yet to be considered: that is, to combine the theory of mixing based on 'average total strain' as a function of the ratio of back pressure flow to drag flow (see later)[14] and the Stokes–Einstein theory for self-diffusion $(D_{AA}\eta_A/T = \text{constant}$, where $D_{AA} = \text{self-diffusivity}$, $\eta_A = $ viscosity of the melt and $T = \text{absolute temperature})$.[15]

It is instructive to note that the universal solution adopted by extruder operators to achieve stability in the metering zone and, therefore, to eliminate 'surging' or pulsating flow through the die by increasing the head pressure is concordant with the above suggestion.

This, in effect, reduces the length of the transition zone by increasing both the rate of molecular diffusion from the melt phase to the rubbery particles and the ratio of the back pressure flow to the drag flow. With regard to the function of the metering zone as a melt-conveying unit, the theory has reached a very high degree of sophistication, despite the doubts that may arise from some of the underlying assumptions:

Assumptions	*Limiting factors*
(1) The flow is in a steady state (i.e. independent of time).	(1) Applicable only to a limited section of the metering zone.
(2) The flow is fully developed (i.e. the velocity gradient across the channels and in the axial direction is zero).	(2) Applicable only to a limited section of the metering zone.
(3) There is no slip at the walls of the screw and barrel.	(3) It would not be possible to extrude easily degradable polymers like PVC.
(4) The melt is incompressible.	(4) It is difficult to estimate with accuracy the effects of hydrostatic stresses on viscosity.

Basically, the analysis starts with an expression for the conservation of momentum of a laminar Newtonian flow in a rectangular channel, assuming that the barrel acts as an infinite plate sliding over the flights of the screw. The theory then introduces appropriate modifications for curvatures at the edges of the channel and for the variations of viscosity as a function of temperature rise and pressure gradient, etc. Normal forces are not considered, but these would only introduce errors in the calculation of power consumption, in addition to those possibly arising from slip.

The shear stresses developed by the moving surface create a linear velocity gradient across the channel and promote the so-called 'drag flow', while the pressure gradient, created by the restrictions through breaker plates and die, sets up a parabolic velocity profile in the opposite direction and distorts accordingly the overall velocity profile. The theory also takes into account the circulatory flow across the channel and in the axial direction. These profiles are shown in Fig. 9.8, where V_z refers to the velocity component along the channel, while V_x and V_y are the velocities in the other orthogonal directions. V_b is the velocity of the barrel (i.e. equivalent to that at the flights of the screw), and V_{bz} and V_{bx} are the components in the down channel and across channel directions, respectively. The assumptions of fully developed flow and incompressibility

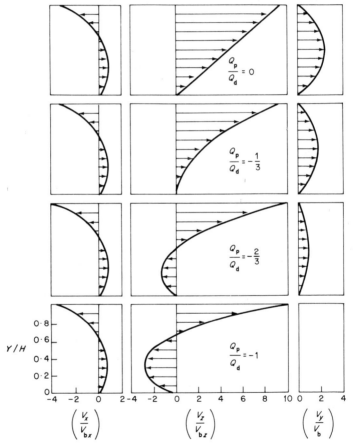

Fig. 9.8. Cross channel, down channel, and axial velocity profiles for various Q_p/Q_d values. (After Tadmor and Klein.[16])

simplify the equation of motion for the flow along the channels and reduce it to

$$\frac{\partial P}{\partial z} = \eta \frac{\partial^2 V_x}{\partial y^2} \tag{9.7}$$

This can be integrated to obtain an expression for the velocity profile along the channel, which can then be used to calculate the flow rate, i.e.

$$Q = \int_0^H \int_0^W V_x \, dy \, dx$$

where H = depth of channels, and W = width of channels.

For isothermal Newtonian flow conditions the flow rate along the channels takes the form of

$$Q = AN - B \frac{\Delta P}{\eta} \tag{9.8}$$

where N = screw speed (number of revolutions per unit time); ΔP = pressure difference from the section of the channel of the metering zone considered and the die; η = viscosity of the melt; and A and B = screw characteristics (constants) which group together all the geometrical parameters of the screw, i.e.

$$A = \pi^2 D^2 H \left(1 - \frac{ne}{t} \right) \sin \Phi \cos \Phi$$

$$B = \pi D H^3 \left(1 - \frac{ne}{t} \right) \sin^2 \Phi / 12L$$

where D = screw flight diameter, H = channel depth, n = number of flights starts (parallel channels), t = channels lead, e = land width of channels, Φ = helix angle, and L = length of the metering zone considered.

Note that the first term of the flow rate equation corresponds to the 'drag flow', which in the absence of slip is independent of the nature of the melt, while the second term represents the 'back pressure flow', i.e. the quantity by which the drag flow rate is reduced by forcing the melt against a back pressure; the back pressure flow is inversely related to the viscosity. If slip takes place, the first term is reduced and the second is increased; while if the effects of temperature rise and the positive pressure gradient on viscosity are considered, the slope of the flow rate curve as a function of the head pressure may no longer be linear, owing to the relative variations of viscosity with temperature and pressure. Note that if the suggestion of the equivalence of the Arrhenius type of equation for the viscosity as a function of temperature to the Doolittle equation for the variation of viscosity as a function of pressure was applicable (see Chapt. 7, p. 318), only the slope would change and the Q versus ΔP relationship would still be a straight line. However, because the viscosity decreases with shear rate as the output increases, the value of η would decrease according to the power law or sinh functions given in Chapt. 7, p. 312. The curve is therefore expected to bow downwards.

Eventually at very high pressures and especially if slip occurs, the pressure flow term could even become of the same order of magnitude as the drag flow and the flow would cease altogether. This phenomenon is not at all uncommon in practice when the polymer is 'overlubricated'.

Illmann[17] noted, in fact, that for over-lubricated materials the curve of output as a function of speed showed a maximum, while those formulations with the correct degree of lubrication gave almost a straight line, as predicted by the theory.

There is another hidden material parameter in the flow rate equation: the effect of temperature at the surface of the screw, a phenomenon which cannot be considered simply from the point of view of a local increase in viscosity. Screw cooling, in fact, can easily bring some of the material at the wall into the rubbery state, which means that the viscosity goes to infinity and, since the material will not flow, the effective depth of the channel is reduced. On the basis of eqn. (9.8) one expects a large decrease in the slope of the flow rate curve as a function of the pressure drop, due to both a reduction in the effective channel depth (which appears as a cubic term) and the local increase in viscosity, and a concomitant reduction in drag flow. This has been confirmed experimentally by Colwell and Nicholls,[18] who plotted the along-channel velocity (which is proportional to the flow rate) as a function of the pressure drop (Fig. 9.9).

It must be noted that although in practice it is preferred to cool the screw only in the feed zone, controlled cooling of the metering zone can be an effective way of accelerating the melting of the broken solid bed and thereby increasing flow stability. To what extent the 'solid' layer of the material on the screw surface is actually stationary or moves slowly through slip and diffuses into the melt at the exit channel portions of the screw has not been verified experimentally.

In considering the theory of flow through the metering zone, each term can be multiplied by a geometrical correction factor to allow for the curvature at the edges of the channels, i.e.

$$Q = f_D Q_D - f_P Q_P \qquad (9.9)$$

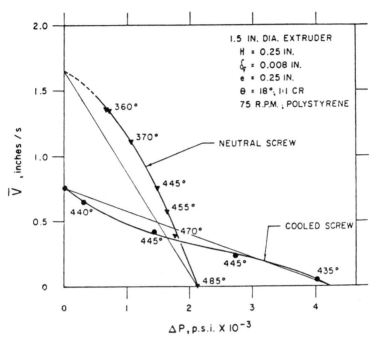

Fig. 9.9. Average velocity in the channel (proportional to flow rate) versus pressure drop for a case of cooled screw and neutral screw. The reduction of flow rate due to a cooled screw is noticeable. (After Colwell and Nicholls.[18])

The values of f_D and f_P are functions of the aspect ratio of the channels H/W and are well documented in the literature.[19] Furthermore, the theory also allows for an additional pressure flow component over the flights of the screw, which is known as the 'leakage flow'; so that the overall flow rate equation becomes:

$$Q = f_D Q_D - f_P Q_P - Q_L \qquad (9.10)$$

Assuming that the 'clearance' annulus between the flights of the screw and the barrel can be approximated by two parallel plates, the flow rate equation for the leakage flow becomes:

$$Q_L = \frac{\pi D_b \delta_c \Delta P_c}{12 \eta_c l_f} \qquad (9.11)$$

where δ_c = clearance, ΔP = pressure drop over the flights (approximately equal to the total pressure drop divided by the number of turns),

η_c = viscosity of melt in the clearance regions, and l_f = total land length of the flights.

Since the pressure gradient in the metering zone is created by the die, it is necessary to consider how the events in these two sections are interrelated. A generalised flow rate equation for the flow through the die can be cast in the following terms:

$$Q_D = k\,\frac{\Delta P}{\eta_D} \tag{9.12}$$

where ΔP = overall pressure differential between the metering zone and the die exit; η_D = shear viscosity of the melt in the die at the particular output shear rate; and k = 'die characteristic', i.e. a geometry factor (for circular dies $k = \pi r^4/8L$, r = radius, L = die land length).

The combination of the flow rate equations for the metering zone of the screw and for the die respectively (ignoring the leakage flow), yields

$$Q = \frac{A\,N}{1 + \left(\dfrac{B}{K}\right)\left(\dfrac{\eta_D}{\eta_b}\right)} \tag{9.13}$$

which is the interception point of the two flow rate curves expressed as a function of ΔP (Fig. 9.10). It can be seen that a range of outputs is possible by changing the screw speed and the ratio η_D/η_b or, in other words, the temperature differential or ratio (T_D/T_b) between the barrel and the die.

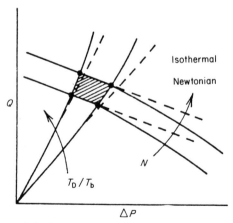

Fig. 9.10.   Effects of die/barrel temperature ratio on flow rate as a function of pressure.

When plotted as a function of the screw speed, on the other hand, eqn. (9.13) yields a series of straight lines whose slope, for a given screw–die combination, depends solely on the ratio of the viscosity in the die and barrel. The feed rate curve for the conveying of the solid granules is also linear, and both screw design and extrusion conditions must be such that the feed rate curve coincides exactly with the flow rate curve for the given screw–die combination. Failure to achieve this balance of flow rates results in a 'starved' flow through the metering zone, which produces a cyclic pulsating flow for every screw revolution (which is unlike the surging resulting from the flow instability set up by the breaking up of the solid profile). Alternatively it may 'choke' the feed zone by displacing the melting zone too much to the rear, thereby preventing the development of a sufficiently high pressure by the conveying action of the solid bed.

The curve for the melt rate as a function of screw speed, on the other hand, is not necessarily a straight line, but follows a power law relationship, whose power index varies from $\frac{1}{2}$ when the heat of melting is supplied entirely by conduction from the barrel to $\frac{3}{2}$ when it is generated purely through viscous dissipation.[19] Therefore, at a given screw speed, the melting rate could well be above or below the flow rate curves. In either case, flow instability would result since an excessive melting rate (e.g. when a high compression ratio screw is used for amorphous materials) would mean that the channel between the last few flights must be filled with melt flowing back from the metering zone. In other words, there will be variations along the screw in the balance of drag flow to pressure flow.

When the melting rate is below the flow rate capacity of the metering zone (e.g. low compression ratio screws used with crystalline materials), instability will result from the solid bed extending too far into the metering zone. To a large extent, however, the balance of conduction heating to viscous heating and, therefore, the matching of melting rate to flow rates through feed zone and metering zone respectively can be accomplished by adjusting the barrel temperature, screw temperature (when screw cooling channels are available) and back pressure by means of suitable screen packs, pressure valve regulation, etc.

9.2.2 Factors Affecting Product Quality

In general, the quality of an extrudate can be expressed in the following terms:

(1)   It must be consistent in dimensions along its length.
(2)   It must be 'homogeneous' through its cross-section.
(3)   It must be free of surface irregularities.

All the required characteristics that are not associated with the items above are considered to be functions of the nature of the material and of the particular die, and other events occurring after the extrudate has left the die.

Item (1) implies that the flow rate through the die must be uniform; i.e. there should be no surging and it should be free of gross distortions, such as melt fracture or draw down resonance. The occurrence of these phenomena determines the maximum output of the machine; surging has already been discussed, while melt fracture phenomena have to be related to the specific dies considered.

Item (3) can be deduced directly from the discussions in Chapt. 7, while other surface irregularities such as grooves along the surface of the extrudate, normally resulting from depositions on the die land (plate-out), are considered to be a general material feature and/or a problem of equipment maintenance.

Item (2), on the other hand, implies that the extrudate should be thoroughly mixed to be free of visible or otherwise detrimental weld lines, and free of voids.

(a) *Effects of Die/Screw Interactions on Efficiency of Mixing of Extruded Products*

The importance of mixing in extrusion processes has already been shown when dealing with the melting of granules or powder in the transition zone. In addition, mixing is an important requirement for the dispersion or dissolution of additives. Obviously, considerable economic savings are achieved when compounding operations are minimised or eliminated altogether. The incorporation of additives in an extrusion process can normally be carried out by means of pre-tumbling of granules (or powder), either directly with additives or with masterbatches and powder concentrates. More recently a technique of pre-dispersing (with partial or total dissolution) of additives in a liquid phase (compatible with the polymer) has been introduced.

For the purpose of this analysis it is possible to distinguish two extreme cases of dispersion of additives: (i) incompatible particles breaking up from agglomerates and taking up 'equidistant' positions in the polymer matrix (e.g. pigments, fillers, etc.) and (ii) compatible 'liquids' which diffuse and dissolve in the polymer matrix. Masterbatches and liquid 'pre-packs' are intermediate cases.

Case (i), Incompatible Particulate Additives. Consider the agglomerate (Fig. 9.11) in a shear field imposed by the flowing melt. The breaking up of

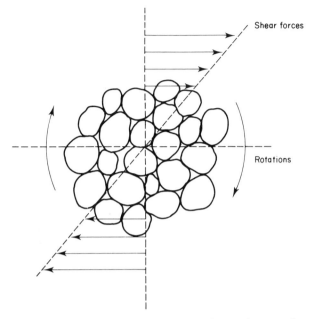

Shear forces

Rotations

Fig. 9.11. Filler particles agglomerate in a polymer melt.

the cluster relies on minimising the rotation of the agglomerate and increasing the level of 'nominal' strain or deformation imposed by the shear stresses transferred to it from the surrounding melt. In terms of orbits described by the agglomerates at given 'net' flow rate, it is to be expected that the longer the path followed by the particles, the greater the time available for each individual particle to reach a state of uniform dispersion. This is why mixing heads at the screw tip or intermeshing multiple screw machines are preferred when mixing is a prime requirement.

In the analysis by Tadmor and Klein[20] the average residence time and average strain have been combined into a parameter called the 'weighted average total strain', $\bar{\bar{\gamma}}$. It has been shown that in the metering zone this increases very rapidly when the $-Q_P/Q_D$ ratio exceeds the value of 0·5. Increasing the back flow obviously lengthens the path of the orbit described by the particles. The beneficial effect of screw cooling on mixing can be interpreted also in terms of longer residence time, since it automatically increases the $-Q_P/Q_D$ ratio.

The recovery characteristics of polymer melts also play an important role in the mixing process, and it can be demonstrated that the higher the recovery compliance, the more efficient the resultant dispersion in laminar shear flow situations.

If slip flow takes place, at a given flow rate, the total strain (shear + elongational) is reduced and, therefore, mixing deteriorates. Consequently, care must be exercised in the choice of lubricants; while lubricants may assist particle separation by reducing the level of stresses required to overcome interparticle friction and adhesion, they may promote slip at the interface between the melt and the surface of the equipment. Ideally, therefore, the lubricant used on the surface of fillers or pigments should be an internal type (i.e. should be compatible with the polymer to minimise the slip flow component).

Case (ii), Compatible Additives. It can be demonstrated that the diffusion rate of compatible additives depends both on the diffusivity coefficient (D_{AB}) of the additive in the polymer matrix and on the weighted average total strain, $\bar{\bar{\gamma}}$. In other words, since

$$\frac{dc}{dt} = D_{AB}\frac{d^2c}{dx^2} \tag{9.14}$$

where dc/dt is the rate of diffusion (i.e. the rate of change of concentration of additive in the polymer matrix) and d^2c/dx^2 is the second derivative of the concentration as a function of the distance from the additive/polymer interface. Increasing the value of $\bar{\bar{\gamma}}$ means, in effect, reducing the distance of the diffusion path (x) of the additive, which results in a higher diffusion rate.

(b) Effects of Die/Screw Interaction on Particle Memory and Voids

It can happen sometimes that extrudates produced from powders, rigid PVC in particular, exhibit a 'grainy' structure, that is to say, that individual particles retain their original identity. As a result of this, the products are very brittle since cracks can easily propagate through the particle boundaries.

The major cause of this problem is the high recovery compliance of the melt (i.e. the material is being processed under conditions which are too near its rubbery state); so that insufficient flow through molecular diffusion takes place within the residence time of the material in the process. Overlubrication of the material, low compression ratio of the screw and insufficient back-pressure flow are additional causes, since all these factors reduce the 'effective' shear stresses on the particles, i.e. those which cause distortion of the particles, as opposed to rotations, and give rise to forward motions. As a result of this, air may also become entrapped at the particle interface and within the polymer particles, when these have a high surface roughness and porosity.

(c) Effects of Die/Barrel Assembly on Weld Lines

On leaving the barrel, the melt is invariably separated, first through the orifices of the breaker plate, and second (when hollow extrudates are produced), through the fins which connect the mandrel to the die/barrel assembly. At this stage the flow is laminar and free of circulatory motions; therefore, the separated streams must be rejoined and their identities destroyed by the time the melt reaches the die exit.

This is achieved by forcing the melt to converge so that the shear stresses at the interface are developed, and sufficient residence time is allowed (i.e. by means of sufficiently long die lands) to enable molecular diffusion to take place across the interface. It is expected, therefore, that the lower the viscosity of the melt at the die, the lower the tendency to produce weld lines in extruded products.

9.2.3 Formalisation of a Global 'Extrudability Index'

From the detailed discussions in the foregoing sections it is clear that there are numerous and often opposing requirements, both with respect to material characteristics and processing conditions, to take into account in defining a processability index for a general extrusion process. Hence it is convenient to use separate indices in relation to specific requirements, in order to avoid the large errors that may arise from the difference in the magnitude of various effects, in a global expression for the extrudability index. These can be cast in the manner below.

(a) Feed Rate Index of Solid Granules

$$\alpha_{FR} = f\left[\frac{1}{\Delta H_{(\mu)}, \phi(S/V, l/d)}\right] \tag{9.15}$$

where $\Delta H_{(\mu)} =$ the activation energy of the coefficient of friction, and $\phi(S/V, l/d) =$ geometrical factor of the feed stock, which is a function of S/V (surface area to volume ratio of the granules) and l/d (the aspect ratio of the granules).

(b) Melting Rate Index

This has already been formulated in eqn. (9.6) and has been cast in the following terms:

$$\alpha_{er} \equiv f\left\{g(\bar{\mu}, \bar{\phi})\left(\frac{\rho_m\left[K_m(T_b - T_m) + \dfrac{\eta}{2}\right]}{2[C_p(T_m - T_s) + \Delta H_m]}\right)^{1/2}\right\}$$

(c) Flow Stabilisation Index
This can be simply expressed as

$$\alpha_{FS} = f(D_{AA}, \eta_A) \qquad (9.16)$$

(d) Flow Rate Index
This can be derived directly from eqn. (9.13) and modified to take into account the coefficient of friction. Since both viscosity and coefficient of friction enter in the flow rate equations for the die and the metering zone and their effects tend to cancel out in the global flow rate equation, the only way the flow rate index can be increased is through their relative activation energies. In other words, if the activation energies for the viscosity ($\Delta H_{(\eta)}$) and for the coefficient of friction ($\Delta H_{(\mu)}$) are high, only small increases in the die temperature are necessary to obtain large increases in flow rate. Therefore,

$$\alpha_{FR} = f(\Delta H_{(\mu)}, \Delta H_{(\eta)}) \qquad (9.17)$$

(e) Quality Index
This can be defined by combining the parameters that affect mixing efficiency and melt welding respectively, i.e.

$$\text{Mixing parameters} = J_{(rec)}, D_{AB}, \mu, \eta$$
$$\text{Welding parameters} = D_{AA}, J_{(rec)}, \eta$$

In the latter, however, the terms $J_{(rec)}$ and η would appear at the denominator and cancel out with those in the mixing term. So, once more the activation energies are the parameters which help to optimise their opposing effects. Hence the quality index can be cast in the following form:

$$\alpha_{Q} = f(D_{AA}, D_{AB}, \Delta H_{(\eta)}, \Delta H_{(J)}, \Delta H_{(\mu)}) \qquad (9.18)$$

where D_{AB} is the diffusivity coefficient of the particular additive in question. If a mixture of additives is considered (as is normally the case), it may be possible to compute a weighted average coefficient, i.e.

$$D_{AX} = \phi_A D_{AA} + \phi_B D_{AB} + \phi_C D_{AC} + \cdots$$

9.3 ANALYSIS OF INJECTION MOULDING PROCESSES

As already mentioned on p. 366, an injection moulding operation occurs in three consecutive steps:

(1)   Feeding and thermally plasticising a stock of granules by means of

an extrusion-type screw that rotates until the required amount of melt has been conveyed to the front of the barrel.

(2) Injecting a metered amount of stock into a cold cavity.

(3) Cooling and ejecting the moulded components.

9.3.1 Analysis of the Feeding and Thermal Plasticisation Steps of the Injection Moulding Process

Feeding and melting of the granules in an injection moulding machine differ from the procedures in extrusion with respect to the following:

(1) The rotation of the screw is intermittent and may take place only over a fraction of the total cycle time. Consequently, while the melt cools in the cavity of the mould, the melting process in the channels of the stationary screw continues through heat transfer by conduction from the barrel and from the melt surrounding the granules. Experiments have shown, in fact, that the melt zone in injection moulding screws is shorter with intermittent rotations than with continuous extrusion plasticisation.[21]

(2) The melt travels through the metering zone of the screw at zero discharge rate (i.e. without any flow through the nozzle), so that the pressure developed at the front of the barrel displaces the screw backwards and continually shortens the length of the feed zone. The pressure that is developed at the front of the barrel depends on the resistance exerted on the screw, i.e. the sum of the weight of the screw, the frictional forces exerted by the screw against the barrel and by the drive gear assembly, and the hydraulic back pressure opposing the return of the screw (which can normally be adjusted).

Consequently, increasing the speed of rotation of the screw mainly produces a faster return without increasing very much the melting rate by viscous heating. In fact, faster rotations and shorter screw return times tend to reduce the length of the melt zone.[21] Only by increasing the back pressure onto the screw will the melt rate increase, as a result of the higher pressure built up at the front of the barrel. In other words, this has the effect of increasing the $-Q_P/Q_D$ ratio in the metering zone, while achieving at the same time a better mixing of the melt in view of the greater weighted average total strain, $\bar{\bar{\gamma}}$.

When the melt is injected into the cavity, the pressure at the front end of the barrel increases to very high values (typically $50\text{--}100 \, \text{MN/m}^2$) and is prevented from producing a back flow along the channels of the metering zone of the screw by means of a non-return valve at the screw tip. This is not the case for PVC and thermosetting materials, which would decompose and 'harden' in the 'dead' spots of the valve; in these cases the melt viscosity in the metering zone and the frictional forces in the feed/compression zone are sufficiently high to prevent appreciable back flow.

9.3.2 Analysis of the Injection, Cooling and Ejection Steps of the Injection Moulding Process

These steps are by far the most complicated in the injection moulding process, in view of the irregular path that the melt must follow, the differential cooling rates and the complex stress patterns set up both during cooling and on ejection. The analysis of these stages of the injection moulding process mainly concerns the following:

(1)   Pressure requirements for mould filling and packing.
(2)   Cavity filling mechanisms.
(3)   Cooling rates and temperature distribution within the material in the mould from the time the cavity has been filled to the time it is being ejected.
(4)   State of the material at ejection and post-ejection events.

All the events in these steps of the process are interdependent and, therefore, cannot be considered in isolation. Furthermore, apart from the additional aspects of mixing, these events control both the quality of the products and the production rate (i.e. cycle time).

Feeding and thermal plasticisation in the screw/barrel assembly must take place while the 'moulding' (jargon for the component being moulded) is cooling and being ejected from the mould; hence they have no effect on the cycle time. Naturally, the screw/barrel assembly has to be designed to provide the required feeding and melting rate so that the above can be achieved; or, in other words, the size of the machine, normally rated on the basis of their swept volume and melting capacity, is selected according to the dimensions of the component(s) to be produced.

9.3.3 Pressure Requirements for Cavity Filling

For the purpose of the analysis two basically different processes must be distinguished: the high pressure (or external pressure) cavity filling process, and the low pressure (or internally generated pressure) process for cellular mouldings. The low pressure extrusion filling process will be regarded as synonymous with the mould packing stage of a conventional high pressure process.

(a) *High Pressure Cavity Filling*

The pressure required to fill the cavity of a mould depends on the flow mechanism, flow path ratio (flow length/height of flow channel) and flow path geometry for the complete channel components of the mould, i.e. sprues, runners, gates and cavities. In general, these are far too complex and variable to be amenable to quantitative treatments, and consequently only qualitative predictions can be made.

As a first approximation one can make the generalised statement that the total pressure required is the sum of the melt delivery pressure (i.e. the pressure required to fill the cavity against atmospheric pressure resistance only) and the pressure that must be exerted onto the melt to compress it and reduce the free volume to a level determined by the allowable volumetric shrinkage of the moulding.

$$\bar{P}_{im} = \sum \Delta p + f(V_m - \bar{V}_s) \tag{9.19}$$

where $\sum \Delta p$ = sum of pressure drop to fill the cavity via sprue, runners and gate channels; and $f(V_m - \bar{V}_s)$ = packing pressure at sprue entry (V_m = specific volume of the polymer at atmospheric pressure, $\bar{V}_s$ = average specific volume of the melt in all channels and cavity on completion of the packing stage).

The delivery pressure $\sum \Delta p$ depends on the viscosity characteristics of the melt (i.e. viscosity as a function of shear rate, temperature, pressure, etc.) and the geometry of the channels. A generalised expression for this pressure drop can be obtained from the momentum equation, which for the case of a unidirectional laminar shear flow reduces to:

$$\frac{\partial^2 V_2}{dy^2} = \frac{2}{\eta}\left(\frac{\partial p}{\partial z}\right)$$

Hence,

$$\frac{\partial V_2}{dy} = \frac{y}{\eta_a}\left(\frac{\partial p}{\partial z}\right) + C_1 \tag{9.20}$$

where $\partial V_2/dy$ = velocity gradient across the section of the channel of height y, and is equal to the shear rate; $\partial p/\partial z$ = pressure drop along the flow path; and C_1 = constant of integration, which depends on the geometry of the flow channels and the boundary conditions.

The pressure drop, therefore, increases with the rate of shear and the apparent viscosity, η_a, and is inversely related to the cross-section height of the flow channels. However, the conditions of unidirectional laminar shear rarely occur in injection moulding; the boundary conditions for the estimation of C_1 are difficult to define, and both the apparent viscosity and the effective height y of the channels change with time as a result of the 'freezing' of the outer layers in contact with the cavity walls. Furthermore the pressure drop for filling the cavity depends on the flow mechanism imposed by the geometry of the gate and the velocity of the emergent jet. Three basic mould filling mechanisms can be stipulated, as follows.

(*i*) *Jet Filling of the Mould Cavity.* When the size of the gate is such that the emerging melt does not touch or glances off the walls of the cavity (typically of thick sections), the 'jet' will reach the opposite end of the cavity and will build layers in a zig-zag fashion (Fig. 9.12).

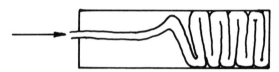

Fig. 9.12.   Jet filling mechanism.

Since the pressure drop associated with the momentum of a free-flowing jet is quite small, the total delivery pressure is that required to overcome the frictional drag in passing through the sprue, runners and gates. It is to be expected, therefore, that this filling mechanism makes the smallest demands on pressure requirements for a given shear rate (or speed of injection) and melt temperature at the gate.

From the materials characteristics viewpoint, one would predict that jet filling is inversely related to the swell ratio of the material and is more likely to take place when melt fracture occurs at the gate. The latter, in fact, would not produce further expansion of the melt, since the normal stresses would relax in the release of energy in the melt fracture process (Fig. 9.13).

To what extent jet filling of cavities is desirable simply on a pressure requirements basis is difficult to ascertain, although it would seem obvious that more of the available pressure can be utilised for packing. The more obvious advantages associated with randomisation of fibres in the moulding of filled compositions will be discussed later.

(*ii*) *Turbulent Flow Filling of Mould Cavities.* When the flow front undergoes frequent changes in direction upon emerging from the gate and upon impinging against obstacles, such as inserts, in the actual cavity, it is difficult to achieve fully developed flow conditions. This mechanism is

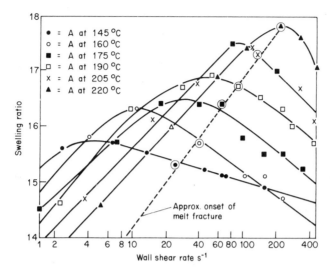

Fig. 9.13. Swelling ratio against shear rate for a low density polyethylene at six temperatures. (After D. L. T. Beynon and B. S. Glyde, *Br. Plast.*, **33** (1960) 416.)

sometimes deliberately produced to avoid 'jetting', which may cause the formation of characteristic splay marks on the surface of moulded products especially with high viscosity materials, and to prevent the occurrence of flow lines on the surface of the component. Figure 9.14 shows the change in filling mechanism from jetting to turbulent flow through altering the position of the gate relative to the cavity and introducing a small tab at the side of the cavity. A typical turbulent filling mechanism is also produced by means of tunnel or submarine gates.

Turbulent flow filling can also occur with gates normal to the plane of the moulding (e.g. pin gates) and can often be recognised from periodic imprints across the surface of the moulding and perpendicular to the flow direction. These imprints are caused by the localised 'freezing' of the melt front in its sinuous path along the cavity. By its very nature, it is expected

Fig. 9.14. Effect of the use of a side-gate on the flow mechanism of the melt in the cavity.

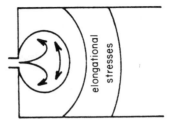

Fig. 9.15.   Mechanism for the setting up of elongational stresses within the melt
flowing through a cavity.

that turbulent flow in the cavities makes considerably higher demands on
pressure than 'jet' filling.

(*iii*) *Divergent Laminar Flow Filling of Mould Cavities.* With side-gated
cavities quite frequently the height of the gate channels is only marginally
smaller than that of the cavity. This means that the swelling of the polymer
melt emerging from the gate will cause the flow front to diverge into a
circular flat profile. As the melt advances into the cavity, the radius of
curvature of the melt front gradually increases (see Figs. 9.15 and 9.16) in
view of the elongational stresses set up as the front is forced to stretch out
laterally. This is also a direct consequence of the surface tension at the front

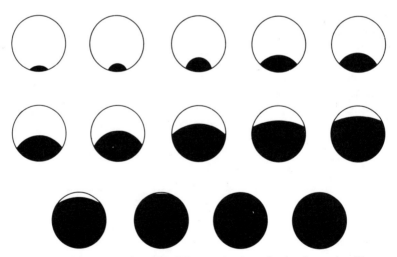

Fig. 9.16.   Divergent laminar flow filling mechanism of a circular cavity. (Retraced
from original photographs by Gilmore and Spencer.[22])

wall of the fluid and is probably accentuated by the normal stress difference, i.e. the elongational stresses in the transverse direction being greater than those parallel to the line of flow emerging from the gate.

This situation also arises in the case of centre-gated (sprue or direct feed gates), i.e. the gate axis is normal to the direction of the flow in the mould cavity. As discussed earlier, in this case the flow may start as turbulent at the entry point; however, the deceleration effect produced by the divergence (spreading) can quickly dampen the oscillatory velocity component in the plane of the cavity and establish laminar flow conditions over a short distance.

The net consequences of laminar divergent flow during mould filling are two-fold.

First, the polymer at the wall 'freezes'; that is to say, it reaches the rubbery state conditions. Consequently the effective height of the flow channel decreases with time, forming a tapered channel for the incoming melt. This, together with the elongational stresses' effect on deformable (rubbery) inclusion is shown in Fig. 9.17. Eventually, at some distance from

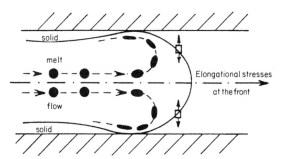

Fig. 9.17. Freezing mechanism of a melt at the cavity walls. (After Z. Tadmor, *J. Appl. Polym. Sci.*, **18** (1974) 1753.)

the gate, the effective height of the flow channel becomes so small that flow may cease altogether, producing 'short' mouldings.

In mould design practice the ratio of the maximum flow length of a material to the cavity height (i.e. the flow path ratio) under 'normal' injection moulding conditions in terms of pressure, speed of injection and average mould and barrel temperatures, is used as a parameter to decide the position of the gate or number of gates required to fill the cavity. Recommended values for the maximum flow path ratio for the most common thermoplastics materials are shown in Fig. 9.18.

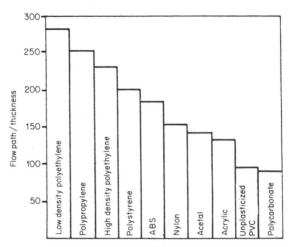

Fig. 9.18. Suggested maximum values for flow length/thickness ratios of some commonly used polymers. (After Jarret.[23])

These values suggest that while at the end of the scale it is the high melt viscosity of the material that limits its 'natural' flow length, the important parameter is the temperature drop required for the onset of rubbery state conditions. Materials such as Nylon 6 or 6,6 and acetals, for instance, despite their low viscosity will afford only slightly higher flow length values than the most viscous materials, such as PVC, acrylics and polycarbonates.

Second, the divergent flow situation creates additional and substantial pressure losses resulting from the much larger values of the elongational viscosity in relation to the shear viscosity of polymer melts. Calculations of the pressure requirements for mould filling of a disc-shaped cavity from a centre gate have been attempted by Barrie,[24] who assumes that the total pressure drop can be taken as the sum of one component resulting from laminar shear flow and one from elongational flow. The calculations were performed on the basis that in shear the melt follows a power law relationship, while the tensile viscosity is independent of the shear rate over the range of practical interest.

9.3.4 Pressure Requirements for the Packing of the Melt in the Mould

That polymer melts are compressible has been recognised for a long time by both processing technologists and rheologists. (See discussions in Chapt. 7.) For instance, in the *TS Note G103* of ICI Ltd (p. 16), one finds that the natural volumetric shrinkage (i.e. at atmospheric pressure) exhibited by common thermoplastics when cooled from typical injection moulding temperatures to room temperature is as shown in Table 9.2.

On the other hand, typical values quoted for the linear mould shrinkage for unfilled thermoplastic are usually in the range of 0·5 to 0·8 % for amorphous polymers and 1–3 % for crystalline polymers. Although the volumetric shrinkage values are expected to be approximately three times higher, the values in Table 9.2 suggest that something like 6–8 % more material has entered the mould cavities than would have been the case if the melt were discharged against atmospheric pressure only (i.e. without melt packing).

Table 9.2

Materials	Cooled		Approximate volume contraction
	From	To	
Acrylics	150 °C	20 °C	7 %
Nylon 6,6	285 °C	20 °C	14 %
Nylon 6	260 °C	20 °C	13 %
Low density polyethylene	190 °C	20 °C	18 %
Polystyrene	195 °C	20 °C	7 %

Obviously, if a melt could be compressed to achieve the same density as it would have at room temperature (neglecting the complications arising from frozen orientation, which will be discussed later) then the cavity could be made to the exact dimensions as those specified for the component to be produced. This would seem, therefore, an ideal solution to the problem of tolerances, and consequently it would be necessary to calculate the follow-up pressure (i.e. the level of pressure that follows the mould filling stage) required to compress the melt by a pre-determined amount.

A theoretical approach to this problem was first attempted by Spencer and Gilmore,[25] who proposed the use of the equation of state of real gases to predict the relationship between pressure, temperature and volume, i.e.

$$(p + \pi_i)(V - a) = RT \qquad (9.21)$$

where π_i = internal pressure of the polymer (i.e. isotropic internal stresses corresponding to the cohesive energy); a = incompressible specific volume, which is proportional to the molecular weight of the monomeric units and is a function of their stereo configuration; p = applied external pressure;

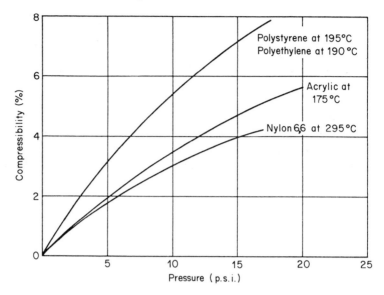

Fig. 9.19.   Compressibility of molten thermoplastics.[26]

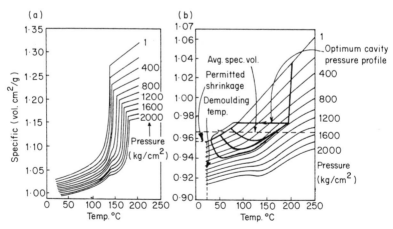

Fig. 9.20. P–V–T diagrams for typical crystalline (a, polyethylene) and amorphous polymers (b, polystyrene).[27]

$V = $ specific volume; $R = $ universal gas constant; and $T = $ absolute temperature.

Although this equation predicts linear relationships between the three stated variables, which are neither expected to hold over an unlimited range of conditions on a theoretical basis (in view of the possible chain/chain interactions) *nor* actually confirmed experimentally (see Figs. 9.19 and 9.20), it does enable the mould designer and processor to make approximate estimates of pressure requirements when the full P–V–T diagram for the particular material is not available.

The experimental values of π_i and a for common thermoplastics materials as obtained by Spencer and Gilmore[25] and Sagalaev et al.[28] are shown in Table 9.3.

Table 9.3: Experimental Values of π_i and a for Common Thermoplastics

Material	π (MN/m^2)	a (cm^3/g)
Polyethylene (branched)	329	0·88
Polypropylene	247	0·83
Poly(4-methylpentene 1)	105	0·83
Polystyrene	187	0·82
Polymethyl methacrylate	218	0·73
Nylon 6	54	0·77
Polycarbonate	46	0·56

The recent introduction of programmers for the pressure and speed of injection controllers on injection moulding machines makes it possible, therefore, to select and control the required changes in these two variables throughout the entire injection cycle. By means of a feedback from pressure transducers and thermocouples in the cavity, the hydraulic pressure and the rate of displacement of the screw are adjusted to the predetermined values. The selection of the rate of change of pressure as a function of the temperature of the melt in the mould is shown in Fig. 9.20(b).

The major problems that arise in the optimisation of pressure and speed of injection profiles in the filling and packing stage of the injection moulding process are as follows.

First, the differential cooling rates across the section of the moulding (see Fig. 9.21) bring the outer layers into the rubbery state prior to achieving the desired level of packing. There can also be a differential cooling rate, both across and along the cavity, as a result of uneven fluxes through the walls of

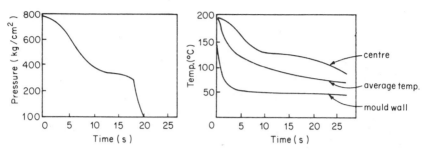

Fig. 9.21.   Pressure and temperature variations in a moulding cycle. (Reproduced from ref. 29.)

the mould according to position and design of the water cooling channels and thermal sink capacity of the various mould components.

Second, variations in cross-sections produce differences in velocity profiles (and associated strain rates) which can also affect the rate of cooling (and/or viscous heating) during the mould filling and packing stage.

Third, the pressure drop along the flow path will produce a gradual variation in the level of cavity packing, and is responsible for the gradual increase in across-section shrinkage away from the gate.

At various time intervals $(t_1, t_2, t_3, t_4,$ etc.) from the application of pressure at the gate one can have a situation similar to that depicted in Fig. 9.22, where the difference in depth of shading in the various regions shown is proportional to the pressure and the temperature, and hence, the time sequence and state of the material during the packing of the cavity. This denotes that the material in the core regions will be compressed to a higher level than the outer layers and, as a consequence, will display a lower degree of shrinkage during cooling.

The moulding will be ejected after the material in the gate channels has

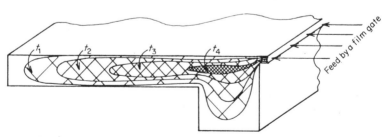

Fig. 9.22.   State of thermoplastics at different stages of mould packing in injection moulding.

reached the rubbery state (so that flow through the gate will cease) and the outer layers will have reached a temperature sufficiently low to resist any distortion or imprints by the ejector pins.

The situation depicted in Fig. 9.22 will produce internal stresses; i.e. the outer layers will be in tension and the core in compression. These are complex stresses and (unless the mouldings are substantially thick) are likely to be plane stresses.

Because of the viscoelastic nature of thermoplastics these stresses will tend to relax at a rate which depends on temperature, and readjust themselves towards a state of equilibrium, i.e. zero internal stresses. If the relaxation process continues after the moulding has been ejected from the cavity so that there will be no external forces acting on the component, then distortions, sinks, bulges, etc., may result, depending on the constraints imposed by the geometrical features of the component.

In order to avoid distortions it may be necessary either to introduce deliberately geometrical features (such as ribs) which can resist the bending deformations arising from the relaxation of internal stress, or to keep the component in the mould (or even quickly transfer it to fixtures which hold it into the same shape as in the mould) until the material reaches its glassy state.

In many cases, however (e.g. LDPE, EVA, ionomers, flexible PVC, etc.) this temperature is either below ambient temperature, or the rate of relaxation to reach an acceptable minimum level at ambient temperatures is too low and therefore uneconomical. This means that the component must be 'annealed', i.e. must ideally be brought to its rubbery state conditions under external constraints, for a sufficient length of time to allow complete relaxation of the internal stresses.

If, on the other hand, prevention of distortions is the only important consideration, and the material has a glass transition temperature sufficiently higher than the ambient temperature (e.g. polystyrene, SAN, acrylics, polycarbonates, etc.), then it suffices to mould the component under high cooling rate conditions (e.g. low mould temperature) and very high speed of injection. This makes it possible for the moulding to be ejected, at an economical cooling cycle, at temperatures substantially below the glass transition temperature so that stress relaxations after moulding are completely suppressed.

In the case of crystalline polymers the natural volumetric shrinkage of the material is far too high to produce an adequate level of melt compaction during the mould packing stage and to reduce the volumetric shrinkage to the level that is achievable with amorphous polymers. Furthermore, the level of crystallinity developed is inversely related to the cooling rate so that the lower cooling rate and higher pressure in the core regions tend to produce a level of crystallinity and, therefore, density which is higher than in the outer regions. It can also be easily appreciated that, if the cooling rates of the outer surfaces of a flat moulding are different, then the difference in density across its section will create a bending moment which results in the bowing of the component. It can be inferred from the above discussion that, while with amorphous materials the core is always likely to be in compression, with crystalline polymers the situation may be reversed. In other words the greater crystallisation shrinkage in the core will 'pull inwards' the outer layers and cause a sink (depression) on the surface by an amount depending on the rigidity of the outer layers and the difference in cooling rates between outer and inner layers. This is obviously likely to be much more pronounced in the thick sections of the moulding.

With the higher modulus materials the shrinkage stresses in the core are often not sufficient to cause further bowing of the outer layers. Therefore, voids will be formed instead within the core to accommodate the natural contraction of volume of the material. This inversion of mechanism is largely responsible for the much greater shrinkage of crystalline polymers than amorphous polymers. In general, one can conclude that mould shrinkage can be minimised through the following four expedients:

First, reducing the level of natural shrinkage by selecting polymers with a lower degree of crystallinity and/or by incorporating inorganic fillers.

Second, accelerating the natural rate of crystallisation in order to minimise differential levels of crystallinity developed through variations in cooling rate. This can be done by selecting polymers with a narrower molecular-weight distribution and/or by incorporating nucleating agents.

Third, promoting very fast cavity filling to minimise the temperature drop across the moulding during the packing stage, thereby increasing the amount of compaction. This is normally achieved by means of a pressure overshoot at the gate during the filling stage and by programming the pressure profile thereafter, or by simply reducing it to a lower level (sometimes called the 'second stage') when a programmer is not available on the injection moulding machine. From the material characteristics aspects this can be achieved by using polymers with a low viscosity, and which are highly pseudo-plastic, but at the same time possess a low recovery compliance to promote a jet filling mechanism. (Referring to the discussions in Chapt. 7, this means that those polymers more likely to possess these characteristics are highly branched.) Fillers should also promote jet filling in view of their ability to decrease the compliance and to increase the shear rate sensitivity of the viscosity.

A higher thermal diffusivity would increase the 'freezing' tendency of the polymer in the filling stage, but this is a property which cannot be greatly changed either through synthesis or by means of additives. Furthermore, a decrease in thermal diffusivity would also increase the global cooling time, which is not desirable on economic grounds. A relationship between thermal diffusivity and cooling cycle in injection moulding has been obtained by Ballman and Shusman,[30] i.e.

$$t_c = -\frac{D^2}{\pi^2 \alpha} \ln \left\{ \frac{\pi}{4} \left(\frac{T_x - T_m}{T_c - T_m} \right) \right\} \qquad (9.22)$$

where t_c = cooling time, D = moulding thickness (maximum), α = thermal diffusivity, T_x = ejection temperature, T_c = barrel temperature, and T_m = mould temperature. Note: $\alpha = K/\rho |C_p|_v$, where K = thermal conductivity, $|C_p|_v$ = specific heat at constant volume, and ρ = density.

Fourth, selecting polymers that only require small pressure increments to produce large reductions in specific volume. In terms of the equation of state this means that small values of π_i and a are required. This can be inferred from the definition of compressibility when the appropriate substitutions from the equation of state are made, i.e.

$$\beta = -\frac{1}{V} \left(\frac{\partial V}{\partial P} \right)_T = \frac{1}{(P + \pi_i) + \left[\frac{a}{R} \times (P + \pi_i)^2 \right]} \qquad (9.23)$$

There seems to be some disagreement however, between the predictions obtained when substituting the values of a and π_i reported in Table 9.3 and the results presented in Fig. 9.19. This discrepancy may be due, however, to experimental errors arising from crystallisation phenomena which cannot be detected from simple observations of volume as a function of pressure.

9.3.5 Formalisation of an 'Index for the Ease of Mould Filling and Packing'
On the basis that predictions from the equation of state are approximately correct an expression for the 'ease of mould filling and packing' can be cast in the following terms, i.e.

$$\alpha_{fp} = f \left(\frac{1}{(\chi_p), (\eta), J_{(rec)}, \beta(\pi_i, a)} \right) \qquad (9.24)$$

where χ_p denotes the degree of crystallinity and η, $J_{(rec)}$, and $\beta(\pi_i, a)$

represent the properties of the melt already discussed. Obviously, to take into account the freezing effect at the cavity wall, one could add at the denominator the thermal diffusivity term. However, as already mentioned, the relatively small difference in thermal diffusivity among the various polymers is unlikely to have a substantial effect on mould filling and packing predictions, except for the case of very slow injection rates, e.g. large-area mouldings.

9.3.6   Problems Arising from Mould Packing

If a constant pressure profile is maintained throughout the entire injection stage, it is likely that, in the gate regions (particularly when the gate is large so that its 'freezing' is being delayed), the pressure will be higher than the level necessary to bring the specific volume to the desired value (the value it would have at ambient temperature under atmospheric pressure only). On the basis of earlier discussions this means that there can be a substantial level of residual stresses in these regions (i.e. the core will be in compression while the outer layers will be subjected to tensile forces) if the moulding is cooled quickly and the glass transition temperature is substantially above ambient temperature.

This situation is very damaging to the performance of the component, especially with respect to its resistance to fracture. A crack that propagates through the outer regions will be subjected to stress intensification substantially greater than that expected from external stresses alone. The susceptibility of the component to brittle fracture is accentuated by the likely 'scar' left behind when separating the moulding from the small plug in the gate channels, since this produces a rather acute external notch.

In the case of crystalline polymers there can be an even greater difference in specific volume between elements of material at the gate and those away from it, since the cooling rate in the gate regions can be considerably lower (because they are subjected for longer periods to the incoming melt and are further away from the cooling channels) and, therefore, a higher degree of crystallinity is developed. If the glass transition temperature of the amorphous phase is below room temperature (e.g. polyolefins), so that the internal pressure relaxes through bulging of the outer skin after the moulding has been ejected from the cavity, then the material in the gate regions will be subjected to tensile (plane) stresses created by the shrinkage gradient along the line of flow. This arises particularly if the gate feeds the cavity from a centre point, rather than from the side.

As for the case discussed above, this situation increases the susceptibility to brittle fractures to the extent that, with very brittle polymers, e.g.

polypropylene (homopolymers) and poly-4-methyl-pentene, fracture at the gate can occur on subsequent cooling and in the absence of external forces.

Furthermore, the decay in packing pressure along the flow path would certainly cause an equivalent reduction in mould shrinkage, making it difficult, therefore, to maintain the tolerances in dimensions in the finished component. With crystalline polymers it can also happen that the shrinkage rate due to crystallisation is greater than the packing rate of the melt through the core. As a result, the melt, during the subsequent follow-up pressure stages, may find a lower resistance to flow through the gap between the cavity wall and the surface of the moulding. This gives rise to the formation of a skin which does not 'weld well' with the surface of the moulding previously formed (owing to the low diffusion rate through the frozen skin) and can peel away during the subsequent service life of the article. This is an effect which is often known in the trade as the 'onion peel'.

9.3.7 Orientation in Injection Moulded Components

Orientation has a predominant effect on the 'quality' of injection moulded components, i.e. with respect to tolerances, dimensional stability, warping, and all of the other performance characteristics affected by anisotropy (see Chapt. 7). Orientation in injection moulded components arises from cooling effects which bring the polymer to its glassy/rubbery state before the molecular relaxation (or recovery) process is completed.

Two aspects of orientation must be distinguished in injection moulding: one arising from the effect of freezing of the outer layers and the other from elongational flow. Both phenomena retard the rate of recoiling of molecular chains after mould filling and packing.

(a) Orientation Resulting from Melt Freezing at the Cavity Walls

On the basis of the discussions on mould filling and packing mechanisms it is evident that the 'type and extent of orientation in injection moulded components can be quite complex. For the same component and material, orientation may vary considerably according to the gate geometry and its position with respect to the cavity, cooling rates, levels of pressures, rates of injection, temperature of discharge of the melt, etc.

By far the most complex situation arises when cavity filling takes place by a 'jet mechanism'; while for the case of a 'turbulent flow' this complexity may only extend over the entrance regions of the cavity, since outside these areas the flow is of the laminar type. On the assumption that, with jet filling, most of the orientation introduced in the passage through the gate can subsequently relax in the cavity (except at the folds, where the streams touch the cavity walls) it can be expected that the net residual orientation in the component is rather small. Furthermore, during mould packing, the shear strains introduced in the regions adjacent to the cavity walls are also rather small and only produce marginal increments in the orientation pattern set up during the filling stage. A similar situation exists also in the entrance regions of the cavity when a turbulent flow discharge operates.

For the case of unidirectional laminar shear flow, the orientation pattern which develops in the component can be inferred directly from the flow mechanism illustrated in Fig. 9.17. The elongational stresses set up at the front of the advancing melt increase the effective λ_U/λ_R ratio in view of the local increase in strain and the greater λ_U value acquired through the change from shear stresses (in the core) to tensile stresses. This results in a net increase in orientation from that occurring in the core regions away from the melt front. When the front regions of the melt impinge against the walls of the cavity, this orientation will be largely retained through the rapid cooling.

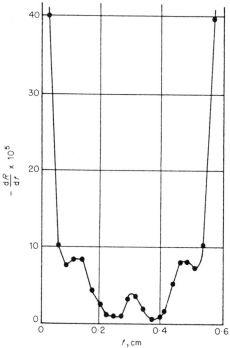

Fig. 9.23.   Birefringence distribution across polystyrene impact bar test piece. (After Murphy.[31])

The freezing of the melt in the outer layers reduces the temperature of the 'layer' beneath and brings it towards the rubbery state conditions. Consequently, at some distance away from the cavity wall, a substantial amount of orientation is expected in view of the shearing action from the incoming melt. However, the temperature in these regions is substantially higher than at the wall and, in view of the shear nature of the stresses, the polymer can relax faster than at the surface. Furthermore, as the thickness of the frozen layers increases, the actual (isotropic) pressure (*not* the pressure gradient) acting on the melt increases from the onset of the packing stage. This may counteract, to a large extent, the tendency toward greater relaxation rates resulting from the high temperature, and may produce further peaks in the residual orientation pattern.

Typical orientation patterns developed in injection moulding for amorphous and crystalline materials respectively are shown in Figs. 9.23 and 9.24.[31] It is interesting to note that although in both cases the flow through the cavity is of the unidirectional laminar shear type (i.e. without divergence of direction which gives rise to elongational flow), some

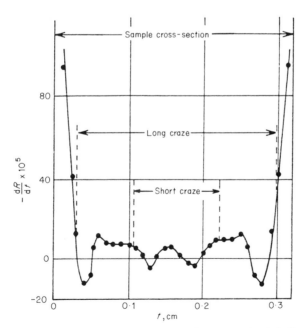

Fig. 9.24.   Birefringence distribution across polystyrene dumb-bell test piece.
(After Murphy.[31])

transverse orientation can take place just beneath the surface of the moulding. Whereas this phenomenon can be easily understood for the case of crystalline polymers on the basis of the discussions in Chapt. 7, it is more difficult to explain in the case of amorphous materials. The only interpretation that can be given in the latter case is that such orientation takes place during the packing stage, when the pressure in the core of the moulding causes some lateral stretching of the material in its rubbery state just beneath the surface.

Note also that, in either case, the true details of the orientation pattern at the surface may have been obscured by the thickness effect of the samples examined and, therefore, the values shown are merely averages of some much more complex peaks. It would seem, for instance, that for the case of crystalline polymers, orientation is actually at a maximum at the surface[32] (Fig. 9.25) and follows a pattern different to that of amorphous polymers (Figs. 9.23 and 9.24). Furthermore, the rapid freezing of the polymer at the melt front, subjected to tensile stresses, impedes the growth of spherulites and, therefore, a fibrillar crystal structure develops, instead, at the surface.

As already mentioned, the details of the orientation pattern in a moulding depend on moulding conditions and types of gates used. For the

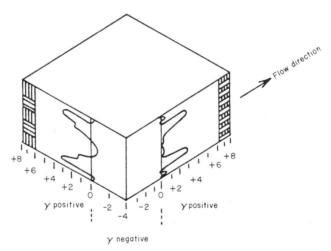

Fig. 9.25.   Isometric diagram of the layers of differently birefringent LDPE in a flash-gated plaque. (Birefringence $\times$ 10^{-3}.) The magnitude of γ indicates the degree of orientation which is different in the two directions. The sign of γ indicates the direction of the polymer chain axes which are also shown diagrammatically in the hatched area. (After Cole.[32])

case of polystyrene, these effects have been studied in detail by Ballman and Toor;[33] some of their findings are shown in Fig. 9.26.

It is understood that, since the pressure drop decreases along the flow path, both the strain rate at the front of the advancing melt and the shear stresses between the frozen skin layers and the flowing melt in the core will decrease. As a result, there will be a gradual reduction in degree of

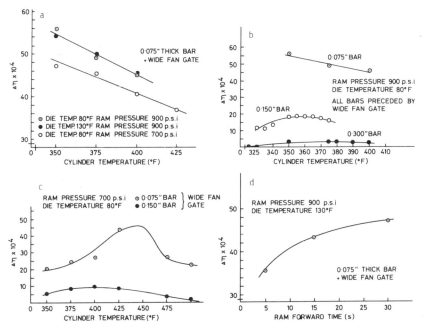

Fig. 9.26. Effect of processing conditions on birefringence of injection moulded polystyrene specimens. (After Ballmann and Toor.[33]) (a) Effect of injection moulding cylinder temperature on orientation. (b) Effect of cylinder temperature on orientation of injection moulded polystyrene—comparison of several cavity thicknesses. (c) Effect of injection moulding cylinder temperature on orientation. (d) Effect of injection moulding ram forward time on orientation.

orientation. This can be seen directly from the birefringence pattern exhibited by an injection moulded rectangular bar. (See Fig. 9.27(a) and (b).)

In interpreting the fringe patterns in the diagrams in Fig. 9.27 note that each shaded area represents a certain amount of orientation. Since white light is polychromatic when it disperses into its component wavelengths,

Cavity entrance   (a)   Wave front

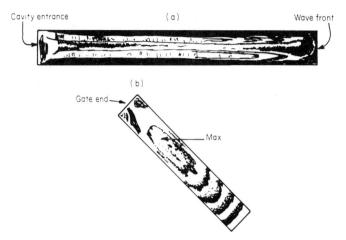

(b)

Gate end

Max

Fig. 9.27.   Birefringence of an injection moulded polystyrene bar. (a) View across the thickness of the bar: maximum orientation is displayed at the skin. (b) View from the flat side of the bar: maximum orientation is displayed at the gate end. (After Ballmann and Toor.[33])

the retardation at any point will be just sufficient to cancel out a particular wavelength from the spectrum. This causes the eye to see a combination of the main colours or the complement of the cancelled colour.

(b) Orientation Arising from Elongational (Divergent) Flow
As already mentioned, divergent or radial flow develops biaxial stresses during mould filling and packing, i.e. elongational (normal) stresses at the melt front and transverse shear stresses in the layers beneath the surface. This is shown in Fig. 9.28 (left). The elongational stresses in the melt

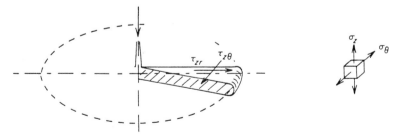

Fig. 9.28.   Biaxial stresses in divergent (radial) flow. Left: Shear stresses at wall and surface layers beneath. Right: Elongational stresses at the melt front.

reaching the cavity wall are the same (right) except that the element considered would undergo a 90° rotation (see Fig. 9.17).

It is expected, therefore, that the above stresses will develop biaxial orientation in the moulding, and that the extent of orientation will be greater in the radial direction than in the transverse direction. The qualitative orientation pattern both across the thickness and along the direction of flow will be similar to that for unidirectional laminar flow.

Although very little attention has, so far, been given to the question of biaxial orientation in injection moulded components, the results of Ogorkiewicz and Weidmann[34] seem to support the above suggestions. The plots of the 100 s isochronous tensile creep modulus for polypropylene copolymer mouldings (Fig. 9.29), produced with a central sprue gate, in fact, deviate from those expected from uniaxial orientation only with respect to the asymmetry. Note that the difference between the A- and B-type mouldings is with respect to moulding conditions only: the A-type mouldings were produced at lower barrel and mould temperature than the B-type mouldings.

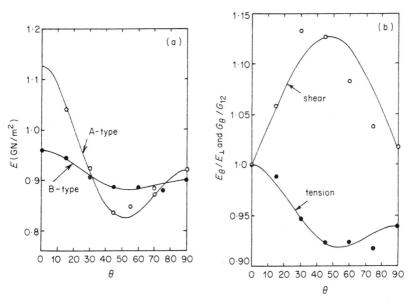

Fig. 9.29. (a) Variation with angle-to-flow direction of experimental values of the 100-s tensile creep modulus at 0·2 % strain. (b) Relative variation with angle-to-flow direction of the 100-s creep modulus in shear (upper curve) and in tension (lower curve) for B-type mouldings. (After Ogorkiewicz and Weidmann.[34])

(c) Problems Arising from Radial Flow Orientation in Injection Moulded Components

In addition to the problems of anisotropy and embrittlement, orientation gives rise to dimensional instabilities during the service life of the component and also to warping on ejection from the cavity. There are two main types of warping problems that arise in injection moulding: bowing and hyperbolic–paraboloid ('figure of eight') distortions, both of which arise from the differential shrinkage between the flow direction and that transverse to it during the cooling stage. The rate of shrinkage in the flow direction is greater than in the transverse direction. The two types of warping can be illustrated by reference to the moulding of a disc, using a standard side gate and a central sprue gate respectively (Fig. 9.30).

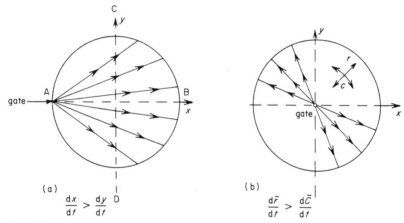

Fig. 9.30. Typical warping in injection moulding. (a)—(1) After packing, $CD = AB$; (2) at ejection time, $CD > AB$, hence the disc will be bowed along the CD line. (b)—(1) After packing, $\bar{c} = 2\pi\bar{r}$; (2) after cooling, i.e. at ejection time, $\bar{c} > 2\pi\bar{r}$, hence a 'figure of eight' distortion has to take place.

Anomalies can arise, however, when the melt contains reinforcing glass fibres. In general, the fibres tend to follow the orientation pattern of the melt, unless these are very long. By acquiring a preferential orientation in the radial direction, however, the fibres will reduce the shrinkage in this direction, in view of the transfer of the shrinkage stresses from the polymer matrix onto the fibres.

Consequently, while fibrous fillers will generally reduce the net shrinkage and also, therefore, the extent of warp, the direction of the warp may be more difficult to predict. For instance, in the case of glass-reinforced polypropylene the amount of shrinkage resulting from relaxation (or recovery) of the polymer orientation can offset the reduction in shrinkage caused by the fibres, so that while the extent of warp will be reduced, the type of warp, i.e. whether bowing or distortional, will remain the same. This is not the case with rapidly crystallising polymers such as nylons or polytetramethylene terephthalate, where the amount of shrinkage due to orientation of the polymer is relatively small and, consequently, the effect of the fibres predominates. Hence, in the radial direction, one could have a shrinkage lower than in the transverse direction. This would reverse the direction of bowing for a side gate feed and would convert the 'figure of eight' distortion for a central gate feed to a 'bowl' distortion (where $\bar{c} < 2\pi\bar{r}$).

It is understood that in considering mould shrinkage, both the effects of pressure and other moulding conditions on mould packing and orientation must be taken into account. The separation of these two effects is often quite difficult and, consequently, one has to rely on trial-and-error procedures for the solution of warping and tolerance problems. In mould design, however, one also relies on the introduction of geometrical features in the component to create obstacles to the natural shrinkage of the polymer, but care must be taken insofar as this may set up harmful residual stresses.

(d) Formalisation of an 'Injection Moulding Index'
In conclusion, since orientation in injection moulding creates problems which outweigh the small advantages resulting from increased stiffness, it must be considered an undesirable feature of the process. Furthermore since the amount of frozen orientation is a function of the compliance, an overall expression for the 'injection moulding index' of polymers will be exactly the same as that given in eqn. (9.24), i.e.

$$\alpha_{im} = f\left(\frac{1}{\chi_p, \eta, J_{(rec)}, \beta(\pi_i, a)}\right)$$

REFERENCES

1. B. S. GLYDE, *Polymer Rheology and Plastics Processing Conference, Loughborough, Reprints*, 17–19 September 1975, p. 186.
2. Z. TADMOR and I. KLEIN, *Engineering Principles of Plasticating Extrusion*, Van Nostrand Reinhold, New York, 1970, pp. 49–78.
3. Z. TADMOR and I. KLEIN, *Engineering Principles of Plasticating Extrusion*, Van Nostrand Reinhold, New York, 1970, p. 77.
4. E. C. BERNHARDT, *Processing of Thermoplastic Materials*, Krieger, New York, 1959, p. 235.

5. Z. TADMOR, I. J. DUVDEVANI and I. KLEIN, *Polym. Eng. Sci.*, **7** (1967) 202.
6. Z. TADMOR and I. KLEIN, *Engineering Principles of Plasticity Extrusion*, Van Nostrand Reinhold, New York, 1970, pp. 83–4.
7. Z. TADMOR and I. KLEIN, *Engineering Principles of Plasticating Extrusion*, Van Nostrand Reinhold, New York, 1970, pp. 124–5.
8. R. M. OGORKIEWICZ (ed.), *Thermoplastics, Properties and Design*, Wiley, New York, 1974, p. 185.
9. Z. TADMOR and I. KLEIN, *Polym. Eng. Sci.*, **9** (1969) 5.
10. M. GALE, Ph.D. Thesis, University of Aston, 1971.
11. H. D. BASSET, A. M. FAZZARI and R. B. STAUB, *Plastics Technology*, **11** (1965) 49.
12. H. D. BASSET, A. M. FAZZARI and R. B. STAUB, Union Carbide Literature, Reprint from *Plastics Technology*, **11** (1965) 49.
13. H. B. KESSLER, A. M. FAZZARI and R. B. STAUB, *Soc. Plast. Eng. J.*, **16** (1960) 267.
14. Z. TADMOR and I. KLEIN, *Engineering Principles of Plasticating Extrusion*, Van Nostrand Reinhold, New York, 1970, pp. 332–58.
15. B. R. BIRD, W. E. STEWART and E. N. LIGHTFOOT, *Transport Phenomena*, Wiley, New York, 1960, pp. 513–15.
16. Z. TADMOR and I. KLEIN, *Engineering Principles of Plasticating Extrusion*, Van Nostrand Reinhold, New York, 1970, p. 201.
17. G. ILLMAN, *SPE Journal*, **23** (1967) 12.
18. R. E. COLWELL and K. R. NICHOLLS, *Industrial Engineer Chemical Journal*, **51** (1959) 841.
19. Z. TADMOR and I. KLEIN, *Engineering Principles of Plasticating Extrusion*, Van Nostrand Reinhold, New York, 1970, pp. 217 and 285.
20. Z. TADMOR and I. KLEIN, *Engineering Principles of Plasticating Extrusion*, Van Nostrand Reinhold, New York, 1970, p. 356.
21. R. C. DONOVAN, D. E. THOMAS and L. D. LEVERSEN, *Polym. Eng. Sci.*, **11** (1971) 353–60.
22. G. D. GILMORE and R. S. SPENCER, *Modern Plastics*, **28** (1951) 117–18, 120–4, 180–7.
23. J. JARRET, *British Plastics*, **39** (1966) 291.
24. I. T. BARRIE, *Plastics and Polymers*, **38** (1970) 47–51.
25. R. S. SPENCER and G. D. GILMORE, *J. Appl. Phys.*, **21** (1950) 523.
26. ICI Ltd, Plastics Division, *Technical Service Note G 103*, p. 15.
27. ANON., *Modern Plastics International*, **3** (1973) 22.
28. G. V. SAGALAEV et al., *Int. Polym. Sci. Tech.*, **1** (1971) 76.
29. ANON., *Modern Plastics International*, **3** (1973) 22.
30. R. L. BALLMAN and T. SHUSMAN, *Ind. Eng. Chem.*, **51** (1959) 847–50.
31. B. M. MURPHY, *Chem. and Ind.*, **8** (1969) 290–1.
32. E. A. COLE, *Plastics and Polymers*, **43** (1975) 183.
33. R. L. BALLMAN and H. L. TOOR, *Plastic Flow as Interpreted by Birefringence*, Paper given at the 15th Annual Technical Conference of Society of Plastics Engineers, New York, 1959, pp. 51–3, Fig. 8.
34. R. M. OGORKIEWICZ and G. W. WEIDMANN, *Plastics and Polymers*, **40** (1972) 337–42.

10

Processing Performance of Thermoplastics in the Viscous State

The principal characteristic of these processes is that flow invariably occurs under the effects of gravitational forces only and that the shaping operation takes place by diffusion of molecules across an interface. They can be divided into thermofusion and solvent fusion processes.

10.1 THERMOFUSION PROCESSES

These comprise the following:

(1) Powder fusion processes, e.g. powder sintering, rotational moulding, powder coating and, as already mentioned, include also the melting processes in the compression zone of an extruder. It can also be argued that, in general, compression moulding processes fall into this class.
(2) Welding processes.

10.1.1 Fusibility of Polymers in Thermofusion Processes

(a) Fusion of Similar Materials

'Perfect fusion' of two components brought into contact with one another takes place if the overall adhesive forces acting across the interface become of the same order of magnitude as the cohesion within the bulk of the component. If the chemical nature of the two components is the same, perfect adhesion can be achieved by removing all surface irregularities at the interface so that a continuous and homogeneous structure is formed; i.e. the interface vanishes altogether.

Hence in the case of cross-linked thermosets perfect adhesion will never be achieved, since the surface energy is always lower than the cohesive energy; i.e. there is no possibility of forming a structure at the interface which is the same as in the bulk of the material.

In the case of thermoplastics, on the other hand, molecular diffusion at the interface is possible, and for the case of crystalline polymers co-crystallisation and spherulitic growth at the interface can take place, so that the original interface regions completely lose their identity. To achieve this, however, it is necessary that the polymer be in its viscous state; otherwise, some orientation will develop at the interface. Recovery would tend to occur upon subsequent service life of the component and break up the bonding established during processing. Particularly dangerous is the possibility of the component's being brought into contact with swelling agents, which would accelerate the recovery of the orientation.

(b) *Fusion of Dissimilar Materials*

If two components are brought together in their viscous state, diffusion across the interface will take place if the solubility parameters are the same. However, for the case of crystalline polymers, even if the solubility parameters are the same, e.g. polypropylene and polyethylene, but the crystal unit cells for the two polymers are different, phase separation takes place during subsequent cooling when the two polymers crystallise independently of one another. Although the amorphous phase will not be distinguishable at the interface, a weak boundary will be performed between the two crystal and spherulitic domains. This is not the case with amorphous polymers. For instance, polymethyl methacrylate, polyvinyl chloride and styrene–acrylonitrile copolymers, which have solubility parameters in the range of 9·3 to 9·7 $(cal/cm^3)^{1/2}$, are compatible and can be easily fused.

It must be borne in mind, however, that one is dealing with interfacial phenomena, and, therefore, two polymers with considerably different solubility parameters, e.g. polystyrene ($\delta \simeq 8·0$) and styrene–acrylonitrile copolymers ($\delta \simeq 9·5$), can be made to bond by oxidising the material of lower solubility parameter, so that at the interface the two materials will have the same solubility parameter. Furthermore, it must be realised that entropic separation of the less-compatible additives, such as lubricants, can also take place and form weak boundary layers. Adhesion promoters could, on the other hand, be used advantageously for their ability to migrate to the interphase and equalise the solubility parameters.

10.1.2 Analysis of Powder Sintering Processes

The processes which involve powder sintering are as follows.

(a) PTFE Sintering

This is carried out by making a preform of the powder by 'cold compaction' and subsequently bringing it to above the melting point for a sufficient length of time to allow diffusion through the interface of the powder particles. Upon cooling, co-crystallisation of the polymer at the interface takes place, thereby consolidating the interfacial bonding. Obviously, in this particular case the 'strength' of the moulding produced depends not only on the sintering conditions, i.e. the extent of interfacial diffusion, etc., but on the quality of contact between the powder particles during the pre-forming operation. Hence it is to be expected that factors such as particle size and distribution of particles play an important role. Small particles and a multimodal distribution of particles give the highest packing density and, therefore, the best performance in sintering. The degree of compaction depends also on the interfacial friction of the particles; consequently, smooth particles are preferred to those with an irregular surface, while volatile lubricants (e.g. low boiling paraffins) can be used to lubricate the flow of particles, as in the case of PTFE pastes for the production of thin mouldings.

(b) Expanded Polystyrene Block Moulding

This is a good example of solvent-assisted thermofusion processes. The residual blowing agent (a low boiling solvent, e.g. pentane) reduces the rubber/melt transition temperature of the material and assists the transport phenomena at the interface when the temperature is raised, typically to 110–120 °C in this case. The expandable beads are first delivered to a mould, where the temperature is raised by the injection of steam. This causes further expansion of the particles which assume a cuboidal geometry and fill the interstitial spaces, while the interfacial diffusion ensures that a good bond is established between the various particles.

(c) Rotational Moulding

While there are some variants to this process (such as 'slush moulding' in which there is no rotation of the mould), the fusion mechanism is basically similar to all rotational moulding processes. One feeds a quantity of powder into a pre-heated mould which is subsequently rotated by a single axis or double axis rotation action to distribute the powder uniformly over the internal walls of the cavity.

Shear stresses resulting from centrifugal action are negligible, owing to the low rotation speed of the mould and, therefore, the only forces acting on the polymer melt are those due to gravity. Consequently, the temperature must be very high in order to reduce the viscosity and accelerate diffusion through the powder interface so that proper fusion and flow over the wall surfaces can take place.

The rotational moulding process occurs in two stages; first the densification stage, when the powder melts and compacts under gravitational forces, and second the mould filling stage, when the melt flows down the walls of the rotating mould, thereby filling all channels, undercuts, etc., not accessible to the powder. Because of this, both the usual geometrical factors for the powder (e.g. particle size, distribution, coefficient of friction, etc.), and the rheological characteristics of the melt at zero shear stress (e.g. Newtonian viscosity and surface tension), are very important.

Furthermore, since the temperature continues to rise during rotation and flow of the melt, the temperature coefficient for both viscosity and surface tension must be considered. According to Rao and Throne,[1] the distance that the polymer melt will flow due to the driving action of the surface tension is given by:

$$\Delta \bar{z}^2 = \frac{R}{2} \int_0^{t_1} \frac{\gamma_{(t)}}{\eta_{(t)}} \, dt \qquad (10.1)$$

where R = radius of the capillary, and $\gamma_{(t)}$ and $\eta_{(t)}$ are respectively the surface tension and the viscosity functions. These are functions of time since their values will change in time, presumably as a result of the temperature rise. In fact, the above equation shows that γ is the driving force while η is the resistance-to-flow factor; hence, the polymer should ideally have a high surface tension against the walls of the mould but a low viscosity.

From this it can be deduced that mould release agents and external lubricants in the polymer composition assist the removal of the component from the mould but they can hinder the reproduction of fine details on the surface of the cavity, owing to the reduction in surface tension between the polymer melt and the mould cavity wall.

According to the above-mentioned authors, however, this inconvenience can be overcome by increasing the temperature to the maximum limit determined by the degradation characteristics of the polymer, since the surface tension decreases linearly with temperature while the viscosity (as already seen) decreases exponentially.

Formalisation of a 'Processability Index for Rotational Moulding'. In conclusion, the processability index for rotational moulding can be expressed in the following terms:

$$\alpha_{rm} = f\left(\frac{\gamma_0, \Delta H_{(\eta)}, D_{AA}, T_{0x}}{|\eta|_0, \dfrac{d\gamma}{dt}, \phi(r, g)}\right) \tag{10.2}$$

where γ_0 = surface tension at a reference temperature T, $|\eta|_0$ = zero shear rate viscosity at a reference temperature T, $\Delta H_{(\eta)}$ = activation energy for the viscosity, D_{AA} = molecular diffusion coefficient for the polymer melt at the maximum processing temperature T_{0x}, $\phi(r, g)$ = a function of the geometrical characteristics of the polymer particles, $d\gamma/dt$ = temperature coefficient of the surface tension (i.e. not a thermally activated phenomenon), and T_{0x} = maximum temperature to which the polymer can be exposed within the heating cycle of the process.

10.1.3 Analysis of Powder Coating Processes
These are mainly used to coat metal components and occasionally to protect the surface of glass objects (e.g. bottles) from scratching, etc. In rotational moulding, external lubricants are used in the formulation of the polymer, and mould release agents are applied on the surface of the cavity of the mould. However, in this case, one uses adhesion promoters in the polymer formulation and 'primers' on the surface of the substrate as a means of assisting adhesion.

The process is carried out in two stages; i.e. the application of the powder on the surface of the substrate is followed by sintering of the powder by thermofusion.

(a) Electrostatic Spraying
This is probably the most important process commercially since it provides uniform coatings. It exploits the dielectric properties of polymers so that the powder can be made to acquire electrostatic charges under the influence of an external field and be attracted to the surface of the substrate.[2] This process is carried out in four basic stages.

First Stage, Electrification of the Powder Particles. The particles acquire charges mostly by interfacial polarisation, whose relaxation time, λ, is of the order of 10^{-2} s, and also by dipolar polarisation, for which $\lambda = 10^{-8}$ s. (See discussions in later sections.)

The polarisation (P) is, of course, an intrinsic property of the material, defined as

$$P = Q_{(vac)} - Q_{(diel)} \tag{10.3}$$

where $Q_{(vac)}$ is the charge density on the plate of a condenser when the system is completely evacuated, and $Q_{(diel)}$ is the charge density when the material is placed between the electrodes of the condenser.

Polarisation is a time-dependent phenomenon, and the rate at which the charge is built up depends on the relaxation time, λ, by the usual exponential relationship

$$P = P_0(1 - e^{-t/\lambda})$$ (10.4)

The rate of charge decay, when the external electric field is removed, follows the relationship

$$P = P_0(1 - e^{-t/\lambda})e^{-[(t - t_1)/\lambda]}$$ (10.5)

Therefore, interfacial polarisation is the most important event in view of its having the longest relaxation time; this means that the attractive forces between the deposited polymer particles and the surface of the substrate persist for a much longer time than for other polarisations, thereby enabling the sintering process that follows to be carried out without the risk of the powder dropping off. In addition, charges can be acquired by means of a double layer polarisation mechanism, i.e. at the solid/gas interface, there will be some ionic species, mainly water and ionised air, adsorbed on the surface of the polymer particles. These will orient themselves in the direction of the applied field, causing the electrification of the particles. This mechanism differs from the interfacial polarisation mentioned earlier, in that it is strictly a surface phenomenon, while the interfacial polarisation occurs within the bulk of the polymer powder as a result of the polarisation of 'dissolved' ionic species. Since we are interested in surface phenomena it can easily be inferred that double layer polarisation is also a very important phenomenon.

Second Stage, Dispersion of Particles (Fig. 10.1). The rotating head of the spray gun is charged by direct current. Dispersion of the particles occurs by virtue of their repulsion (since they have the same charge) and by the momentum acquired from the rotating nozzle.

Third Stage, Deposition of the Powder Particles. The deposition of the powder particles on the object to be coated, which is being earthed, is determined by the attraction forces between the charged particles and the earth, i.e.

$$F = \frac{Q_1 Q_2}{\varepsilon_a r^2}$$

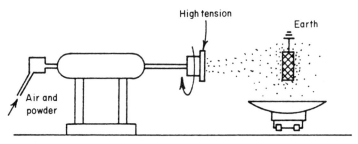

Fig. 10.1.   Dispersion of powder particles in electrostatic spraying.

where ε_a is the permittivity of the medium separating the powder and the object (i.e. air), r is the distance between the particle and the object, and Q_1 and Q_2 are the respective charges.

When a certain layer of powder has been deposited there will be a repulsion between the incoming particles and those on the surface of the object. Hence, a uniform layer of powder will be deposited, while the excess falls off. To increase the thickness one can reverse the polarity of the nozzle so that the powder will acquire charges of opposite sign and will be attracted by the charges on the surface of the object. The maximum layer that can be built up depends, therefore, on the relaxation time of the polarisation of the powders and on the density of the powder, since the powder coating can eventually fall off due to gravitational effects.

Fourth Stage, Sintering of Powder Coatings. This is usually carried out in ovens at high temperatures to ensure rapid melting and diffusion, as in the case of rotational moulding.

Formalisation of a 'sinterability index': The processability index for the sintering stage of powder coating processes is the same as for rotational moulding. The only difference is with respect to the electrification and deposition stage. The processability index for these two parts of the process can be simply expressed as follows:

$$\alpha_{ed} = f\left(\frac{\lambda, P_0}{\rho}\right) \tag{10.6}$$

where λ = the relaxation time of the material, which depends primarily on the polarity of the structural units and on the ionisability and mobility of additives present on the surface; P_0 = the polarisation at infinite time (i.e. the maximum polarisation), which is primarily dependent on the total amount of dipoles and ionic species present in the polymer; and $\rho =$ the density of the polymer particles.

10.1.4 Other Powder Coating Processes

Other important powder coating processes include 'flame spraying' and 'fluidised bed coating'. These differ substantially from electrostatic spray coating insofar as the deposition of the powder on the substrate is achieved by thermal fusion. In other words, the sintering process is carried out in two stages. The powder is deposited by virtue of the immobilisation achieved through fusion against the hot object and/or against itself. As far as the processability is concerned, it can be expressed in exactly the same manner as in previous cases.

10.2   WELDING PROCESSES

These differ from powder processes mainly with respect to the fact that fusion is restricted to certain regions of the component to be produced. The major processes used for welding thermoplastics include: heated tool welding, hot gas welding, high frequency welding and ultrasonic welding.

10.2.1 Analysis of Dielectric/High Frequency Welding Processes

As mentioned earlier, the polarisation of a polymer subjected to an electric field occurs primarily through interfacial (or ionic) polarisation and dipolar polarisation. The contributions from atomic polarisation (infrared regions) and electronic polarisation (optical regions) are negligible in view of the very short relaxation times involved (10^{-13}–10^{15} s) and the low polarisability of the atoms.

If an alternating current is applied to the polymer, there will be a delay in the response of the entities involved in the polarisation (i.e. ions and dipoles). This will obviously be due to obstacles involved: inertial effects and interactions with the surrounding entities, which create internal frictions in an attempt to follow the direction of the current. This causes internal 'losses', which manifest themselves as a temperature rise according to the frequency of the current and the strength of the electric field. The ionic species will probably travel longer distances in following the direction of the current. At the same time they cause much greater losses and are more effective in raising the temperature of the polymer than dipolar oscillations.

The essence of dielectric welding is, therefore, to heat the material internally to the temperature at which molecular diffusion can take place and then apply shear stresses by pressing the two surfaces concerned against each other.[3]

The energy loss, which corresponds to the enthalpy rise of the polymer in the absence of heat losses to the surroundings, is given by the expression:

$$N = 0.555 \times 10^{-12} \times f \times k \times \tan \theta \times F^2 \qquad (10.7)$$

where N = electric energy loss (W/cm^3), f = current frequency (c/s), F = field strength (V/cm), k = dielectric constant, θ = loss angle (rads), and $\tan \theta$ = loss tangent.

Note that k, like the modulus of the material, undergoes a major drop around the glass transition temperature and in the region of the rubbery/viscous state transition, while $\tan \theta$ follows the usual peaks in these regions. Consequently, a drop in k is compensated by a rise in $\tan \theta$. From the above expression it can be directly deduced that the more polar materials, such as nylons, PVC, etc., and those formulations containing the highest levels of ionic additives, e.g. flexible PVC, can be dielectrically welded much more easily than the low polarity materials, such as polyolefins.

As already mentioned, since it is often desirable to weld different materials together, it is important that the difference in their solubility parameters should be as small as possible, and that the zero shear rate viscosity should also be very low.

10.2.2 Formalisation of a 'Processability Index' by Dielectric Welding

An overall expression for the dielectric-weldability index of polymers can therefore be as follows:

$$\alpha_{dw} = f \left(\frac{\bar{k}, \overline{\tan} \, \theta, \Delta H_{(\eta)}, D_{AB}}{|\eta|_0, (\delta_A - \delta_B)|_{\delta_A > \delta_B}} \right) \qquad (10.8)$$

where $\bar{k}$ = the mean dielectric constant from room temperature to that corresponding to the viscous state; $\overline{\tan} \, \theta$ = the mean loss factor; $\Delta H_{(\eta)}$ = the activation energy of the viscosity; D_{AB} = the diffusibility of polymer A with polymer B (for the same polymer this becomes D_{AA}); and $\delta_A - \delta_B$ = the difference in solubility parameters, which will be omitted when considering welding of the same polymer, accordingly, the expression could be modified by a factor which makes this term zero when $\delta_A = \delta_B$.

10.2.3 Formalisation of a 'Processability Index' by Ultrasonic Welding

Internal heating of the material can be accomplished by the application of alternating mechanical stresses. The two welding processes that utilise this property of the material are:

(1)   Friction (or spin) welding
(2)   Ultrasonic welding

Only the analysis of the latter will be considered in this case.

(a) *Ultrasonic Welding Processes*
These processes use dynamic mechanical stresses at very high frequencies ($\simeq 20$ kilocycles/s), i.e. just above the audible range, by means of transducers which convert electrical energy into mechanical pulses. These pulses are amplified by means of so-called 'horns' which also have the function of producing stresses at the interface to enhance molecular diffusion between the two surfaces in contact. There is obviously a very strong analogy between the theoretical concepts underlying mechanical and dielectric losses.

The mechanisms are similar, the only difference being that, under mechanical stresses, the losses occur by virtue of segmental motions of the polymer chains. Also, the only contribution by the net displacement of structural entities, such as low-molecular-weight species, is quite small in comparison with the ionic polarisation for dielectric heating. There are two basic methods used in ultrasonic welding.[4]

Near-Field (or Contact) Welding. This is used with materials having a large damping peak at low temperatures, e.g. low density polyethylene, polypropylene copolymers, toughened polystyrene, etc., which tend to attenuate the applied stresses and, therefore, prevent the stress waves from reaching the interface concerned. At room temperature, the attenuation factor for low density polyethylene is 142×10^{-3}, while, for polymethyl methacrylate, it is only 6×10^{-3}.

Far-Field (or Transmission) Welding. This is used with materials having low attenuation factors, so that the pulses can be applied at a distance of up to 20–25 cm away from the area of welding. This is highly desirable when welding two halves of a hollow component, e.g. ducts, etc. (Fig. 10.2). In general, it can be said that these far-field transmission characteristics are highly desirable.

The processability index for ultrasonic welding can be expressed in the same way as that for dielectric welding, except that the loss factor is placed

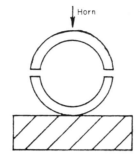

Fig. 10.2.   Far-field ultrasonic welding.

in the denominator to indicate the desirability of far-field transmission characteristics, i.e.

$$\alpha_{uw} = f\left(\frac{E^*, \Delta H_{(\eta)}, D_{AB}}{|\eta_0|, \tan\theta, (\delta_A - \delta_B)|_{\delta_A > \delta_B}}\right) \tag{10.9}$$

where E^* is the dynamic complex modulus.

10.3   SOLVENT-ASSISTED FUSION PROCESSES

These can be subdivided as follows:

(1)   Fusion processes with permanent solvents, e.g. PVC paste processes.
(2)   Fusion processes with migratory solvents, e.g. cements, solution coating, film casting, latex coatings, etc.
(3)   Fusion processes with partial migration of solvents, e.g. PVC 'organosol' processes, coatings with plasticised polymers and assisted by solvents.

The techniques for these processes are numerous and it is not within the scope of this book to describe them.

10.3.1 Formalisation of a 'Processability Index' for Solvent Fusion Processes

From the processability point of view, one can use exactly the same arguments as for rotational moulding and welding. The exception is in the case of migratory solvents where one must include the latent heat of vaporisation of the solvent in the denominator. For mixed migratory/ permanent solvents, the analysis with respect to diffusibility coefficients and solubility parameters becomes much more complicated.

REFERENCES

1. M. A. RAO and J. L. THRONE, *Polym. Eng. Sci.*, **12** (1972) 237.
2. Society of Manufacturing Engineers, *Fundamentals of Powder Coating*, Dearborn, Michigan, p. 17.
3. H. P. ZADE, *Heat Sealing and High Frequency Welding of Plastics*, Wiley, New York, 1959.
4. R. H. THOMAS, SR., *Ultrasonics in Packaging and Plastics Fabrication*, Cahners Books, New York, 1974.

Index